OPTICAL ELECTRONICS

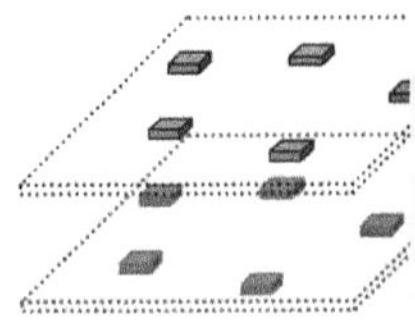

SELF-ORGANIZED INTEGRATION AND APPLICATIONS

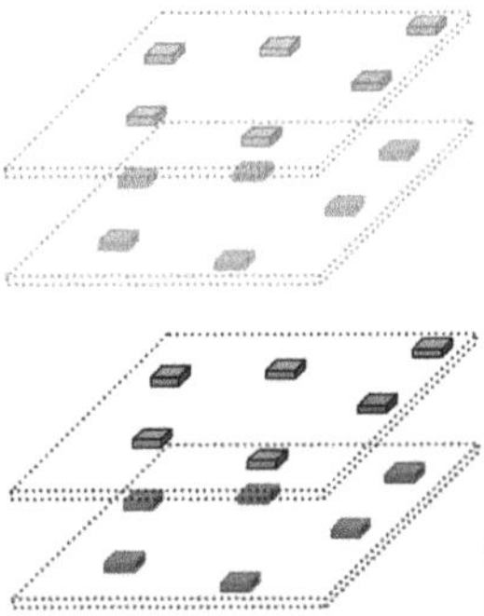

OPTICAL ELECTRONICS

SELF-ORGANIZED INTEGRATION AND APPLICATIONS

Tetsuzo Yoshimura

PAN STANFORD PUBLISHING

Published by

Pan Stanford Publishing Pte. Ltd.
Penthouse Level, Suntec Tower 3
8 Temasek Boulevard
Singapore 038988

Email: editorial@panstanford.com
Web: www.panstanford.com

British Library Cataloguing-in-Publication Data
A catalogue record for this book is available from the British Library.

Optical Electronics: Self-Organized Integration and Applications

ISBN 978-981-4310-90-1 (Hardcover)
ISBN 978-981-4364-08-9 (eBook)

Printed in the USA

To Yoriko

Contents

Preface

The implementation of optics into electronic systems has improved the performance of computers and communication systems. In future, the role of optics in electronics will increase more and more, giving rise to a situation where enormous optical components are involved with optical wiring within electronic systems. In such cases, optics should be embedded into the systems without excess spaces, fabrication costs, and system design efforts, allowing us to treat optics unconsciously.

This book intends to propose and review comprehensive strategies for implementation of optics into electronic systems with minimized optics excess. We define "simply-optics-implemented electronics" as "optical electronics."

The first core technology of optical electronics is the self-organized optical waveguide based on the self-organized lightwave network (SOLNET). SOLNET is regarded as "optical solder," which enables self-aligned optical couplings between misaligned optical devices and resolves the most serious problem in optics, namely, "assembling with micron or sub-micron accuracy," which is not necessary in electronics.

The second core technology is the three-dimensional (3-D) optical circuit, which is built by stacking optical waveguide films. SOLNET contributes to construct vertical waveguides between the optical waveguide films.

The third core technology is the material-saving heterogeneous thin-film device integration process "photolithographic packaging with selectively occupied repeated transfer (PL-Pack with SORT)" which replaces the packaging process based on flip-chip bonding with the all-photolithographic semiconductor process. This achieves integration of a lot of optical and electronic devices into films efficiently to provide optoelectronic (OE) system films.

The fourth core technology is the high-speed/small-size light modulators and optical switches, which are mandatory to build high-performance OE systems.

These core technologies have potentiality to realize the self-organized 3-D integrated optical circuits, which are the fundamental platform for the optical interconnect within computers, the optical switching system, the integrated solar energy conversion system for future photo-voltaics and photo-synthesis, and the bio/medical photonics. I hope that readers will be interested in the optical electronics and take some hints from it for their research work.

I would like to thank Professor Kunihiko Asama of Tokyo University of Technology for his encouragement and support, General Managers Isao Fudemoto and Osamu Takahashi of Fujitsu Laboratories Ltd. for giving me an opportunity to work in Silicon Valley. I would also like to thank colleagues of Fujitsu Computer Packaging Technologies, Inc. (FCPT), San Jose, California, and Fujitsu Laboratories, Ltd., and students who graduated from Yoshimura Laboratory in Tokyo University of Technology for their collaboration in research work that contributes to the accomplishment of this book. Finally, I would like to express sincere gratitude to Mr. Stanford Chong, director and publisher, Pan Stanford Publishing Pte. Ltd., for giving me a great opportunity to write this book and Mr. Arvind Kanswal for his great efforts and sincere help in completing this book.

Tetsuzo Yoshimura

Chapter 1

Introduction

Electronics is currently one of the fundamental technologies. During last several tens of years, optics has gradually penetrated into electronics. Figure 1.1 illustrates the implementation of optics into electronics. Optical communication systems, optical interconnects, optical switching systems, optical storage systems, and solar energy conversion systems are involved. Recently, bio/medical photonics has appeared as another important field.

The typical example is the optical communication systems. As shown in Fig. 1.2, since optics was introduced into long-haul systems, with the passage of time, it has been implemented into shorter-distance systems, such as local-area network (LAN) and "inter-box" interconnects and is ready to come into "within box" interconnects. As optical networking has penetrated into systems with short wiring distances, the line counts have been increased.

In computers, increases in clock frequencies and wiring density cause problems such as RC delay, signal attenuation, power dissipation (or heat generation), and cross-talk. Optics implementation will be required to solve these problems within boxes, including board and chip-level wiring. Optics implementation will also be required for massive optical switching systems to treat rapidly increasing data rate in networking. Optical-wiring-implemented

Optical Electronics: Self-Organized Integration and Applications
Tetsuzo Yoshimura

ISBN 978-981-4310-90-1 (Hardcover), 978-981-4364-08-9 (eBook)
www.panstanford.com

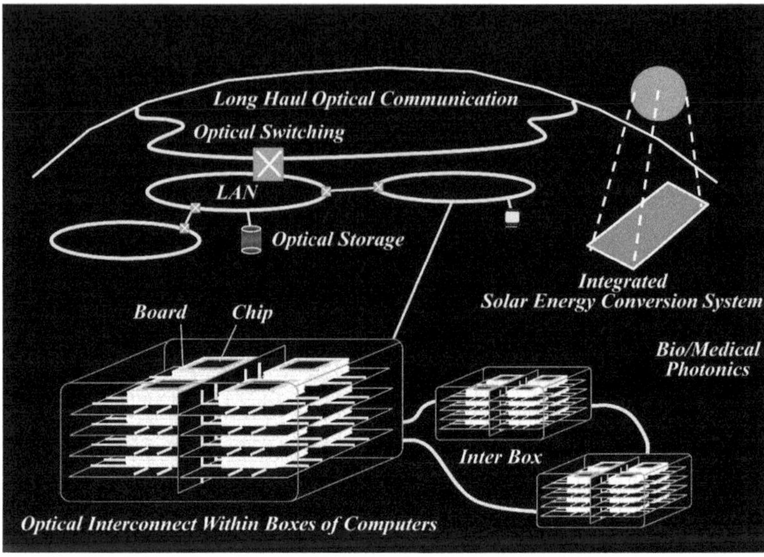

Figure 1.1. Optics implementation into electronics. See also Color Insert.

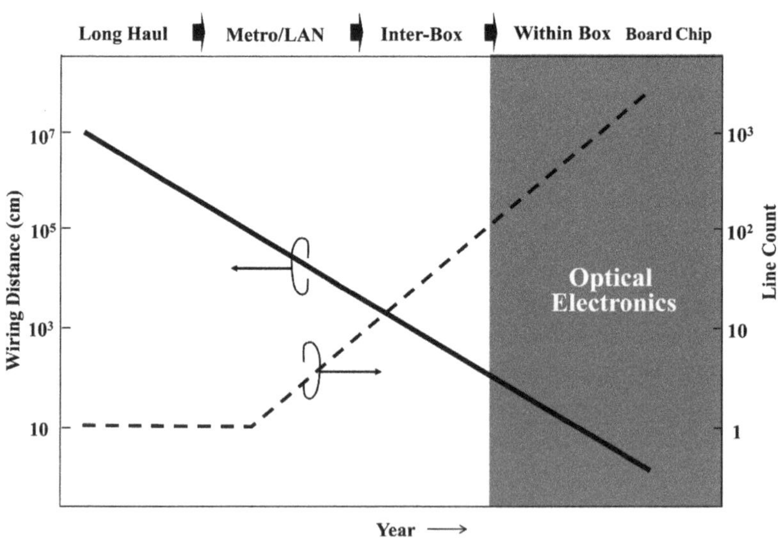

Figure 1.2. Optical networking penetration into electronic systems.

films for the optical interconnects within boxes can be applied to film-based integrated solar energy conversion systems. Thus, the optics penetration into various kinds of electronic systems will be promoted more and more in future to improve their performance.

Two remarkable features exist in the optics within boxes, being distinguished from LAN and box-to-box data transmission; "very short line distance of cm order" and "very large channel count of hundreds to thousands." This means that hundreds to thousands of channels should be distributed on a board for electrical-to-optical (E-O) and optical-to-electrical (O-E) signal conversion. Thus, optics requires excess components, space, fabrication processes, and design efforts when it is implemented into electronic systems, causing size and cost issues.

Currently, for the optics implementation, packaging processes based on the flip-chip bonding are used for assembling discrete devices such as chips. The discrete devices are space/material-consuming and the packaging processes are time-consuming. Therefore, we should consider the following two issues:

- How to reduce the size and the material consumption of the optical devices?
- How to simplify the device assembling process?

In this book, strategies for providing solutions to the above-mentioned issues are described. The main objective is to show how optics will be implemented into electronic systems with reasonable sizes and costs to realize optical electronics. "Optical electronics" is the concept of "simply-optics-implemented electronics."

In Chapter 2, merits of optics implementation into electronics are summarized. Then scalable film optical link module (S-FOLM), which is the backbone of optical electronics, is reviewed, and the final goal "self-organized three-dimensional (3-D) integrated optical circuit" is proposed. The circuit is based on S-FOLM and consists of stacked waveguide films, where thin-film devices are embedded, and the self-organized lightwave network (SOLNET).

In Chapter 3, as fundamentals of the analysis of optical circuits, the beam propagation method (BPM) and the finite difference time domain (FDTD) method are explained.

Then, concepts and theoretical/experimental demonstrations of core technologies in optical electronics as well as their current status are described in Chapters 4 to 9, including SOLNET, which forms self-aligned coupling waveguides between misaligned devices, optical waveguide films with vertical mirrors and 3-D optical circuits, material-saving heterogeneous thin-film device integration processes, and light modulators/optical switches.

Chapter 10 presents optoelectronic (OE) hardware built by optical electronics, such as optical solder for alignment-free optical couplings, optical wiring in free spaces, integrated optical interconnects within boxes, and 3-D optical switching systems. Waveguide-based integrated solar energy conversion systems are presented in Chapter 11. In addition, in Chapter 12, as future technologies for optical electronics, the molecular layer deposition (MLD), and artificial electro-optic materials and devices fabricated by MLD are described. Subjects related to bio/medical photonics are briefly presented.

Chapter 2

From Electronics to Optical Electronics

In this chapter, problems in current electronic systems and merits of optics-implemented electronic systems for overcoming the problems are described. The optics-implemented electronics should be simple, namely, the structure, the fabrication process, and the design/operation procedure should be just like pure "electronics." Such system simplification is achieved by optical electronics. Various types of optics implementation methods for optical electronics are compared, and the concept of S-FOLM and the self-organized 3-D integrated optical circuit are described. Core technologies in optical electronics are summarized.

2.1 Merits of Optics Implementation Into Electronics

High-performance computers are typically built from modules with the hierarchy shown in Fig. 2.1(a). Large-scale integrated circuits (LSIs) are placed on multi-chip modules (MCMs). The MCMs are placed on printed circuit boards (PCBs). The PCBs are placed on backplanes (BPs). The approximate sizes of LSI, MCM, PCB, and BP are, respectively, ~1 cm, ~10 cm, several tens of centimeters, and ~100 cm.

Optical Electronics: Self-Organized Integration and Applications
Tetsuzo Yoshimura

ISBN 978-981-4310-90-1 (Hardcover), 978-981-4364-08-9 (eBook)
www.panstanford.com

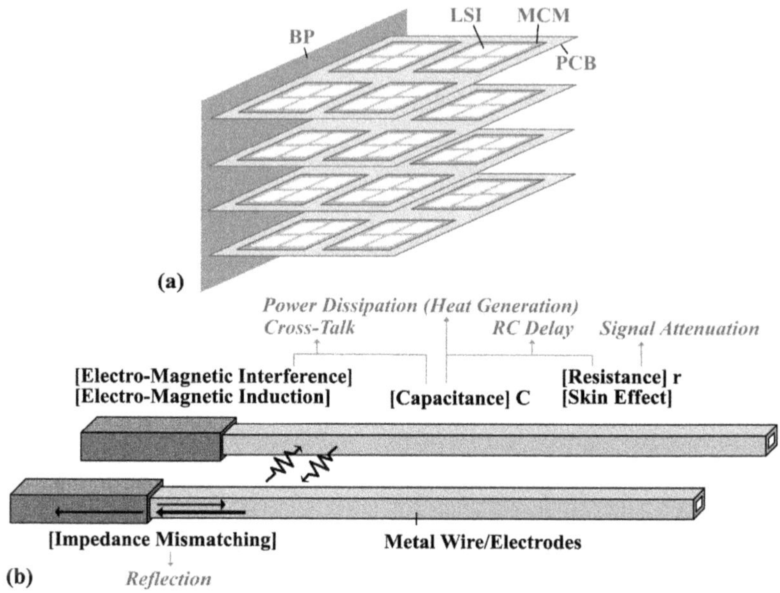

Figure 2.1. (a) Structural hierarchy in high-performance computers. (b) Problems in the electrical wiring.

As shown in Fig. 2.1(b), increases in clock frequency and wiring density in each level in the hierarchy cause problems such as crosstalk, power dissipation (heat generation), RC delay, signal attenuation, and signal reflection. The cross-talk arises from electromagnetic interference, electromagnetic induction, and electrical couplings between metal wires/electrodes due to capacitance. The power dissipation and the RC delay mainly arise from charge-discharge of metal wires/electrodes through electrical resistance of the metal wires/electrodes and transistors. The signal attenuation arises from electrical resistance of the wires/electrodes including the skin effect. The signal reflection arises from impedance mismatching in electrical wires/electrodes. Since the electromagnetic interference, electromagnetic induction, capacitance and electrical resistance tend to increase with increasing clock frequency and wiring density, the problems become serious with increases in clock frequency and wiring density in computers. The same situation occurs in switching systems in communications.

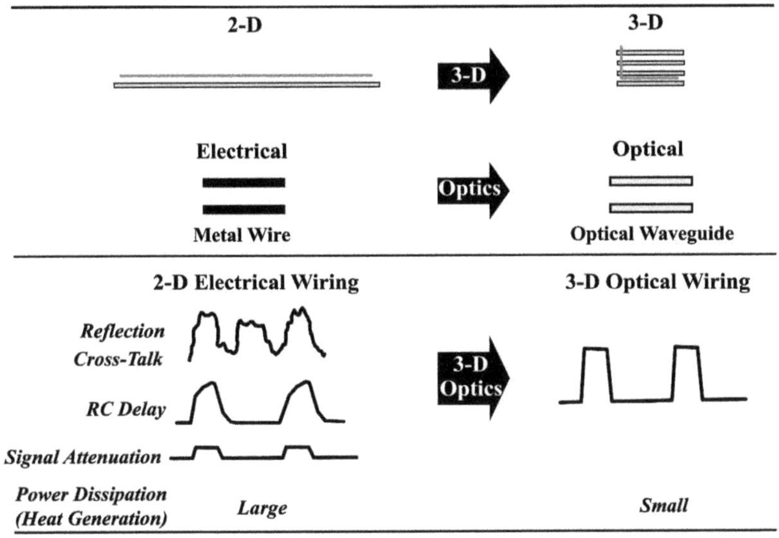

Figure 2.2. Merits of the three-dimensional wiring and the optical wiring.

There are two ways to overcome the problems: three-dimensional (3-D) wiring [1] and optical wiring [2]. Figure 2.2 shows merits of the 3-D wiring and the optical wiring. By introducing 3-D wiring, wiring distances are reduced, resulting in the reduction of the electromagnetic interference, electromagnetic induction, capacitance and electrical resistance. Actually, in many research organizations, studies on the 3-D electrical wiring in LSIs are under way. By replacing metal wires with optical waveguides, the electromagnetic interference, electromagnetic induction, capacitance, electrical resistance, and impedance mismatching can drastically be reduced. The removal of the skin effect leads to frequency-independent loss of propagated signals. Therefore, it is expected that 3-D optical wiring will achieve data transmission with sharp and large signals in high-speed and high-density systems.

The optical wiring has additional merits as shown in Fig. 2.3: "Short OK" and "Open OK" characteristics. When optical waveguides cross each other, almost no cross-talk appears due to straight-propagation characteristics of light waves, enabling an increase in wiring design freedom. By using the "Short OK" characteristics,

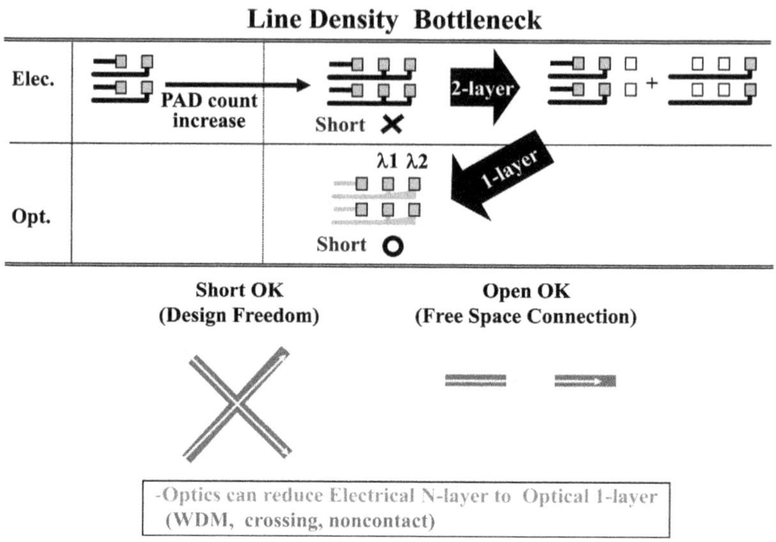

Figure 2.3. Additional merits of the optical wiring.

optics can reduce layer count for wiring; for example, electrical 2-layer to optical 1-layer.

Meanwhile, when optical waveguides are separated with some distances, due to their free-space-propagation characteristics, most of the light waves from one optical waveguide can be transmitted into the other optical waveguide. The "Open OK" characteristics enable "free space connection." This property is useful to construct 3-D optical circuits, in which light wave transmission to vertical directions can be easily carried out without vertical waveguides.

Furthermore, by using the wavelength division multiplexing (WDM), a lot of optical signals with different wavelengths can be transmitted in a single optical waveguide. This realizes drastic increase in transmission data rates as well as reduction of layer count for wiring, for example, electrical N-layer to optical 1-layer.

As described, optics-implemented electronics is expected to provide the next-generation high-performance systems. As shown in Chapter 11, it will also provide integrated solar energy conversion systems.

2.2 Optical Electronics

For the optics implementation, the most important issue is the removal of the optics excess, namely, excess components, spaces, fabrication processes, and design efforts. The optics excess increases the system size and raises the cost unacceptably. In other words, the optics-implemented electronics should be simple in order to be used widely just as the pure electronics. We define the concept for the "simply-optics-implemented electronics" as optical electronics.

Optical electronics, in which "optics can be treated like electronics," is schematically compared with the conventional optics-implemented electronics in Fig. 2.4. For components and spaces, in the conventional optics-implemented electronics, discrete optical devices are used, and spaces both for electronic devices (LSIs) and for optical devices are required. In optical electronics, on the other hand, optical devices are integrated in LSIs or other parts in the systems as thin-film flakes, and spaces only for electronic devices are required. This contributes to space and material consumption reduction.

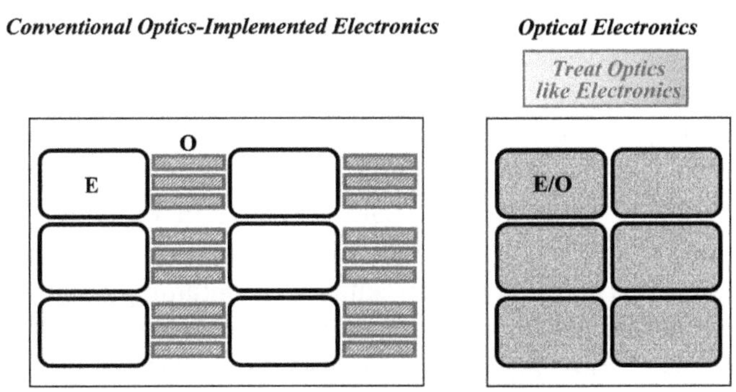

	Conventional Optics-Implemented Electronics	Optical Electronics
Components:	**-LSIs** **-Discrete Optical Devices**	**-LSIs** **-Integrated Thin-Film Flakes of Optical Devices**
Spaces:	**-for LSIs** **-for Optical Devices**	**-for LSIs**
Fabrication Processes:	**-Packaging (Flip-Chip Bonding)** **-Assembling for Alignments**	**-Heterogeneous Integration (Photolithography)** **-Self-Organized Alignments**
Design:	**-Electronics** **-Optics**	**-Electronics**

Figure 2.4. Comparison of optical electronics with the conventional optics-implemented electronics.

For fabrication processes, in the conventional optics-implemented electronics, flip-chip-bonding-based packaging, in which assembling with individual optical device alignment is necessary, is used. In optical electronics, on the other hand, the heterogeneous integration based on the photolithography similar to the LSI fabrication process is used for embedding the thin-film device flakes, and optical devices are coupled by self-organized alignment. For design, it is carried out by considering both electronics and optics in the conventional optics-implemented electronics. In optical electronics, the design is carried out by considering only electronics. Thus, it is expected that simple optics-implemented electronic systems will be fabricated by optical electronics.

2.2.1 *Optics Implementation Methods*

In Fig. 2.5, a typical optical interconnect model is illustrated as an example of optics-implemented electronics. Electrical signals from an LSI output terminal are converted to optical signals by a light modulator or a vertical cavity surface-emitting laser (VCSEL). The optical signals are transmitted through an optical waveguide to a photodetector (PD) to be converted to electrical signals. The electrical signals are provided into an LSI input terminal.

In the optical interconnect model, there are various structures of the E-O and O-E conversion parts. The structures are classified in Fig. 2.6, for example with VCSELs and PDs. In the optoelectronic (OE) module type [3], III-V-compound bulk chips of VCSELs and PDs, and bulk chips of integrated circuits (ICs) containing drivers for VCSELs and amplifiers for PDs, which are indicated as D/A IC in Fig. 2.6, are used. The chips are assembled on a substrate by the flip-chip-bonding-based packaging together with LSI chips. The VCSELs and

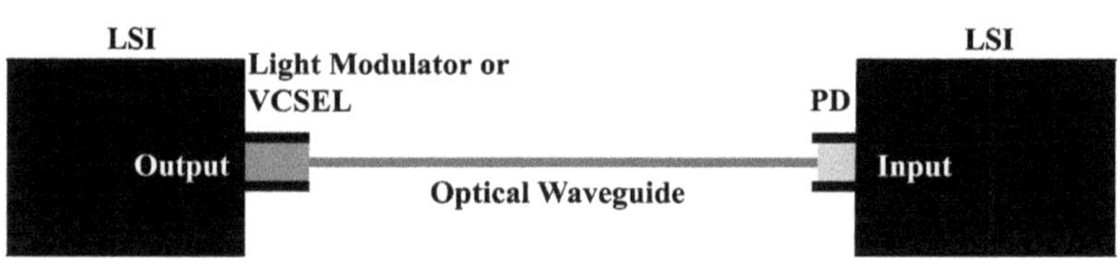

Figure 2.5. Typical optical interconnect model.

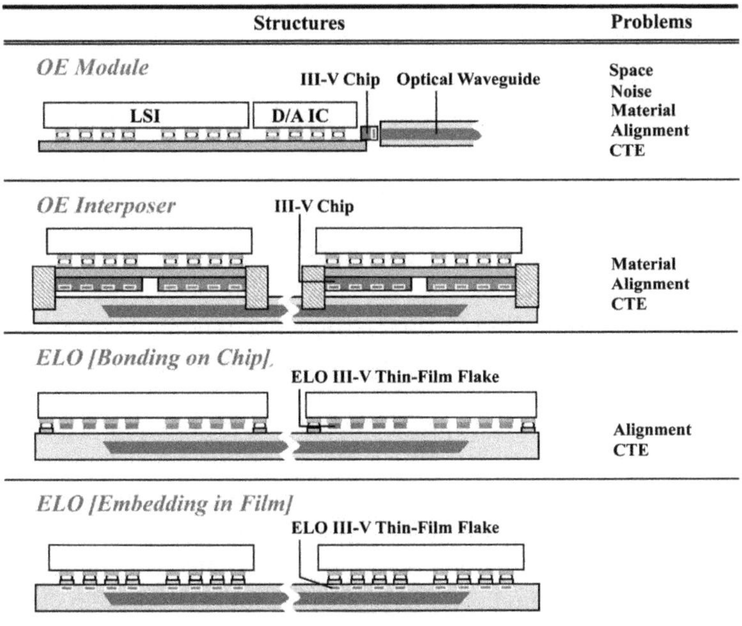

Figure 2.6. Classification of E-O and O-E conversion structures. See also Color Insert.

PDs are coupled to optical waveguides. Although the OE module type is already available, it has several problems:

- Excess space for bulk chips is necessary.
- Large noise is generated due to relatively long electrical lines for driver–VCSEL and PD–amplifier connections.
- Material consumption for the VCSEL/PD bulk chips is huge.
- Huge efforts are required for the alignment of the VCSEL/PD chips and the optical waveguides by the flip-chip-bonding-based packaging.
- Relative position changes between the VCSELs/PDs and the optical waveguides are large because a considerable difference in coefficient of thermal expansion (CTE) exists between the bulk chips and the optical waveguides.

In the OE interposer type [4], the III-V-compound bulk chips are assembled on interposers together with D/A IC chips. The interposers are inserted between LSIs and optical waveguides. The VCSELs and PDs are aligned to vertical mirrors in the optical waveguides to achieve optical couplings. The assembling processes are carried out by the flip-chip-bonding-based packaging. In the OE interposer type, although the problems of the excess spaces and large noise are removed, the problems of material consumption for the VCSEL/PD bulk chips, the efforts for alignment of the VCSEL/PD and the optical waveguides, and the CTE mismatch remain.

In the epitaxial lift-off (ELO) [Bonding on Chip] type [5,6], VCSEL/PD thin-film flakes of III-V-compound epitaxial films are mounted on output and input pads of LSIs. A typical ELO process is shown in Fig. 2.7 [7]. (1) An epitaxial film is grown on a growth substrate. (2) The epitaxial film is separated into flakes in an array by etching. (3) A polymer film is attached on the surface. (4) The substrate is removed to transfer thin-film flakes on the polymer film. (5) The thin-film flakes are transferred onto LSI pads by the pick-and-place method. The LSIs with VCSEL/PD thin-film flakes are mounted on optical waveguides by the flip-chip-bonding-based packaging. In the ELO [Bonding on Chip] type, material consumption for VCSELs and PDs is drastically reduced because the thin-film flakes are used. However, considerable efforts are still required for the alignment of the VCSEL/PD and the optical waveguides because the LSIs are mounted by the flip-chip-bonding-based packaging. There is also the CTE mismatch problem.

In the ELO [Embedding in Film] type [8–20], thin-film flakes of the III-V-compound epitaxial films are embedded into optical waveguide films. By using the photolithographic packaging with the selectively occupied repeated transfer (PL-Pack with SORT) [21–24], the thin-film flakes are embedded by all-photolithographic process similar to the LSI fabrication process. In other words, the flip-chip-bonding-based packaging is not necessary. Consequently, the alignment problem occurring between the VCSEL/PD and the optical waveguides can be solved. Furthermore, the CTE mismatch problem is also solved because individual thin-film flakes of VCSELs and PDs are placed just on the vertical mirror in the optical waveguide films as isolated tiny islands. The detail is described in Chapter 8.

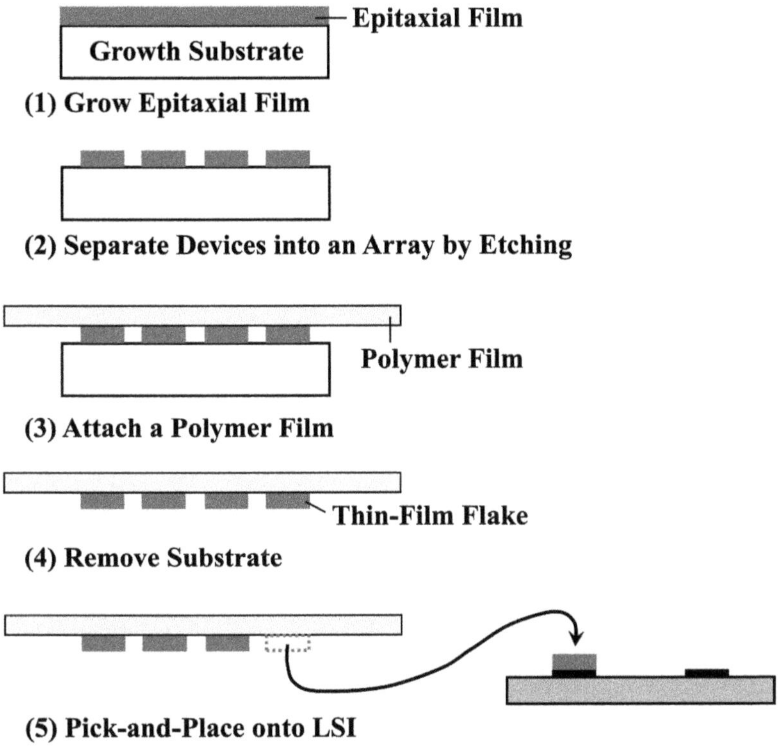

Figure 2.7. Epitaxial lift-off.

From the above-mentioned arguments, it is concluded that ELO [Embedding in Film] type is the best solution to the interfaces of E-O conversion and O-E conversion.

2.2.2 *Three-Dimensional Optoelectronic Platforms Based on Scalable Film Optical Link Module (S-FOLM)*

The future directions of optics implementation methods are summarized in Table 2.1. Optical wiring will be 3-D. E-O and O-E interfaces will be made from thin-film flakes of optical devices. Fabrication process will be the heterogeneous integration using the all-photolithographic process. Thus, we propose the 3-D OE platform as a low-cost solution that satisfies the future directions. The 3-D OE

Table 2.1. Current and future optics implementation methods

Current		Future
2-D	*Optical Wiring*	3-D
Bulk Chip	*E-O/O-E Interface*	Thin-Film Flake
Packaging: Flip-Chip-Bonding	*Fabrication*	Heterogeneous Integration: All-Photolithographic Process

platform is based on the scalable film optical link module (S-FOLM) [10–12, 14–26], which is a new concept of OE hardware consisting of films with embedded thin-film devices. By combining SOLNET [22, 25–28] with S-FOLM, 3-D OE platform consisting of self-organized 3-D optical circuits will be constructed [22].

Figure 2.8 shows generalized concept of S-FOLM. A set of films, in which flakes of thin-film devices are embedded, is prepared. The thin-film devices include optical waveguides, VCSELs, PDs, light modulators/optical switches, wavelength filters, optical amplifiers,

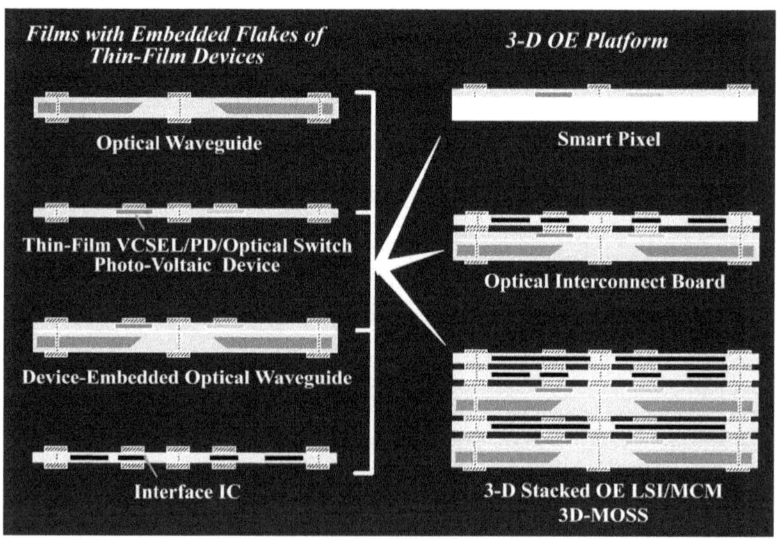

Figure 2.8. Concept of S-FOLM. Reprinted with permission from T. Yoshimura, J. Roman, Y. Takahashi, M. Lee, B. Chou, S. I. Beilin, W. V. Wang, and M. Inao, "Optoelectronic Scalable Substrates Based on Film/Z-Connection and Its Application to Film Optical Link Module (FOLM)," *Proc. SPIE* **3952**, 202–213 (2000). See also Color Insert.

photovoltaic devices, lenses, mirrors, interface ICs, LSIs, and so on. By combining the films in stacked configurations, various kinds of 3-D OE platforms are constructed. By stacking a VCSEL/PD-embedded film on an LSI, a smart pixel will be available. When a VCSEL/PD-embedded optical waveguide film and an interface-IC-embedded film are stacked, an optical interconnect board will be made. If LSI-embedded films are added in the stack, 3-D stacked OE LSI or MCM will be made. If optical-switch-embedded films are used in the stack, the three-dimensional micro optical switching system (3D-MOSS) [22] will be available. Such scalability will contribute to system cost reductions. The stacking may be performed by the building-up or the film/Z-connection [29].

The interface-IC-embedded film, on which LSIs are mounted, acts as an interface film. The ICs may have multiplexing (MUX) and de-multiplexing (DMUX) circuits, error correction circuits, coding/decoding circuits, etc., as well as driver/amplifier circuits. This interface approach permits the system designers to use conventional development tools without a need to understand optics, which is an important feature for introducing optics into a variety of electronic systems.

Thus, S-FOLM constructs maximized kinds of OE platforms by combining minimized kinds of films, realizing all levels of optical networking including WDM backbones, massive parallel optical links, optical interconnect boards, 3-D stacked OE LSIs, and so on. According to customer's requests, any products with any specs can be supplied just by adjusting the stacking combination of the films. This implies that S-FOLM has a great standardized-interface capability and packaging scalability. As a result, S-FOLM enables us to minimize optics excess and keep designers unconscious of optics, contributing to cost reduction of OE systems.

In Fig. 2.9, an optical transceiver utilizing S-FOLM is compared with that utilizing a conventional OE module. In the optical transceiver with the conventional OE module, OE device bulk chip (for example, VCSEL chip) and driver/amplifier chip are mounted on a substrate surface by the flip-chip bonding. In the optical transceiver with S-FOLM, which utilizes the ELO [Embedding in Film] shown in Fig. 2.6 for E-O and O-E conversion, the OE device chip is divided into small/thin flakes, and the flakes are embedded

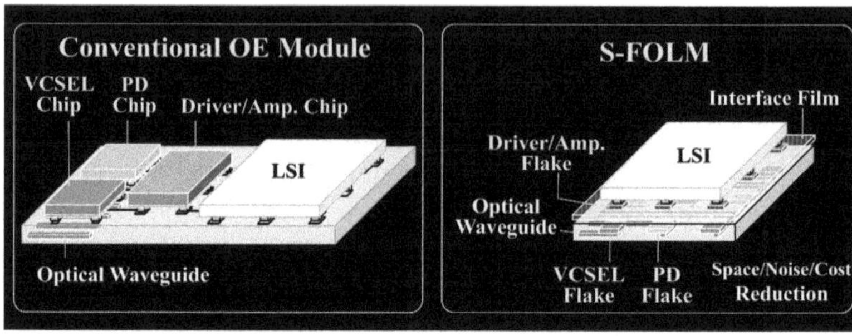

Figure 2.9. Comparison of an optical transceiver utilizing S-FOLM and that utilizing conventional OE module.

just under the LSI corresponding to individual pad positions. An interface film containing thinned driver/amplifier chips is inserted between the LSI and the thin-film OE device flakes. This leads the optical transceiver with S-FOLM to the following advantages:

- Excess space is not necessary on the surface.
- Long metal lines are eliminated, resulting in the reduction of noises, namely, bit error rates, and delay.
- Excess cost is reduced by minimizing epitaxial-material consumption, and by embedding the device flakes by the photolithographic process like PL-Pack with SORT.
- Standardized interface is available by the interface film. This enables us to treat optics as a transparent agent for interconnect-spec improvement.
- In addition, any place/any shape placement capability is available by selective attachment of the films of S-FOLM onto the substrate.

In S-FOLM, semiconductor materials exist only at the places where they are necessary. The other area is occupied by polymer or glass. This is cost effective due to epitaxial-material saving. In a typical case, a cost reduction of one order of magnitude or more is expected, as discussed in Section 8.3.1.

S-FOLM is effective for WDM applications, where OE devices with different wavelength characteristics need to be integrated into a same array. To do this, in the conventional module structure shown in Fig. 2.9, different-wavelength devices should be integrated into one chip with highly sophisticated semiconductor fabrication technologies, raising chip cost. In S-FOLM, on the other hand, a low-cost array containing different-wavelength devices will be easily available by using SORT process, in which the different-wavelength devices are put together into one polymer film at one time from different wafers with individual single-wavelength-devices. The SORT process is discussed in detail in Chapter 8.

In the 3-D OE platform, both 3-D architectures and optical wiring are implemented. So, it is promising for the construction of optical interconnects, optical switching systems, and integrated solar energy conversion systems. The details are presented in Chapters 10 and 11.

2.2.3 *Self-Organized 3-D Integrated Optical Circuits*

In future OE systems such as optical interconnects within boxes, massive optical switching systems, and integrated solar energy conversion systems, a large number of optical couplings in optical circuits will be involved. In such cases, enormous alignment efforts with micron or submicron accuracy are required, raising system fabrication costs. Vertical waveguide formation for the 3-D optical circuits is another issue. In order to solve the problems, a concept of self-organized 3-D integrated optical circuit shown in Fig. 2.10 is proposed. After a precursor with optical devices distributed

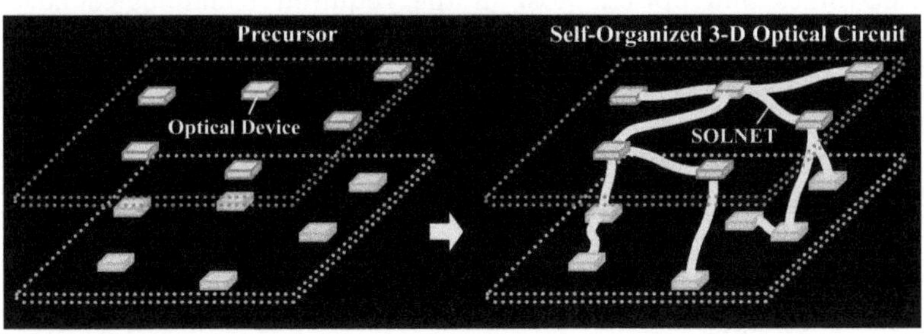

Figure 2.10. Concept of the self-organized 3-D integrated optical circuit. Reprinted with permission from T. Yoshimura and H. Kaburagi, "Self-Organization of Optical Waveguides Between Misaligned Devices Induced by Write-Beam Reflection," *Appl. Phys. Express* **1**, 062007- 1-3 (2008).

three-dimensionally is built, optical waveguides are formed between the devices automatically in a self-aligned manner by write beams emitted from the optical devices to construct the self-organized 3-D optical circuit.

To realize this concept, SOLNET can be used. SOLNET, which is discussed in detail in Chapters 4 and 5, is a technology that self-organizes optical waveguides in photo-induced refractive index increase (PRI) materials such as photo-polymers as follows: Optical devices such as optical waveguide, optical fiber, LD, optical switch, and free space light beam input are placed in the PRI material whose refractive index increases with write beam exposure. By the introduction of write beams into the PRI material from the optical devices, the refractive index distribution of the PRI material changes from time to time. The write beam propagation is affected by the refractive index distribution, which is generated by the write beams themselves. The overlapping write beams are attracted to each other in the PRI material, and they merge into one optical waveguide, constructing self-aligned coupling waveguides connecting the optical devices automatically. Thus, SOLNET is expected to realize self-organized 3-D OE platforms.

2.3 Core Technologies in Optical Electronics

Optical electronics contains a variety of technologies related to electronics and optics. Most of the required technologies have already been established. However, the following core technologies for optical electronics have not been established yet:

- Self-organization of optical waveguides for self-aligned optical couplings and the construction of 3-D optical wiring.
- Optical waveguide films with vertical mirrors and 3-D stacking of the optical waveguide films for 3-D optical circuits.
- Resource-saving heterogeneous integration processes for thin-film device flakes.
- High-speed/small-size light modulators and optical switches.

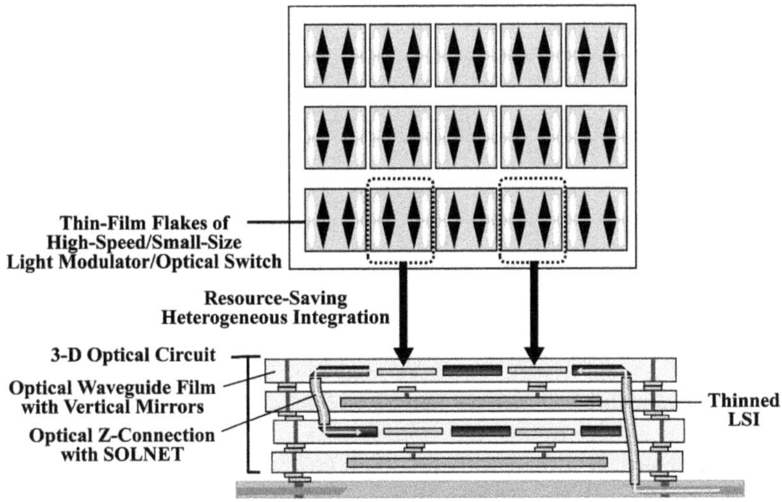

Figure 2.11. Core technologies for the self-organized 3-D OE platform.

As an example of the self-organized 3-D OE platform built by the core technologies described above, a structure of 3-D stacked OE LSI described in Section 10.3 is shown in Fig. 2.11.

Optical waveguide films with vertical mirrors, in which thin-film flakes of high-speed/small-size light modulators and optical switches are embedded together with other devices such as PDs, are stacked with films containing thinned LSIs to construct 3-D architectures. The optical waveguide films can be fabricated by the built-in mask method described in Chapter 6, for example, and stacked each other as described in Chapter 7. The thin-film flakes of the devices are embedded into the films by resource-saving heterogeneous integration using the PL-Pack with SORT described in Chapter 8. Optical Z-connections for inter-film optical link can be fabricated by vertical waveguides of SOLNET to form self-organized 3-D optical circuits (Section 10.2). SOLNET is also used for optical couplings between optical waveguides and embedded optical devices (Section 10.1).

High-speed/small-size light modulators and optical switches consist of optical waveguides with core widths of several-microns or nano-scale. As described in Chapter 9, the light modulator/optical switch performance will be improved by using high-performance electro-optic (EO) materials such as lead lanthanum zirconate titanate (PLZT), quantum dots, and polymer multiple quantum dots (MQDs) made by the molecular layer deposition (MLD) described in Sections 12.1 and 12.2.

References

1. K. Takahashi, "Research and Development on Ultra-High-Density 3-Dimensional LSI-Chip-Stack Packaging Technologies," The 3rd Annual Meeting on Electronics System Integration Technologies Digest (edited and published by the Electronic System Integration Technology Research Department, Association of Super-Advanced Electronics Technologies (ASET)), 43–94 (2002).
2. D.A.B. Miller, "How Large a System Can We Build Without Optics?" Workshop Notes, 8th Annual Workshop on Interconnections Within High Speed Digital Systems, Lecture 1.2 (1997).
3. H. Takahara, "Optoelectronic Multichip Module Packaging Technologies and Optical Input/Output Interface Chip-Level Packages for the Next Generation of Hardware Systems," *IEEE J. Select. Top. Quantum Electron.* **9**, 443–451 (2003).
4. T. Mikawa, M. Kinoshita, K. Hiruma, T. Ishitsuka, M. Okabe, S. Hiramatsu, H. Furuyama, T. Matsui, K. Kumai, O. Ibaragi, and M. Bonkohara, "Implementation of Active Interposer for High-Speed and Low-Cost Chip Level Optical Interconnects," *IEEE J. Select. Top. Quantum Electron.* **9**, 452–459 (2003).
5. R. Pu, C. Duan, and C. Wilmsen, "Hybrid Integration of VCSEL's to CMOS Integrated Circuits," *IEEE J. Select. Top. Quantum Electron.* **5**, 201–208 (1999).
6. N. M. Jokerst, M. A. Brooke, S. Cho, S. Ilkinson, M. Vrazel, S. Fike, J. Tabler, Y. J. Joo, S. Seo, D. S. Wils, and A. Brown, "The Heterogeneous Integration of Optical Interconnections into Integrated Microsystems," *IEEE J. Select. Top. Quantum Electron.* **9**, 350–360 (2003).
7. E. Yablonovitch, T. Gmitter, J.P. Harbison, and R. Bhat, "Extreme Sensitivity in the Lift-Off of Epitaxial GaAs Films," *Appl. Phys. Lett.* **51**, 2222–2224 (1987).

8. T. Yoshimura, J. Roman, Y. Takahashi, S. Beilin, W. Wang, and M. Inao, 4th International Conference on Organic Nonlinear Optics (ICONO'4), Extended Abstracts, Chitose, Japan, 278 (1998).
9. T. Yoshimura, J. Roman, Y. Takahashi, S. I. Beilin, W. V. Wang, and M. Inao, "Optoelectronic Amplifier/Driver-Less Substrate <OE-ADLES> for Polymer-Waveguide-Based Board-Level Interconnection — Calculation of Delay and Power Dissipation," *Nonlinear Opt.* **22**, 453–456 (1999).
10. T. Yoshimura, J. Roman, Y. Takahashi, M. Lee, B. Chou, S. I. Beilin, W. V. Wang, and M. Inao, "Proposal of Optoelectronic Substrate with Film/Z-Connection Based on OE-Film," *Proc. 3rd IEMT/IMC Symposium*, 140–145 (1999).
11. T. Yoshimura, J. Roman, Y. Takahashi, M. Lee, B. Chou, S. I. Beilin, W. V. Wang, and M. Inao, "Optoelectronic Scalable Substrates Based on Film/Z-connection and Its Application to Film Optical Link Module (FOLM)," *Proc. SPIE* **3952**, from Photonics West 2000, San Jose, California, 202–213 (2000).
12. T. Yoshimura, J. Roman, P. Reilly, Y. Takahashi, M. Lee, B. Chou, S. I. Beilin, W. V. Wang, and M. Inao, "Proposal of Film Optical Link Module (FOLM) and Beveled Cut Process for the High Density Waveguide Couplers," *Proc. 1999 IEEE LEOS Annual Meeting Conference*, San Francisco, California, 27–28 (1999).
13. C. Choi, L. Lin, Y. Liu, J. Choi, L. Wang, D. Haas, J. Magera, and R. T. Chen, "Flexible Optical Waveguide Film Fabrications and Optoelectronic Devices Integration for Fully Embedded Board-Level Optical Interconnects," *J. Lightwave Technol.* **22**, 2168–2176 (2004).
14. T. Yoshimura, Y. Takahashi, M. Inao, M. Lee, W. Chou, S. Beilin, W.-C. Wang, J. Roman, T. Massingill, "Systems Based on Opto-Electronic Substrates with Electrical and Optical Interconnections and Methods for Making," U.S. Patent 6,343,171 B1 (2002).
15. T. Yoshimura, Y. Takahashi, M. Inao, M. Lee, W. Chou, S. Beilin, W.-C. Wang, J. Roman, T. Massingill, "Optoelectronic Substrates with Electrical and Optical Interconnections and Method for Making," U.S. Patent 6,611,635 B1 (2003).
16. T. Yoshimura, Y. Takahashi, J. Roman, M. T. McCormack, S. Beilin, W.-C. Wang, M. Inao, "Optical Coupling Structures and the Fabrication Processes," U.S. Patent 6,684,007 B2 (2004).

17. T. Yoshimura, Y. Takahashi, M. Inao, M. Lee, W. Chou, S. Beilin, W.-C. Wang, J. Roman, T. Massingill, "Three-Dimensional Opto-Electronic Modules with Electrical and Optical Interconnections and Methods for Making," U.S. Patent 6,690,845 B1 (2004).
18. T. Yoshimura, Y. Takahashi, K. Kuwabara, S. Beilin, M. Peters, W.-C. Wang, M. Inao, "Optical Reflective Structures and Method for Making," U.S. Patent 6,706,546 B2 (2004).
19. T. Yoshimura, Y. Takahashi, J. Roman, M. T. McCormack, S. Beilin, W.-C. Wang, M. Inao, "Single and Multilayer Waveguides and Fabrication Process," U.S. Patent 6,785,447 B2 (2004).
20. T. Yoshimura, Y. Takahashi, M. Inao, M. Lee, W. Chou, S. Beilin, W.-C. Wang, J. Roman, T. Massingill, "Multi-Layer Opto-Electronic Substrates with Electrical and Optical Interconnections and Methods for Making," U.S. Patent 6,845,184 B1 (2005).
21. T. Yoshimura, K. Kumai, T. Mikawa, O. Ibaragi, and O. Bonkohara, "Photolithographic Packaging with Selectively Occupied Repeated Transfer (PL-Pack with SORT) for Scalable Optical Link Multi-Chip-Module (S-FOLM)," *IEEE Trans. Electron. Packag. Manuf.* **25**, 19–25 (2002).
22. T. Yoshimura, M. Ojima, Y. Arai, and K. Asama, "Three-Dimensional Self-Organized Micro Optoelectronic Systems for Board-Level Reconfigurable Optical Interconnects — Performance Modeling and Simulation," *IEEE J. Select. Top. Quantum Electron.* **9**, 492–511 (2003).
23. T. Yoshimura, J. Roman, W.-C. Wang, M. Inao, M. T. McCormack, "Device Transfer Method," U.S. Patent 6,669,801 B2 (2003).
24. T. Yoshimura, Y. Arai, "3D Optoelectronic Micro System," U.S. Patent 7,387,913 B2 (2008).
25. T. Yoshimura, J. Roman, Y. Takahashi, W. V. Wang, M. Inao, T. Ishitsuka, K. Tsukamoto, K. Motoyoshi, and W. Sotoyama, "Self-Organizing Waveguide Coupling Method 'SOLNET' and Its Application to Film Optical Circuit Substrates," *Proc. 50th Electron Comp. Technol. Conf. (ECTC)* 962–969 (2000).
26. T. Yoshimura, J. Roman, Y. Takahashi, W. V. Wang, M. Inao, T. Ishituka, K. Tsukamoto, K. Motoyoshi, and W. Sotoyama, "Self-Organizing Lightwave Network (SOLNET) and Its Application to Film Optical Circuit Substrates," *IEEE Trans. Comp., Packag. Technol.* **24**, 500–509 (2001).
27. T. Yoshimura, W. Sotoyama, K. Motoyoshi, T. Ishitsuka, K. Tsukamoto, S. Tatsuura, H. Soda, T. Yamamoto, "Method of Producing Optical Waveguide System, Optical Device and Optical Coupler Employing the

Same, Optical Network and Optical Circuit Board," U.S. Patent 6,081,632 (2000).

28. T. Yoshimura, T. Inoguchi, T. Yamamoto, S. Moriya, Y. Teramoto, Y. Arai, T. Namiki, and K. Asama, "Self-Organized Lightwave Network Based on Waveguide Films for Three-Dimensional Optical Wiring Within Boxes," *J. Lightwave Technol.* **22**, 2091–2100 (2004).
29. B. Chou, S. Beilin, H. Jiang, D. Kudzuma, M. Lee, M. McCormack, T. Massingill, M. Peters, J. Roman, Y. Takahashi, and W. V. Wang, "Multilayer high density flex technology," *Proc. 49th Electron. Comp. Technol. Conf. (ECTC)* 1181–1189 (1999).

Chapter 3

Analysis Tools for Optical Circuits

In order to understand and design optical circuits, light beam propagation in optical waveguides that are the basic element of the optical circuits should be analyzed. The waveguide operation can easily be understood by considering the electron confinement in quantum wells. Figures 3.1 (a) and (b), respectively, show the analogy between the electron confinement and the photon confinement, and a schematic illustration of light beam confinement into an optical waveguide. It is known that the wavefunction of electron ψ and the electric field of light waves $\mathbf{E}$ are determined by the following wave equations:

$$\left(-\frac{\hbar^2}{2m}\nabla^2 + V(\mathbf{x})\right)\psi(\mathbf{x}) = E\psi(\mathbf{x}) \tag{3.1}$$

$$\left(\nabla^2 + (n(\mathbf{x})\,k_0)^2\right)\mathbf{E}(\mathbf{x}) = 0. \tag{3.2}$$

Here, m, V and E are mass, potential energy, and total energy of an electron, respectively; $\hbar$ is Plank constant divided by 2π; n is refractive index and k_0 is wavenumber in vacuum. Equations (3.1) and (3.2) can be written as follows:

$$\left.\begin{array}{l} \left(-\frac{\hbar^2}{2m}\nabla^2 - (E - V(\mathbf{x}))\right)\psi(\mathbf{x}) = 0 \\ \left(-\frac{\hbar^2}{2m}\nabla^2 - \frac{\hbar^2 k_0^2}{2m}n^2(\mathbf{x})\right)\mathbf{E}(\mathbf{x}) = 0 \end{array}\right\} \tag{3.3}$$

Optical Electronics: Self-Organized Integration and Applications
Tetsuzo Yoshimura

ISBN 978-981-4310-90-1 (Hardcover), 978-981-4364-08-9 (eBook)
www.panstanford.com

Comparing the two equations in (3.3), it is found that the following relationships exist:

$$\psi(\mathbf{x}) \leftrightarrow \mathbf{E}(\mathbf{x}) \tag{3.4}$$

$$E - V(\mathbf{x}) \leftrightarrow \frac{\hbar^2 k_0^2}{2m} n^2(\mathbf{x}) \tag{3.5}$$

It is known that an electron tends to be confined in a region with small potential energy $V(\mathbf{x})$ as shown in Fig. 3.1(a). The relationship of Eq. (3.5) indicates that "$-V(\mathbf{x})$" corresponds to "$n^2(\mathbf{x})$." This suggests that a photon tends to be confined in a region with high refractive index. Therefore, by constructing line-shaped regions with higher refractive index in a planar substrate, light beams are confined in the line as shown in Fig. 3.1(b). The line is the core of the optical waveguide, and the surrounding area is clad. By constructing higher-refractive-index regions in a medium with designated patterns, light beams can be guided in the designated routing to make optical circuits.

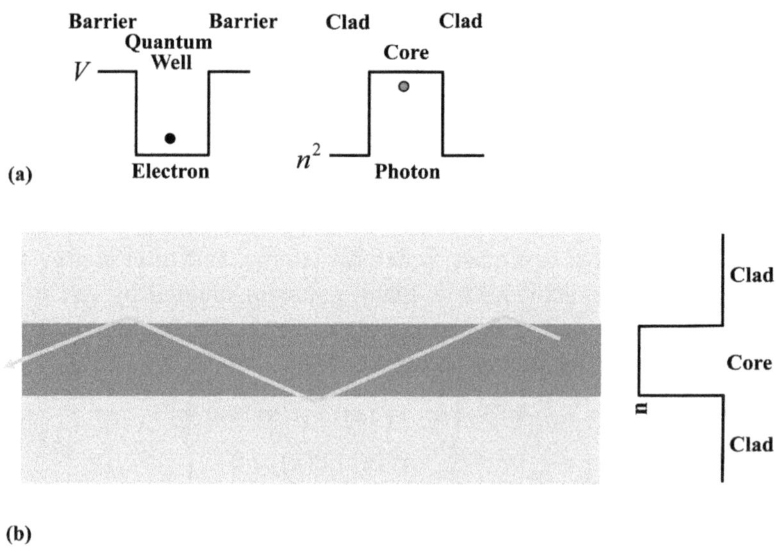

Figure 3.1. (a) Analogy between the electron confinement and the photon confinement, and (b) schematic illustration of light beam confinement into an optical waveguide.

For the analysis of the optical circuits, the beam propagation method (BPM) and the finite difference time domain (FDTD) method are useful. BPM simulates light beam propagation along the propagation axis. Since BPM can perform high-speed calculations with a small memory capacity, it is suitable for large-size models. However, BPM cannot treat off-axis light beam propagation such as reflected light beams and wide-angle light beams. FDTD method, on the other hand, can treat any light beams, including reflected and wide-angle light beams, although it requires a large memory capacity and long calculation time. Therefore, it is necessary for us to use these two methods appropriately depending on the models.

This chapter summarizes BPM using the Fourier transform and the FDTD method.

3.1 Beam Propagation Method (BPM)

In Fig. 3.2, a schematic illustration for simulation of light beam propagation by BPM is shown. Waveguide cores have refractive index (n_1) higher than that of the surrounding clad (n_2). Electric fields of light beams introduced into the calculation region are successively calculated along the propagation axis (z-axis) with a step of Δz to simulate the light beam propagation following the procedure mentioned further in the text.

Figure 3.3(a) shows a model for calculations by BPM using the Fourier transform [1].

The whole calculation region with a width of W is regarded as a virtual waveguide with a width of W. Virtual modes in the virtual waveguide schematically shown in Fig 3.3(b) are expressed as follows:

$$E_\nu(x,z) = F_\nu \sin K_\nu x \cdot e^{-in_2k_0z},$$
$$F_\nu = \sqrt{\frac{4\mu_0\omega}{n_2k_0W}},\ K_\nu = \frac{\pi\nu}{W},\ \nu = 1,\ 2,\ 3,\dots \qquad (3.6)$$

Here, ν represents the mode number and ω angular frequency of the light. μ_0 and k_0 are respectively magnetic permeability and wavenumber in vacuum. The electric field distributions in the calculation region, denoted by E^0, can be expressed by the

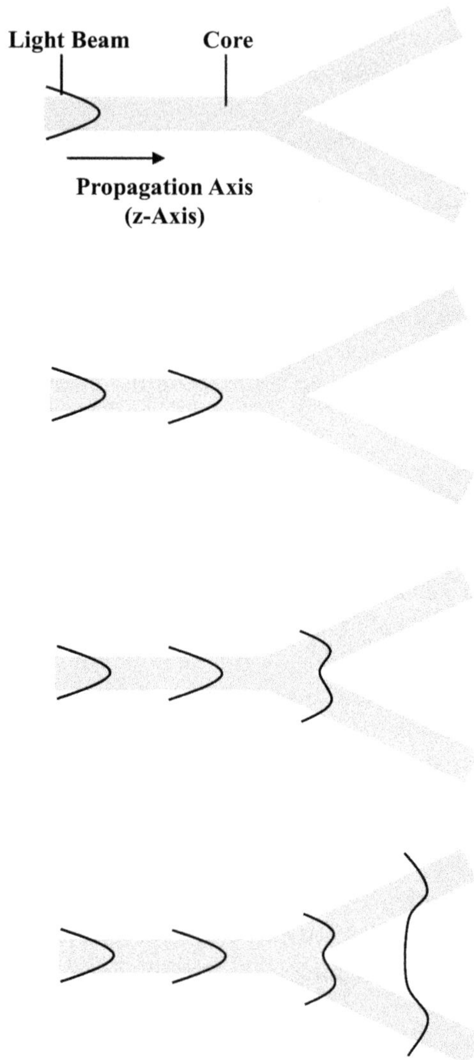

Figure 3.2. Schematic illustration of light beam propagation simulated by BPM.

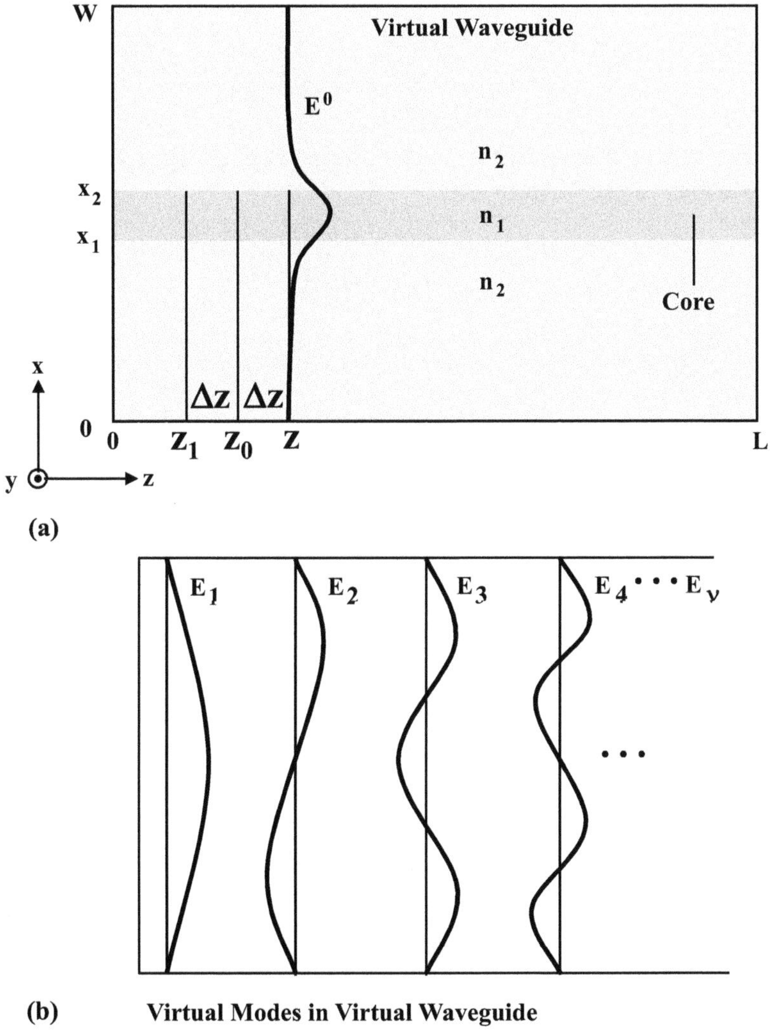

Figure 3.3. (a) A model for calculations by BPM. (b) Virtual modes in a virtual waveguide.

superposition of the electric field distributions of the virtual modes as follows:

$$E^0(x,z) = \sum_{\nu=1}^{\infty} A_\nu(z) E_\nu(x,z) \tag{3.7}$$

Here, $A_\nu(z)$ is the weight coefficient for each $E_\nu(x, z)$. $A_\nu(z)$ and its derivative can be written as

$$A_\nu(z) = \frac{\int_0^W E_\nu^*(x,z)E^0(x,z)dx}{\int_0^W E_\nu^*(x,z)E_\nu(x,z)dx}, \tag{3.8}$$

$$\begin{aligned}\frac{\mathrm{d}A_\mu(z)}{\mathrm{d}z} = &\left[n_2^2k_0^2\sum_{\nu=1}^{M} A_\nu(z)\int_0^W f(x,z)E_\mu^*(x,z)E_\nu(x,z)dx\right.\\ &\left.-K_\mu^2 A_\mu(z)\int_0^W E_\mu^*(x,z)E_\mu(x,z)dx\right]\Big/\\ &\left[i2n_2k_0\int_0^W E_\mu^*(x,z)E_\mu(x,z)dx\right].\end{aligned} \tag{3.9}$$

$f(x,z)$ is given by

$$f(x,z) = \frac{n^2(x,z)}{n_2^2} - 1, \tag{3.10}$$

where, $n(x,z)$ is refractive index in the calculation region.

Figure 3.4 shows the calculation procedure in BPM. The weight coefficient at z_{n+2} is determined by the information derived in the

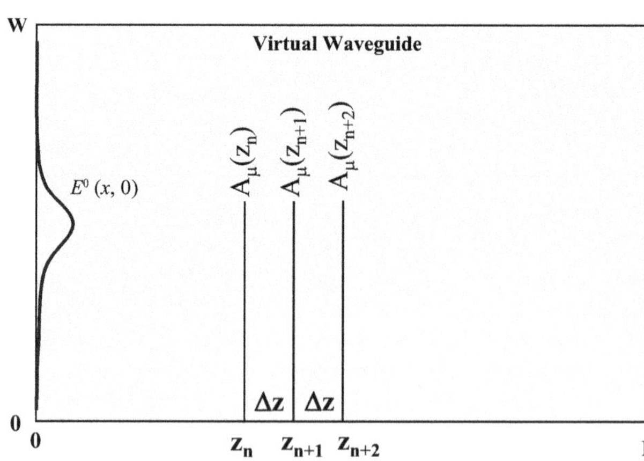

Figure 3.4. Calculation procedure in BPM.

previous steps, namely, the weight coefficient at z_n and slope of the weight coefficient along z-axis at z_{n+1}, by the following equation:

$$A_\mu(z_{n+2}) = A_\mu(z_n) + \left.\frac{\mathrm{d}A_\mu(z)}{\mathrm{d}z}\right|_{z=z_{n+1}} (2\Delta z) \tag{3.11}$$

By repeating the procedure from $z = 0$ to L, which is the end point of the calculation region, $A_\nu(z)$ at $z = 0$ to L is calculated. Finally, by substituting $A_\nu(z)$ into Eq. (3.7), electric field distributions all over the calculation region can be obtained.

3.2 Finite Difference Time Domain (FDTD) Method

In FDTD method, electric fields $\mathbf{E}$ and magnetic fields $\mathbf{H}$ are calculated by solving Maxwell's equations directly. Following are the starting equations:

$$\frac{\partial \mathbf{E}}{\partial t} = -\frac{\sigma}{\varepsilon}\mathbf{E} + \frac{1}{\varepsilon}\nabla \times \mathbf{H} \tag{3.12}$$

$$\frac{\partial \mathbf{H}}{\partial t} = -\frac{1}{\mu}\nabla \times \mathbf{E} \tag{3.13}$$

Here, σ, ε, and μ, are conductivity, dielectric constant, and magnetic permeability, respectively. Difference equations for the differential equations (3.12) and (3.13) can be written as follows:

$$\mathbf{E}^n = \frac{\left(1 - \frac{\sigma\Delta t}{2\varepsilon}\right)}{\left(1 + \frac{\sigma\Delta t}{2\varepsilon}\right)}\mathbf{E}^{n-1} + \frac{\frac{\Delta t}{\varepsilon}}{\left(1 + \frac{\sigma\Delta t}{2\varepsilon}\right)}\nabla \times \mathbf{H}^{n-\frac{1}{2}} \tag{3.14}$$

$$\mathbf{H}^{n+\frac{1}{2}} = \mathbf{H}^{n-\frac{1}{2}} - \frac{\Delta t}{\mu}\nabla \times \mathbf{E}^n \tag{3.15}$$

Here, n represents time steps, meaning that $t = n\Delta t$ when time interval for a step is Δt.

Using Eq. (3.14), electric fields at t are determined by information at previous time, namely, electric fields at $t - \Delta t$ and magnetic fields at $t - (1/2)\Delta t$. Similarly, using Eq. (3.15), magnetic fields at $t + (1/2)\Delta t$ are determined by magnetic fields at $t - (1/2)\Delta t$ and electric fields at t. The calculation sequences in FDTD method is summarized in Fig. 3.5.

In actual calculations, we need to modify Eqs. (3.14) and (3.15) to give difference equations for positions in addition to time. For

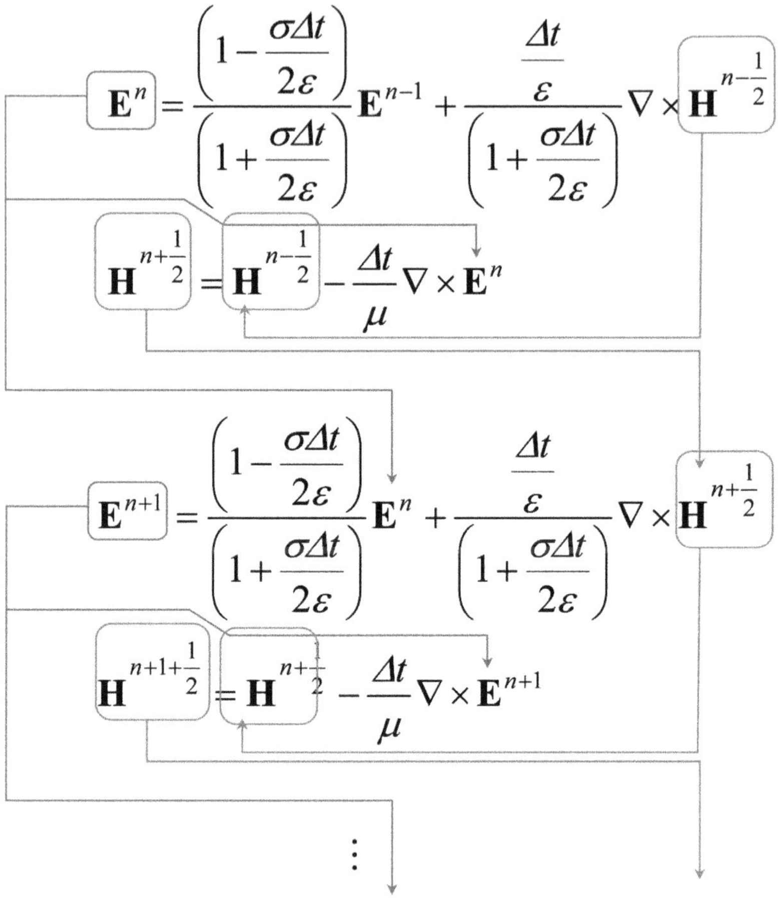

Figure 3.5. Calculation sequences in FDTD method.

two-dimensional models in the x–y plane with polarization //z, the difference equations for E_z^n, $H_x^{n+(1/2)}$, and $H_y^{n+(1/2)}$ are given as follows:

$$
\begin{aligned}
E_z^n(i,j) &= C_{EZ}(i,j)E_z^{n-1} + C_{EZIX}(i,j)\left(H_y^{n-(1/2)}\left(i+\frac{1}{2},j\right)\right. \\
&\quad \left. -H_y^{n-(1/2)}\left(i-\frac{1}{2},j\right)\right) - C_{EZIY}(i,j)\left(H_x^{n-(1/2)}\left(i,j+\frac{1}{2}\right)\right. \\
&\quad \left. -H_x^{n-(1/2)}\left(i,j-\frac{1}{2}\right)\right)
\end{aligned}
\tag{3.16}
$$

$$C_{EZ}(i,j) = \frac{\left(1 - \frac{\sigma(i,j)\Delta t}{2\varepsilon(i,j)}\right)}{\left(1 + \frac{\sigma(i,j)\Delta t}{2\varepsilon(i,j)}\right)} \quad C_{EZIX}(i,j) = \frac{\frac{\Delta t}{\varepsilon(i,j)}}{\left(1 + \frac{\sigma(i,j)\Delta t}{2\varepsilon(i,j)}\right)} \frac{1}{\Delta x}$$

$$C_{EZIY}(i,j) = \frac{\frac{\Delta t}{\varepsilon(i,j)}}{\left(1 + \frac{\sigma(i,j)\Delta t}{2\varepsilon(i,j)}\right)} \frac{1}{\Delta y} \tag{3.17}$$

$$H_x^{n+(1/2)}\left(i, j+\frac{1}{2}\right) = H_x^{n-(1/2)}\left(i, j+\frac{1}{2}\right) - C_{HXIY}\left(i, j+\frac{1}{2}\right) \times \left(E_z^n(i, j+1) - E_z^n(i,j)\right) \tag{3.18}$$

$$C_{HXIY}\left(i, j+\frac{1}{2}\right) = \frac{\Delta t}{\mu\left(i, j+\frac{1}{2}\right)} \frac{1}{\Delta y} \tag{3.19}$$

$$H_y^{n+(1/2)}\left(i+\frac{1}{2}, j\right) = H_y^{n-(1/2)}\left(i+\frac{1}{2}, j\right) + C_{HYIX}\left(i+\frac{1}{2}, j\right) \times \left(E_z^n(i+1, j) - E_z^n(i,j)\right) \tag{3.20}$$

$$C_{HYIX}(i+\frac{1}{2}, j) = \frac{\Delta t}{\mu(i+\frac{1}{2}, j)} \frac{1}{\Delta x} \tag{3.21}$$

Here, i and j represent positional steps along the x-axis and the y-axis, respectively, meaning that $x = i\Delta x$ and $y = j\Delta y$. As shown in Fig. 3.6(a), $E_z\,(i, j)$ at time step n can be determined by surrounding $H_x(i, j - (1/2))$, $H_x(i, j + (1/2))$, $H_y(i - (1/2), j)$ and $H_y(i + (1/2), j)$ as well as $E_z(i, j)$ at previous time steps. For $H_x(i, j + (1/2))$ at time step $n + (1/2)$, as shown in Fig. 3.6(b), it is determined by $E_z(i, j)$ and $E_z(i, j+1)$ as well as $H_x(i, j + (1/2))$ at previous time steps. For $H_y(i + (1/2), j)$ at time step $n + (1/2)$, as shown in Fig. 3.6(c), it is determined by $E_z(i, j)$ and $E_z(i+1, j)$ as well as $H_y(i + (1/2), j)$ at previous time steps. By repeating the procedure, electric field and magnetic field distributions all over the calculation region can be obtained as a function of time.

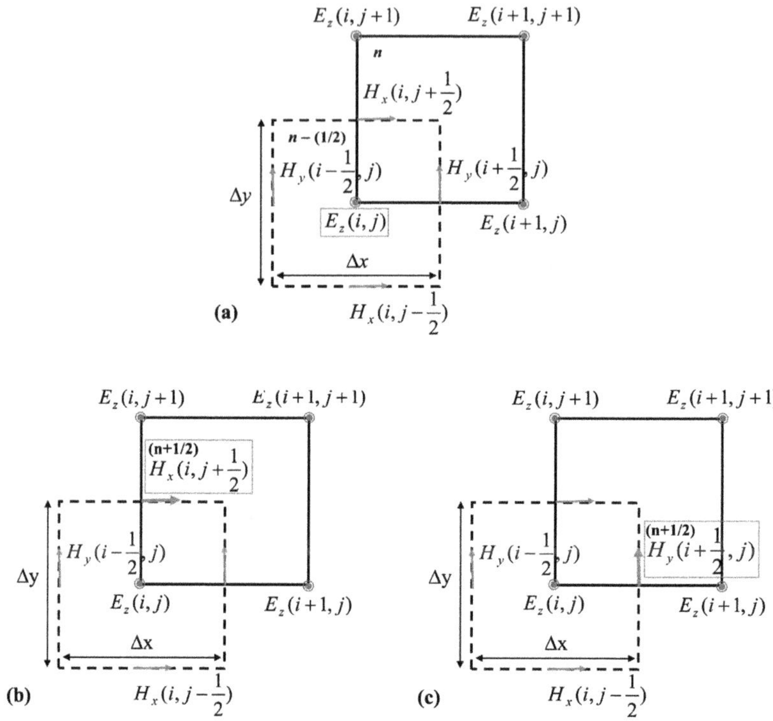

Figure 3.6. Diagrams for calculations of (a) E_z, (b) H_x, and (c) H_y.

Reference

1. M. Seino, H. Nakajima, Y. Daido, I Sawaki, and K. Asama, "Optical Waveguide Analysis Using the Fourie Transform and Its Application to Intersecting Waveguides," *IEICE Ronbunshi* **J-66C**, 732–739 (1985) [in Japanese].

Chapter 4

Self-Organized Optical Waveguides: Theoretical Analysis

Self-organized lightwave network (SOLNET) [1–9] is a novel optical waveguide formation method, which utilizes an attractive force induced between light beams in photo-induced refractive index increase (PRI) materials. SOLNET, which is one of the core technologies in optical electronics, realizes self-aligned optical couplings between optical devices, tolerating the mutual positional misalignment. SOLNET can also form straight and even downtapered optical waveguides in a free space, enabling 3-D optical wiring.

This chapter describes the concept of SOLNET and reviews the theoretical analysis that was carried out by the beam propagation method (BPM) and the the finite difference time domain (FDTD) method.

4.1 Concept of Self-Organized Lightwave Network

The concept of SOLNET is shown in Fig. 4.1. In one-beam-writing SOLNET, a write beam is introduced into a PRI material from an optical device such as an optical waveguide, an optical fiber, a laser diode (LD), etc. In the PRI material, refractive index increases by write beam exposure. When the write beam propagates in the PRI

Optical Electronics: Self-Organized Integration and Applications
Tetsuzo Yoshimura

ISBN 978-981-4310-90-1 (Hardcover), 978-981-4364-08-9 (eBook)
www.panstanford.com

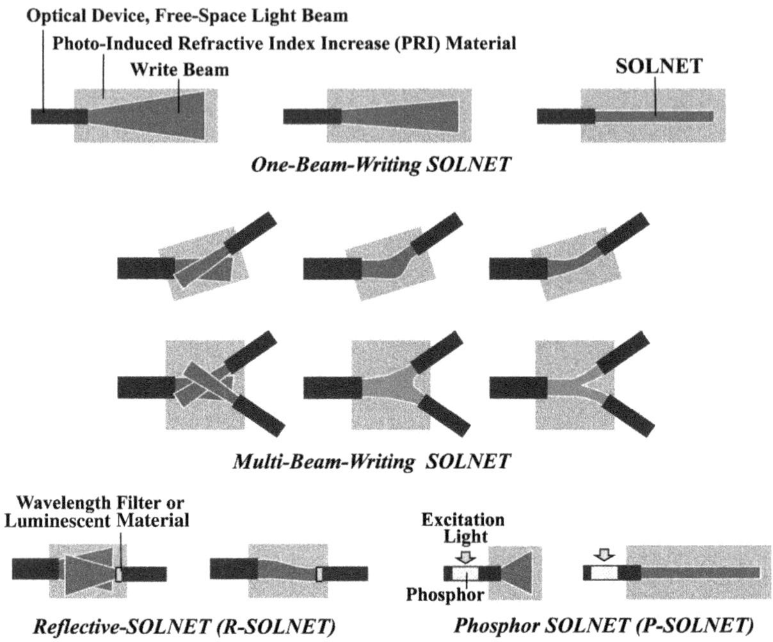

Figure 4.1. Concept of SOLNET. See also Color Insert.

material, the refractive index distribution changes from time to time. Since the write beam propagation is affected by the refractive index distribution, which is generated by the write beam itself, the write beam is concentrated along the propagation axis, that is, the self-focusing is induced. The self-focusing constructs an optical waveguide from the optical device with a mode size close to that of the optical device. The optical waveguide constructed in the PRI material is the SOLNET. Such phenomenon also occurs by the introduction of write beams as free space light beams.

In multi-beam-writing SOLNET, two or more optical devices are put into a PRI material. In Fig. 4.1, the cases of two-beam-writing SOLNET and three-beam-writing SOLNET are drawn. For two-beam-writing SOLNET, two optical devices are put in counter directions and the PRI material is placed between them. Write beams are introduced into the PRI material from both of two optical devices. The write beams are attracted to each other in the PRI material and merge into an optical path by self-focusing to construct a self-aligned coupling waveguide between the optical devices automatically. The self-aligned waveguides can be formed even if the optical devices are misaligned and have different core sizes. The same phenomenon occurs in three-beam-writing SOLNET, connecting three optical devices automatically.

Reflective SOLNET (R-SOLNET) can simplify the multi-beam-writing SOLNET fabrication process. In R-SOLNET, some of the write beams in the multi-beam-writing SOLNET are replaced with reflected write beams from reflective elements such as wavelength filters and mirrors on core facets of the optical devices. A write beam from an optical waveguide and the reflected write beam from a wavelength filter on a facet of the other optical waveguide overlap near the wavelength filter. In the overlap region, the refractive index of the PRI material increases, pulling the write beam to the wavelength filter location more and more. We call this effect the "pulling water" effect. Finally, by self-focusing, a self-aligned coupling waveguide is formed between the optical waveguide and the other optical waveguide.

In R-SOLNET, the wavelength filter can be replaced with an luminescent material. When a write beam from an optical waveguide is introduced onto the luminescent material through the PRI material, the write beam is absorbed by the luminescent material followed by photo-luminescence luminescent from the material. The luminescence acts like a reflected write beam to induce the "pulling water" effect. Consequently, R-SOLNET grows between the optical waveguide and the target site with the luminescent material.

In phosphor SOLNET (P-SOLNET), phosphor is doped in a part of the optical waveguide. By exposing the doped phosphor to excitation light, write beams generated from the phosphor propagate in the optical waveguide to be emitted into the PRI material. P-SOLNET is effective when write beams cannot be introduced from outside; for example, when SOLNET is formed in inner parts of 3-D OE platforms.

4.2 Photo-Induced Refractive Index Increase (PRI) Materials

The PRI materials are, for example, photo-polymers, photo-definable materials, photo-refractive crystals, etc. Figure 4.2(a) shows the mechanism of refractive-index-distribution construction in the

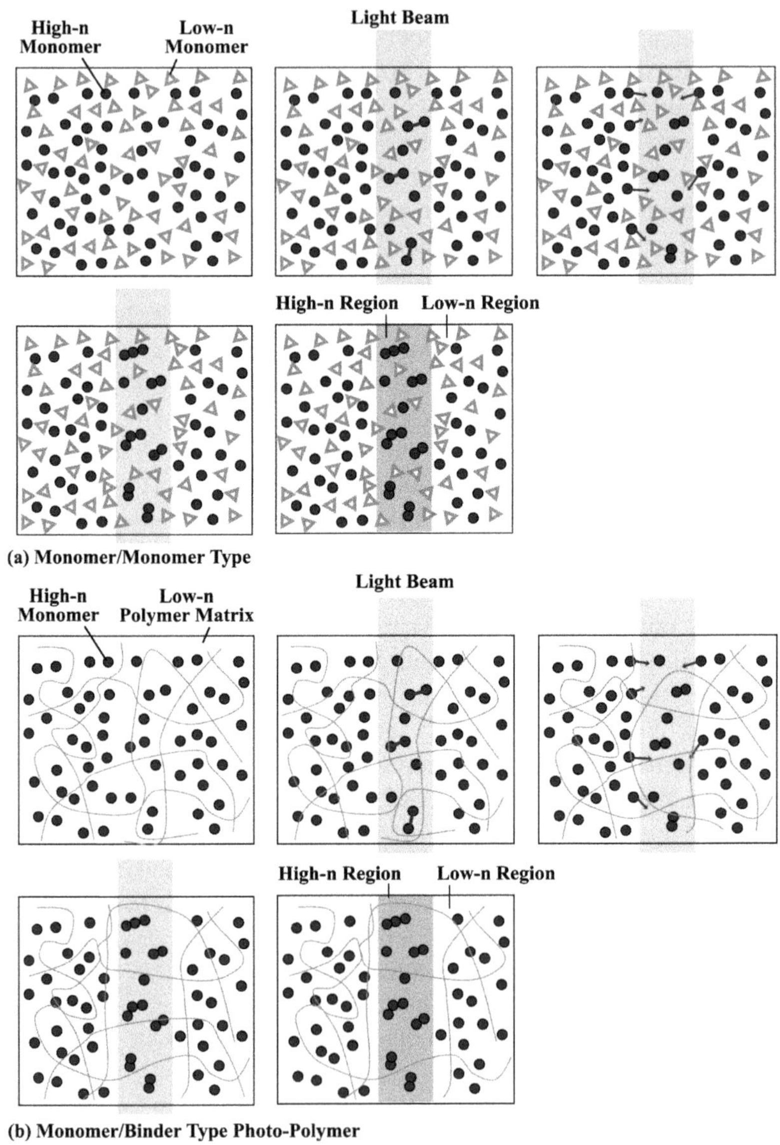

Figure 4.2. Mechanism of refractive-index-distribution construction in photo-polymers: (a) Monomer/monomer type. (b) Monomer/binder type.

monomer/monomer-type photo-polymer. In the material, high-refractive-index monomers and low-refractive-index monomers are mixed. The high-refractive-index monomers have higher photo-reactivity to write beams than the low-refractive-index monomers. When the photo-polymer is exposed to the write beam, the high-refractive-index monomers are combined to make dimers, oligomers, or polymers by photo-chemical reactions. Then, the high-refractive-index monomers diffuse into the exposed region from surrounding regions to compensate for the reduction of the high-refractive-index monomer concentration in the exposed region. This process increases the refractive index of the exposed region, resulting in the PRI effect. The wavelength of the write beams typically ranges from 350 to 800 nm. The spectral response can be adjusted by sensitizers.

Figure 4.2(b) shows the mechanism in the monomer/binder-type photo-polymer. In this case, high-refractive-index monomers are dispersed in a low-refractive-index polymer matrix. Similar to the case of the monomer/monomer-type photo-polymer, refractive index of the exposed region increases, resulting in the PRI effect.

The gamma characteristics, namely, the relationship between the refractive index change and the write beam exposure, can be controlled by adjusting the refractive index, photo-reactivity, and concentration of monomers. Precise control of the gamma characteristics might be possible by mixing two, three, or more kinds of monomers with different refractive index and photo-reactivity into the PRI materials.

4.3 Simulation of SOLNET by BPM

4.3.1 *One-Beam-Writing SOLNET*

4.3.1.1 Procedure

Figure 4.3 shows diagrams of simulation procedure for one-beam-writing SOLNET [1, 2, 4] using two-dimensional BPM with paraxial approximation. Two optical waveguides are arranged in counter directions with photo-polymer filling in between. In step 0, an initial

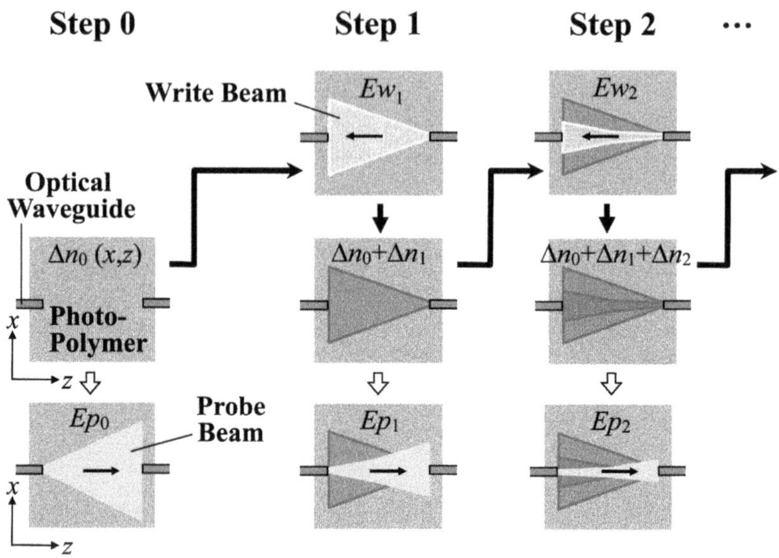

Figure 4.3. BPM simulation procedure for one-beam-writing SOLNET. Reprinted with permission from T. Yoshimura, J. Roman, Y. Takahashi, W. V. Wang, M. Inao, T. Ishitsuka, K. Tsukamoto, K. Motoyoshi, and W. Sotoyama, "Self-Organizing Waveguide Coupling Method "SOLNET" and Its Application to Film Optical Circuit Substrates," *Proc. 50th Electron. Components Technol. Conf. (ECTC)*, 962–969. © 2000 IEEE.

refractive index distribution is given as follows:

$$n(x, z) = n_0 + \Delta n_0(x, z),$$

where n_0 is a constant background refractive index. In Fig 4.3, only varying terms (Δn_i) are noted. A probe beam introduced from the left-side optical waveguide propagates into the photo-polymer with electric field distribution of E_{p_0} (x, z) under an influence of Δn_0. In step 1, a write beam is introduced from the right-side optical waveguide into the photo-polymer. The write beam propagates with electric field distribution of Ew_1 under an influence of Δn_0 to generate additional refractive index change Δn_1, resulting in a new distribution $\Delta n_0 + \Delta n_1$. A probe beam introduced from the left-side optical waveguide propagates with electric field distribution of E_{p_1} under $\Delta n_0 + \Delta n_1$. In step 2, a write beam propagates with electric field distribution of Ew_2 under an influence of refractive index distribution $\Delta n_0+\Delta n_1$, resulting in a distribution $\Delta n_0+\Delta n_1+\Delta n_2$. A probe beam propagates with electric field distribution of E_{p_2} under $\Delta n_0+\Delta n_1+\Delta n_2$. By repeating the procedure, simulations of one-beam-writing SOLNET are accomplished.

It is assumed in the situation that the increase in refractive index upon exposure follows the following relationship:

$$\Delta n = aE^2 \tag{4.1}$$

Here, Δn is a refractive index change induced by write beams with electric field of E [V/m].

Energy density of the write beam exposure on the PRI material over time duration Δt [s] is $(1/2)\varepsilon\ E^2\ v\Delta t$, where ε[J/(mV^2)] is dielectric constant and v [m/s] light velocity. For E of 31 V/m, using ε and v in vacuum, light intensity, that is, energy density per second, is calculated to be 0.13 mW/cm^2. Since Δn is assumed to change linearly in proportion to the energy density of write beams, Δn during Δt can be expressed as

$$\Delta n = \gamma(1/2)\varepsilon E^2 v\Delta t. \tag{4.2}$$

The constant γ [m^2/J] represents sensitivity of the PRI material. Therefore, a in Eq. (4.1) is given by

$$a = \gamma\varepsilon v\Delta t/2 \tag{4.3}$$

Since refractive-index saturation is observed in PRI materials, in the calculation program, we put a sequence, which increases refractive index according to Eq. (4.1) until it reaches the saturation limit, and after that, the refractive index is held at the limit.

4.3.1.2 Results

Parameters used for the simulation are listed in Table 4.1. We assumed that refractive index change is proportional to the write beam intensity with a saturation limit of 0.024 as shown in Fig. 4.4 for simplicity although actual refractive index change versus write beam exposure curves are S-shaped with some γ-values like curves in photographs.

Figure 4.5 shows results of the BPM simulation. The upper and the lower contour diagrams, respectively, represent refractive index distribution and probe beam electric field distribution. The two

Table 4.1. Parameters for BPM simulation

Waveguide width	8 μm
n_0	1.5
Core/Clad refractive index difference	0.01
Gap between optical waveguides	400 μm
Write beam	Gaussian (480 nm, Half width: 3.6 μm)
	Peak electric field: 1 for One-beam writing
	$\sqrt{0.5}$ for Two-beam writing
Probe beam	Gaussian (1.3 μm, Half width: 3.6 μm)
	Peak electric field: 1

Source: T. Yoshimura, J. Roman, Y. Takahashi, W. V. Wang, M. Inao, T. Ishitsuka, K. Tsukamoto, K. Motoyoshi, and W. Sotoyama, "Self-Organizing Waveguide Coupling Method "SOLNET" and Its Application to Film Optical Circuit Substrates," *Proc. 50th Electron. Components Technol. Conf. (ECTC)*, 962–969. © 2000 IEEE.

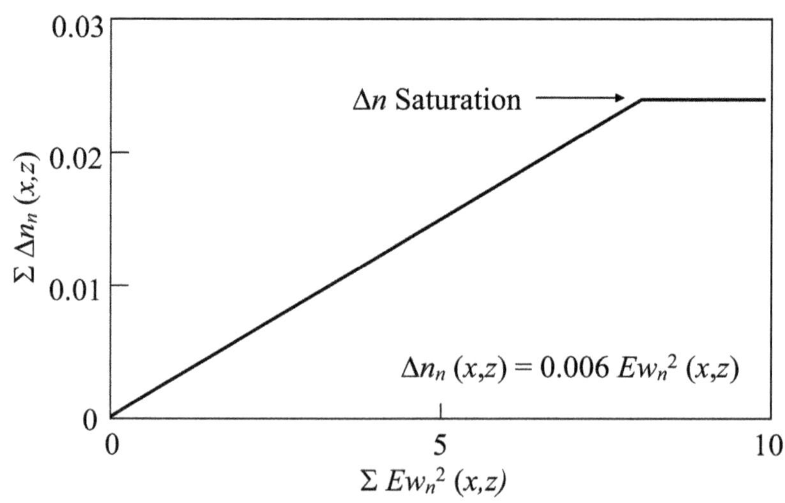

Figure 4.4. Assumed refractive index change characteristics in the PRI material. Reprinted with permission from T. Yoshimura, J. Roman, Y. Takahashi, W. V. Wang, M. Inao, T. Ishitsuka, K. Tsukamoto, K. Motoyoshi, and W. Sotoyama, "Self-Organizing Waveguide Coupling Method "SOLNET" and Its Application to Film Optical Circuit Substrates," *Proc. 50th Electron. Components Technol. Conf. (ECTC)*, 962–969. © 2000 IEEE.

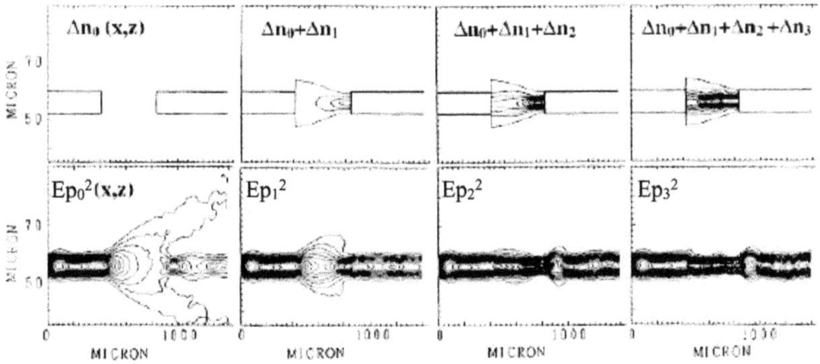

Figure 4.5. One-beam-writing SOLNET with no misalignment. Reprinted with permission from T. Yoshimura, J. Roman, Y. Takahashi, W. V. Wang, M. Inao, T. Ishitsuka, K. Tsukamoto, K. Motoyoshi, and W. Sotoyama, "Self-Organizing Waveguide Coupling Method "SOLNET" and Its Application to Film Optical Circuit Substrates," *Proc. 50th Electron. Components Technol. Conf. (ECTC)*, 962–969. © 2000 IEEE.

rectangles are the left-side and right-side optical waveguides, which are arranged with no misalignment. It can be seen that with write beam exposure progress, a straight optical waveguide is formed gradually, which connects the left-side and the right-side optical waveguides. As a result, a probe beam can efficiently be coupled between the two optical waveguides.

The constructed optical waveguide width is close to the right-side optical waveguide width. The result demonstrates the proof-of-concept for one-beam-writing SOLNET. It is expected that a downtapered optical waveguide might be obtained if wider beams are introduced.

Figure 4.6 shows results of BPM simulation for one-beam-writing SOLNET in the case that lateral misalignment (4 μm) and angular misalignment (2°) between the left-side and the right-side optical waveguides exist. Considerable leakage of probe beams remains even after the optical waveguide of SOLNET is constructed. Therefore, it is concluded that one-beam-writing SOLNET is effective when no misalignment exists between optical devices. When misalignment exists, it does not work. Two-beam-writing SOLNET can resolve this problem as described below.

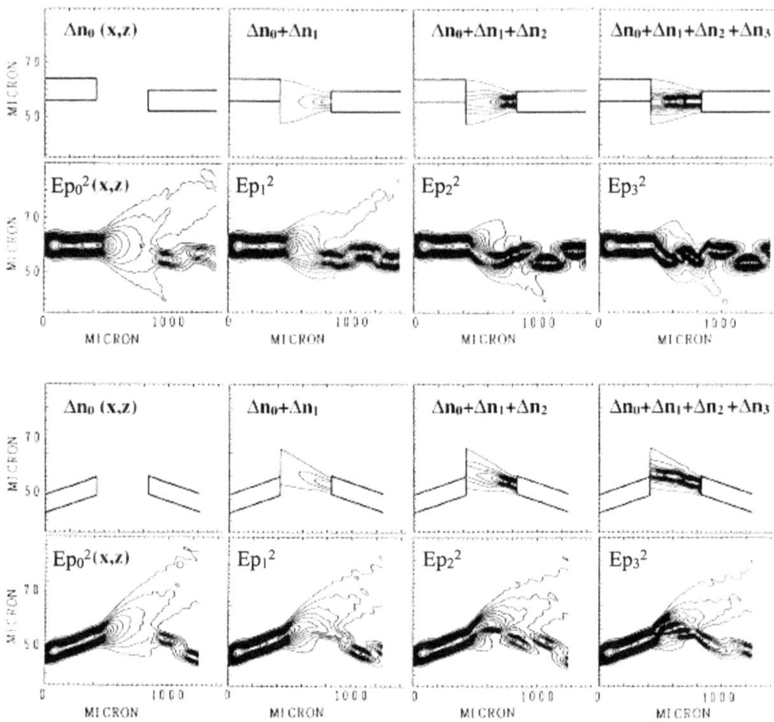

Figure 4.6. One-beam-writing SOLNET with misalignment. Reprinted with permission from T. Yoshimura, J. Roman, Y. Takahashi, W. V. Wang, M. Inao, T. Ishitsuka, K. Tsukamoto, K. Motoyoshi, and W. Sotoyama, "Self-Organizing Waveguide Coupling Method "SOLNET" and Its Application to Film Optical Circuit Substrates," *Proc. 50th Electron. Components Technol. Conf. (ECTC)*, 962–969. © 2000 IEEE.

4.3.2 *Two-Beam-Writing SOLNET*

4.3.2.1 Procedure

Figure 4.7 shows diagrams of simulation procedure for two-beam-writing SOLNET [1, 2, 4]. Basically, the same configuration and calculation procedure as in the case of one-beam-writing SOLNET can be applied to two-beam-writing SOLNET. In step 0, an initial refractive index distribution is given. In step 1, a write beam introduced from the left-side optical waveguide (Ewf_1) and a write beam from the right-side optical waveguide (Ewb_1) propagate under

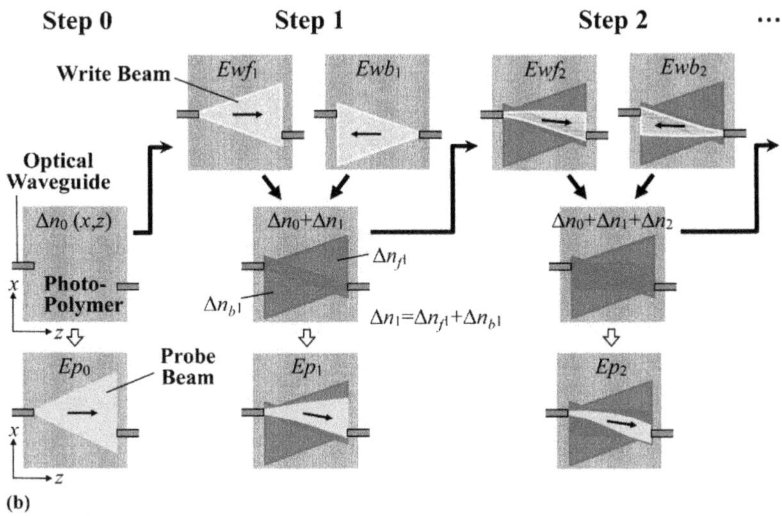

Figure 4.7. BPM simulation procedure for two-beam-writing SOLNET. Reprinted with permission from T. Yoshimura, J. Roman, Y. Takahashi, W. V. Wang, M. Inao, T. Ishitsuka, K. Tsukamoto, K. Motoyoshi, and W. Sotoyama, "Self-Organizing Waveguide Coupling Method "SOLNET" and Its Application to Film Optical Circuit Substrates," *Proc. 50th Electron. Components Technol. Conf. (ECTC)*, 962–969. © 2000 IEEE.

Δn_0 to generate additional refractive index change Δn_{f1} and Δn_{b1}, respectively, resulting in a new distribution $\Delta n_0 + \Delta n_1$. Here, $\Delta n_1 = \Delta n_{f1} + \Delta n_{b1}$. In step 2, write beams of Ewf_2 and Ewb_2 propagate under $\Delta n_0 + \Delta n_1$ to generate additional refractive index change $\Delta n_2 = \Delta n_{f2} + \Delta n_{b2}$, resulting in a distribution $\Delta n_0 + \Delta n_1 + \Delta n_2$. By repeating the procedure, simulations of two-beam-writing SOLNET can be accomplished.

4.3.2.2 Results

Figure 4.8 shows results of BPM simulation for two-beam-writing SOLNET when lateral misalignment of 4 μm and angular misalignment of 2° between the left-side and the right-side optical waveguides exist. By introducing write beams from both of the optical waveguides, a refractive index distribution is gradually formed in the gap region filled with the PRI material. Finally, a

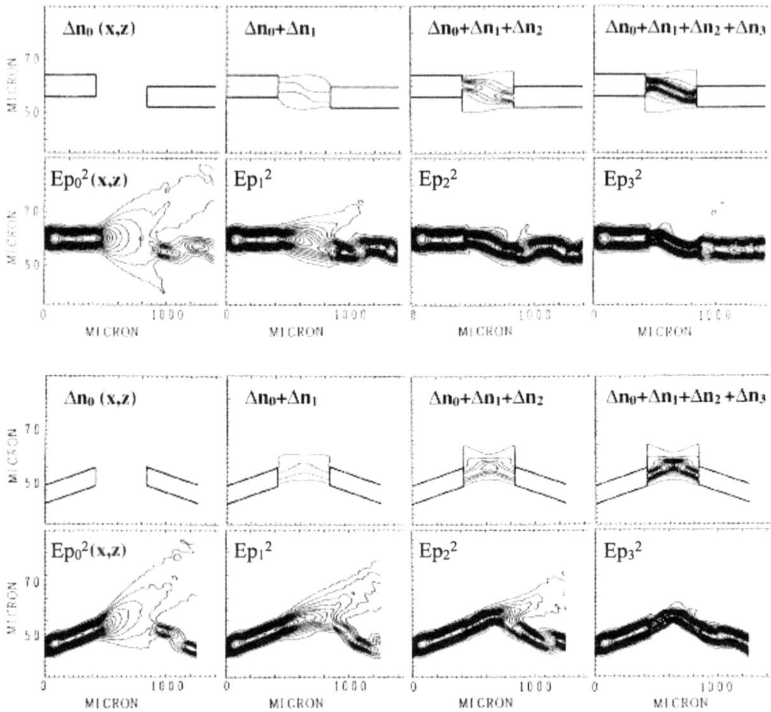

Figure 4.8. Two-beam-writing SOLNET with misalignment. Reprinted with permission from T. Yoshimura, J. Roman, Y. Takahashi, W. V. Wang, M. Inao, T. Ishitsuka, K. Tsukamoto, K. Motoyoshi, and W. Sotoyama, "Self-Organizing Waveguide Coupling Method "SOLNET" and Its Application to Film Optical Circuit Substrates," *Proc. 50th Electronic Components & Technology Conference (ECTC)*, 962–969. © 2000 IEEE.

self-organized coupling waveguide appears which connects the two optical waveguides automatically. This enables reduction of the probe-beam-leakage as well as smooth propagation with small mode disturbance, resulting in a coupling efficiency improvement.

Such phenomena arise from an interaction between the two write beams from the left and from the right in the PRI material. The interaction is an attractive force, so that the two write beams merge together into one optical waveguide.

In Fig. 4.9, coupling efficiency is plotted as a function of exposure step, i.e., writing time, based on the data shown in Fig. 4.8. It is

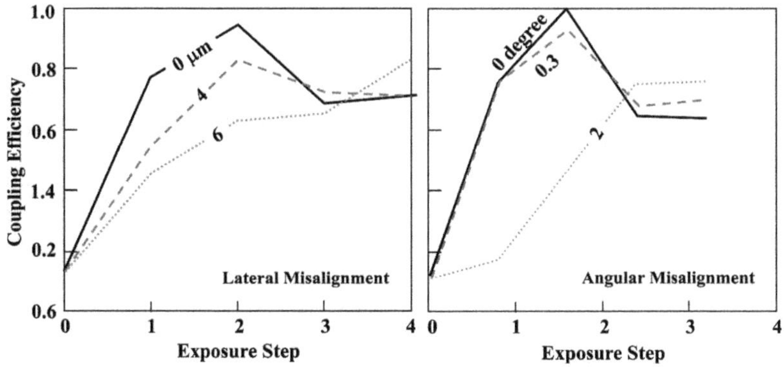

Figure 4.9. Simulated coupling efficiency versus exposure step characteristics in two-beam-writing SOLNET. Reprinted with permission from T. Yoshimura, J. Roman, Y. Takahashi, W. V. Wang, M. Inao, T. Ishitsuka, K. Tsukamoto, K. Motoyoshi, and W. Sotoyama, "Self-Organizing Waveguide Coupling Method "SOLNET" and Its Application to Film Optical Circuit Substrates," *Proc. 50th Electron. Components Technol. Conf. (ECTC)*, 962–969. © 2000 IEEE.

found that with writing time, the coupling efficiency increases up to a peak value, then, slightly decreases, and finally, reaches a saturated value. The writing time required for the coupling efficiency to reach the peak tends to become large with an increase in misalignment. In Fig. 4.10, lateral and angular tolerance in coupling efficiency are compared for one-beam-writing SOLNET, two-beam-writing SOLNET and butt joint couplings. It is found that two-beam-writing SOLNET has excellent tolerance, keeping efficiency around 80% for lateral misalignment of 6 μm, and angular misalignment of 2°. These results demonstrate the proof-of-concept for two-beam-writing SOLNET.

4.3.3 *Reflective SOLNET (R-SOLNET)*

4.3.3.1 Procedure

Figure 4.11 shows a model for simulation of R-SOLNET [5, 6]. Two optical waveguides of 8 μm width are placed in a counter configuration with a 2 μm lateral misalignment and a 240 μm gap that is filled with a PRI material to form a PRI region. The refractive

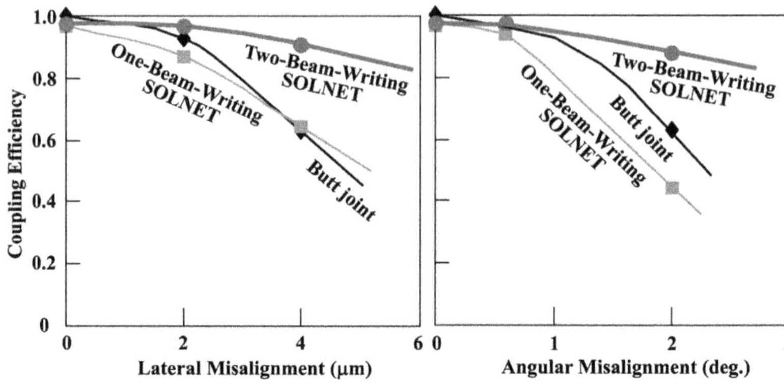

Figure 4.10. Tolerance in coupling efficiency. Reprinted with permission from T. Yoshimura, J. Roman, Y. Takahashi, W. V. Wang, M. Inao, T. Ishitsuka, K. Tsukamoto, K. Motoyoshi, and W. Sotoyama, "Self-Organizing Waveguide Coupling Method "SOLNET" and Its Application to Film Optical Circuit Substrates," *Proc. 50th Electron. Components Technol. Conf. (ECTC)*, 962–969. © 2000 IEEE.

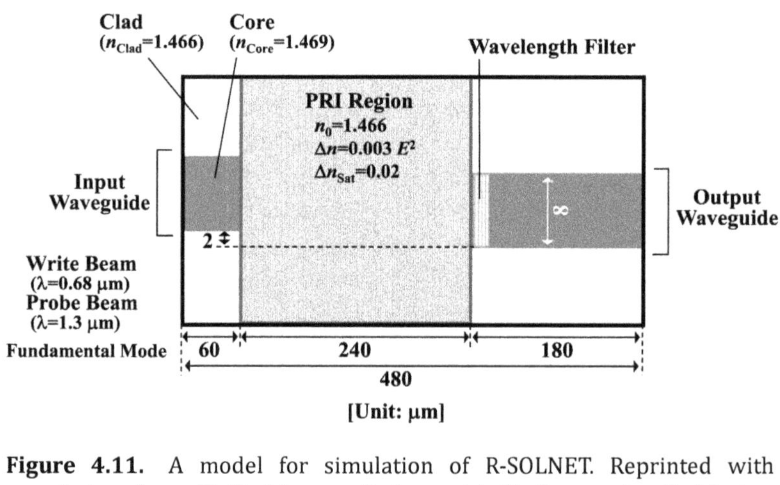

Figure 4.11. A model for simulation of R-SOLNET. Reprinted with permission from T. Yoshimura, T. Inoguchi, T. Yamamoto, S. Moriya, Y. Teramoto, Y. Arai, T. Namiki, and K. Asama, "Self-Organized Lightwave Network Based on Waveguide Films for Three-Dimensional Optical Wiring Within Boxes," *J. Lightwave Technol.* **22**, 2091–2100. © 2004 IEEE.

index of core and clad is, respectively, 1.469 and 1.466. In front of the output waveguide, a wavelength filter, which transmits probe beams with a transmittance of 1 and reflects the write beams with a reflectivity of 1, is placed. Wavelengths of the write beam and the probe beam are, respectively, 0.68 μm and 1.3 μm. The write beam and the probe beam are introduced from the input waveguide. The input beams have fundamental mode profiles. It is assumed that the refractive index of the PRI material is the same as that of clad, namely, 1.466 before exposure.

The calculation procedure is shown in Fig. 4.12. In step 0, refractive index change of the PRI material, Δn_0 , is 0 , which means that the PRI index is uniformly 1.466. A probe beam that does not affect the refractive index of the PRI material is introduced from the input waveguide for calculation of the electric fields, E_{p0} in order to evaluate light beam coupling between the input and the output waveguides.

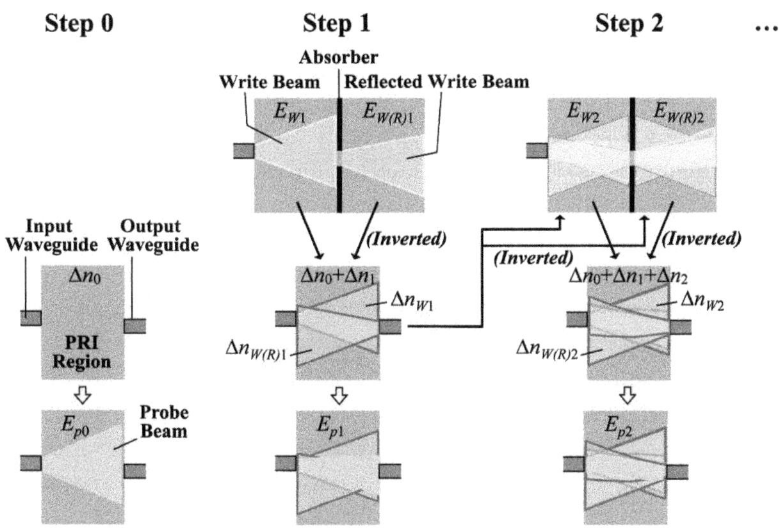

Figure 4.12. BPM simulation procedure for R-SOLNET. Reprinted with permission from T. Yoshimura, T. Inoguchi, T. Yamamoto, S. Moriya, Y. Teramoto, Y. Arai, T. Namiki, and K. Asama, "Self-Organized Lightwave Network Based on Waveguide Films for Three-Dimensional Optical Wiring Within Boxes," *J. Lightwave Technol.* **22**, 2091–2100. © 2004 IEEE.

In step 1, first, electric fields of a write beam, E_{W1}, and electric fields of a reflected write beam from the wavelength filter, $E_{W(R)1}$, are calculated using a configuration shown in Fig. 4.12 (top). In the center of the PRI region, a light absorber is placed with a window corresponding to the wavelength filter aperture. The electric field distribution of the reflected write beam can be obtained by inversing the right-side distribution. Using Eq. (4.1), refractive index change distribution of the PRI material in step 1, Δn_1, can be calculated. Here, Δn_1 is a superposition of refractive index changes induced by the write beam and the reflected write beam. Thus, total refractive index change becomes $\Delta n_0 + \Delta n_1$, which determines electric fields of a probe beam, E_{p1}. In step 2, E_{W2} and $E_{W(R)2}$ are calculated, then, refractive index change distribution in step 2, Δn_2, is obtained, giving total refractive index change of $\Delta n_0 + \Delta n_1 + \Delta n_2$, which determines E_{p2}. By repeating this procedure, simulations of R-SOLNET can be accomplished.

4.3.3.2 Results

Figure 4.13(a) shows simulated refractive index distributions and electric field distributions of probe beams in R-SOLNET for the model shown in Fig. 4.11. The brighter the color is, the larger the refractive index or the larger electric field it represents. In step 0, since refractive index is uniform in the PRI region, a probe beam emitted from the input waveguide expands with propagation due to diffraction, inducing a considerable leakage of the probe beam to outside the output waveguide. In step 2, a coupling waveguide is partially constructed, which reduces the probe beam leakage. In step 3, the self-focusing arising from the superposition of the write beam and the reflected write beam becomes remarkable, resulting in a clear coupling waveguide between the input and the output waveguides, consequently, strong confinement of the probe beam.

When we use a model in which the wavelength filter is removed, one-beam-writing SOLNET is formed instead of R-SOLNET as shown in Fig. 4.13(b). Because no reflected write beam exists, a coupling waveguide stretching straight from the input waveguide is constructed. Calculated coupling efficiency from the input to the output of the one-beam-writing SOLNET for step 0 and step 2, and

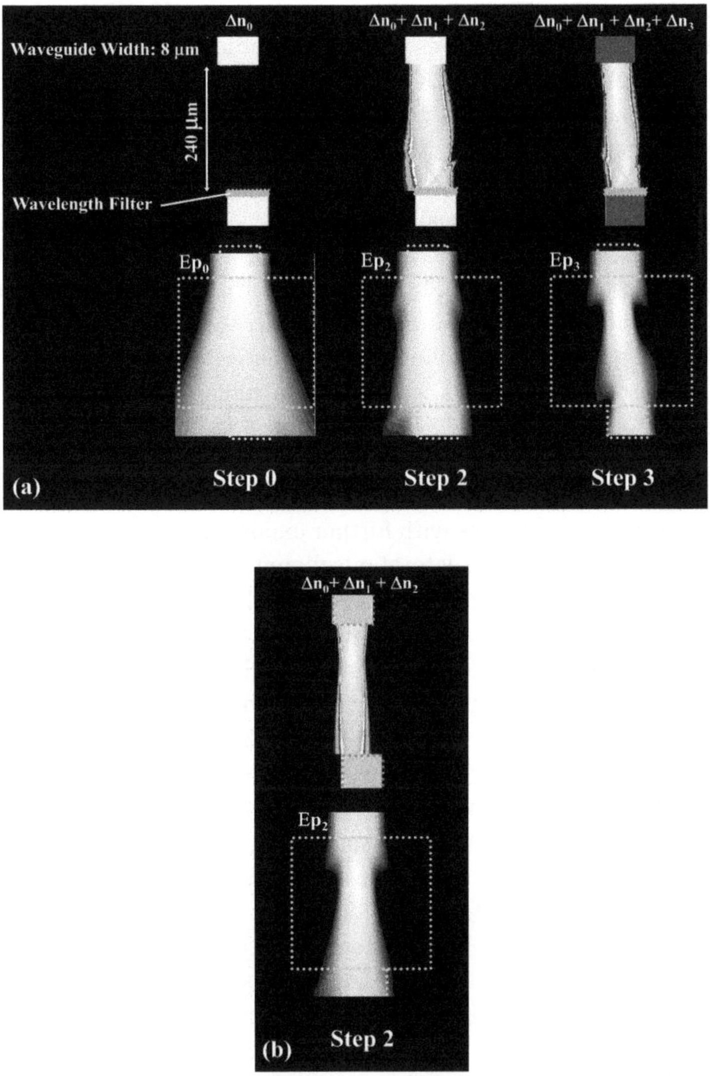

Figure 4.13. (a) R-SOLNET with lateral misalignment of 2 μm for Step 0, Step 2, and Step 3. (b) One-beam-writing SOLNET with lateral misalignment of 2 μm for Step 2. Reprinted with permission from T. Yoshimura, T. Inoguchi, T. Yamamoto, S. Moriya, Y. Teramoto, Y. Arai, T. Namiki, and K. Asama, "Self-Organized Lightwave Network Based on Waveguide Films for Three-Dimensional Optical Wiring Within Boxes," *J. Lightwave Technol.* **22**, 2091–2100. © 2004 IEEE. See also Color Insert.

that of the R-SOLNET for step 2 are, respectively, 0.58, 0.76, and 0.82. Namely, R-SOLNET achieves the highest efficiency, demonstrating the proof-of-concept for R-SOLNET.

From these results, it is concluded that although the intensity of the reflected write beam is smaller than that of the write beam, the reflected write beam from the wavelength filter clearly contributes to pull the coupling waveguide up to the output waveguide. The reflected write beam acts as a trigger to guide the main write beam flow to the reflecting point. In other words, the reflected write beam paves the way for the self-organized optical waveguide to reach the reflecting point by the "pulling water" effect.

Figure 4.14(a) shows coupling efficiency as a function of exposure step for R-SOLNET and for one-beam-writing SOLNET. Step 0 corresponds to the free-space coupling. In both cases, coupling efficiency increases with the exposure step to reach a maximum, and decreases with further exposure.

Figure 4.14(b) shows lateral misalignment dependence and gap dependence of coupling efficiency for R-SOLNET, one-beam-writing SOLNET and the free-space coupling. R-SOLNET exhibits the highest coupling efficiency among the three. It achieves efficiency higher than 0.5 in 4 μm misalignment and exhibits wider tolerance for misalignment than the one-beam-writing SOLNET. The usefulness of R-SOLNET is remarkable for gaps wider than 100 μm. For 240 μm gap, the coupling efficiency in 2 μm misalignment is ~0.35, ~0.65, and ~0.85 for the free-space coupling, one-beam-writing SOLNET and R-SOLNET, respectively. With a decrease in gap, the coupling efficiency in R-SOLNET decreases due to the reduction in beam-overlap between the incident write beam and the reflected write beam, resulting in small differences in coupling efficiency between R-SOLNET and the free-space coupling for small gaps, e.g., 60 μm.

In 3-D optical circuits with 8 μm-wide optical waveguides in stacked OE films, the typical film-to-film distance is ~50 μm; so R-SOLNET may especially be useful for connections to next neighboring or farther films. For 3-D optical circuits with nano-scale waveguides in intra-chip optical interconnects, due to larger diffraction angle, R-SOLNET might be effective in couplings even to the neighboring OE films. The implementation of wavelength filters

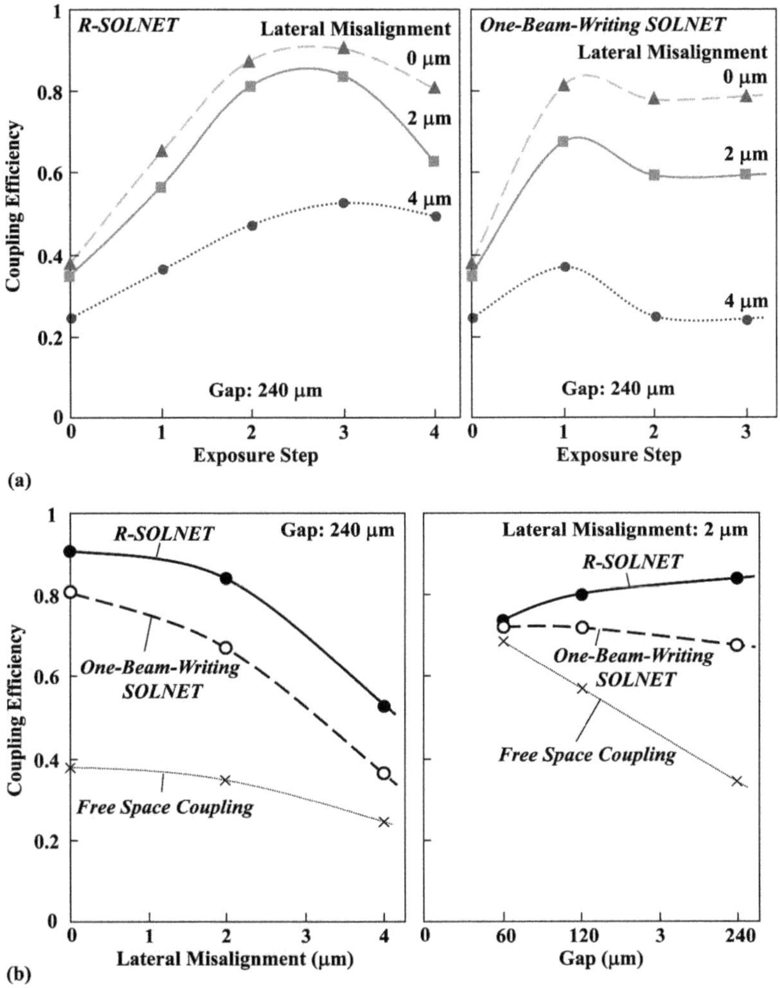

Figure 4.14. Simulation results of coupling efficiency. (a) Coupling efficiency versus exposure step characteristics for R-SOLNET and one-beam-writing SOLNET. (b) Lateral misalignment and gap dependence of coupling efficiency for R-SOLNET, the one-beam-writing SOLNET and the free-space coupling. Reprinted with permission from T. Yoshimura, A. Hori, Y. Yoshida, Y. Arai, H. Kurokawa, T. Namiki, and K. Asama, "Coupling Efficiencies in Reflective Self-Organized Lightwave Network (R-SOLNET) Simulated by the Beam Propagation Method," *IEEE Photon. Technol. Lett.* **17**, 1653–1655. © 2005 IEEE.

deposited on curved end facets of optical waveguides for increasing the overlap between the write beams might improve R-SOLNET performance.

4.4 Simulation of SOLNET by FDTD Method

4.4.1 *SOLNET Simulator*

SOLNET simulator based on the FDTD method enables us to treat SOLNET with reflections , photo-luminescence, and large bending [7, 8, 9]. This means that R-SOLNET can be simulated straightforward. The flow chart for the SOLNET simulator is shown in Fig. 4.15. First, parameters for calculations are set. Initial refractive index distribution in the calculated region is input into data sets. Then, calculations of electric fields and magnetic fields are carried out with absorption boundary conditions.

Since the refractive index of PRI materials changes with time, the refractive index distribution is rewritten at each time step with duration of Δt. The refractive index change Δn during Δt is

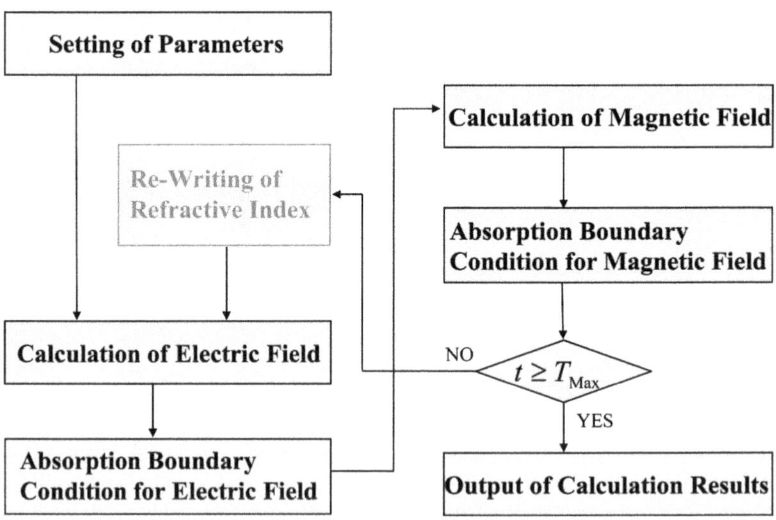

Figure 4.15. Flow chart of calculations in the SOLNET simulator based on the FDTD method.

expressed as Eq. (4.2) with $\gamma\,[\mathrm{m}^2/\mathrm{J}]$ that is a parameter related to the sensitivity of the PRI materials. For simplicity, it is assumed that the refractive index changes in proportion to the write beam exposure until it reaches the saturation limit.

In real systems, typical response time of PRI materials ranges around a few seconds to a few minutes. In the FDTD method, however, it is difficult to carry out calculations with such long time durations since Δt should be sub-ps. So, in the simulations, we rescaled the time parameter as follows:

$$\Delta n = \gamma'(1/2)\varepsilon E^2 v \Delta t' \tag{4.4}$$

where, $\Delta t' = \Delta t \times 10^{12}$.

The electric fields of light sources are expressed as follows:

$$E(t) = (1/2)(1 - \cos(\pi t/T_\omega))E_0 \sin \omega_0 t \quad 0 \leq t \leq T_\omega$$

$$E(t) = E_0 \sin \omega_0 t \qquad t \geq T_\omega$$

Here, ω_0 is the angular frequency of the light wave and $\omega_0 T_\omega = 4\pi$.

The parameters used in the present calculations are as follows. Mesh sizes for the FDTD calculations are $\Delta x = \Delta y = 20$ nm and $\Delta t = 0.0134$ ps. Amplitude of the electric field is $E_0 = 10$ V/m, which corresponds to light intensity of 0.013 mW/cm^2 for ε and v in vacuum. γ' is 1 m^2/J with Δn saturation limit of 0.3. When the time reaches T_{Max} that is set initially, the simulation is terminated to output calculation results.

4.4.2 *L-Shaped One-Beam-Writing SOLNET*

Figure 4.16 shows a model and results of simulation for one-beam-writing SOLNET formed via a 45° mirror. A 45° perfect conductor mirror is inserted in the PRI region. Core width of the input/output waveguides is 0.5 μm and the calculated region size is 20 × 20 μm^2. The refractive index of the clad region and the core are assumed to be 1.0 and 1.5, respectively. The refractive index of the PRI material changes from 1.5 to 1.8 by the PRI effect. Write beam wavelength is 650 nm. The beam source is located near left edge of the input waveguide with a pitch of 20 nm along the y-axis. Polarization is parallel to the z-axis.

When a write beam is introduced from the input waveguide to the PRI region, the write beam increases refractive index in

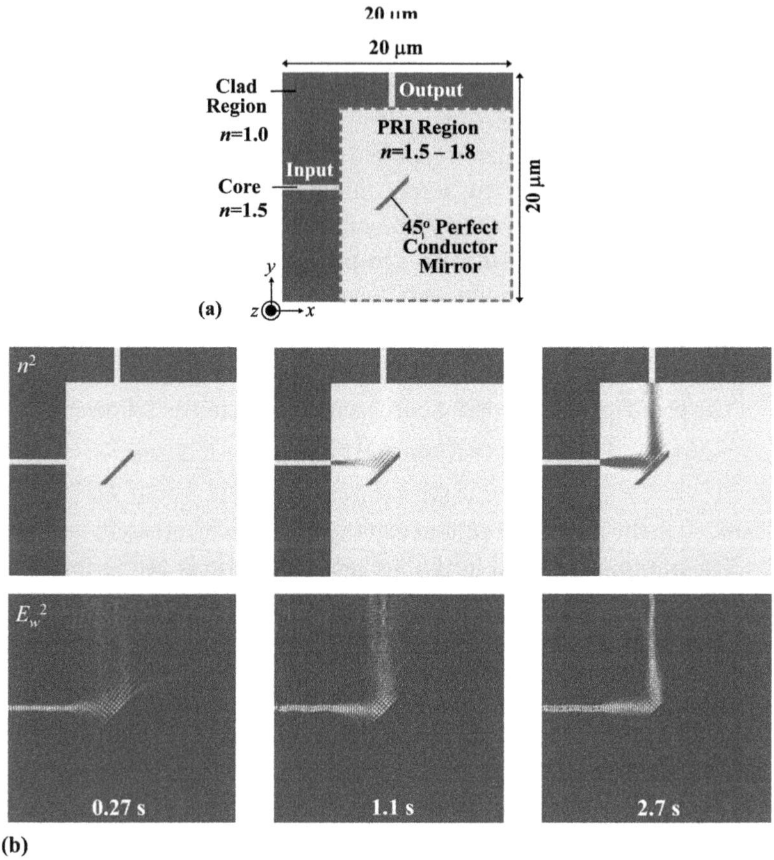

Figure 4.16. Simulation of one-beam-writing SOLNET formed via a 45° mirror: (a) Simulation model. (b) Simulated refractive index distributions and electric field distributions of propagating write beams. Reprinted with permission from T. Yoshimura and H. Kaburagi, "Self-Organization of Coupling Optical Waveguides by the 'Pulling Water' Effect of Write Beam Reflections in Photo-Induced Refractive-Index Increase Media," *Proc. SPIE* **7221**, 722111- 1-10 (2009).

the region. Initially the write beam is reflected to upside by the mirror with diffraction. Then, self-focusing gradually appears in the exposed part. Finally, SOLNET is grown along the center axis of the write beam propagation to construct an L-shaped nano-scale waveguide.

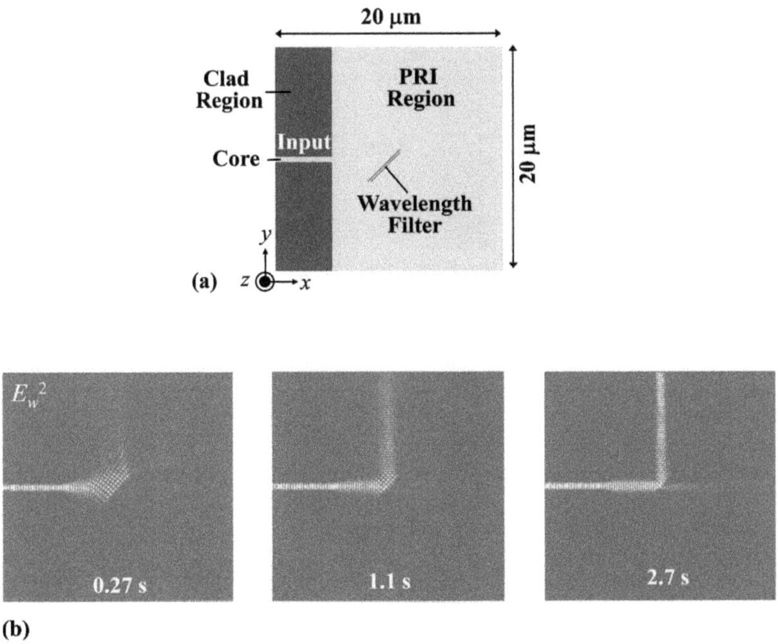

Figure 4.17. Simulation of one-beam-writing SOLNET formed via a 45° wavelength filter: (a) Simulation model. (b) Simulated electric field distributions of propagating write beams. Reprinted from *Opt. Commun.*, **281**, H. Kaburagi and T. Yoshimura, Simulation of Micro/Nano-Scale Self-Organized Lightwave Network (SOLNET) Using the Finite Difference Time Domain Method, 4019–4022, Copyright (2008), with permission from Elsevier.

Figure 4.17 shows a model and results of simulation for one-beam-writing SOLNET formed via a 45° wavelength filter [7]. The wavelength filter consists of four dielectric films. As in the case of the 45° mirror, SOLNET with core width of ~0.5 μm is grown along the center axis of the write beam propagation to construct an L-shaped optical waveguide.

4.4.3 *R-SOLNET Between a Micro-Scale Waveguide and a Nano-Scale Waveguide*

Figure 4.18 shows a model and results of simulation for one-beam-writing SOLNET and R-SOLNET formed between a micro-scale

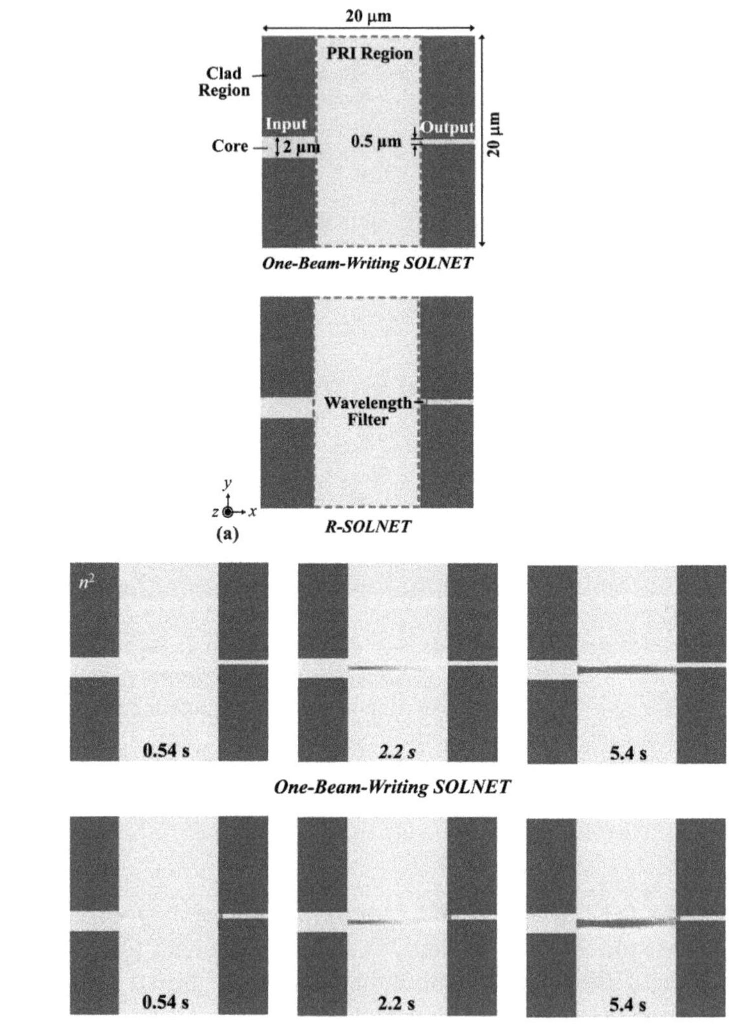

Figure 4.18. Simulation of one-beam-writing SOLNET and R-SOLNET for optical couplings between a micro-scale waveguide and a nano-scale waveguide: (a) Simulation model. (b) Simulated refractive index distributions. Reprinted from *Opt. Commun.*, **281**, H. Kaburagi and T. Yoshimura, Simulation of Micro/Nano-Scale Self-Organized Lightwave Network (SOLNET) Using the Finite Difference Time Domain Method, 4019–4022, Copyright (2008), with permission from Elsevier.

waveguide with core width of 2 μm and a nano-scale waveguide with core width of 0.5 μm [7]. The two optical waveguides are misaligned by 0.5 μm. Refractive index of each part is the same as in the model shown in Fig. 4.16. In R-SOLNET, a dielectric multilayer wavelength filter is placed on the core facet of the nano-scale waveguide. The wavelength filter, which reflects write beams of 650 nm and transmits probe beams of 1.3 μm, consists of 80 nm-thick layer with refractive index of 1.5 and 80 nm-thick layer with refractive index of 2.5. Total layer count is four.

For one-beam-writing SOLNET, refractive index in the PRI region increases with time along the center axis of the write beam propagation to form a straight waveguide. Due to the lateral misalignment of 0.5 μm, misalignment is caused between SOLNET and the output waveguide. For R-SOLNET, on the other hand, SOLNET is guided to the location of the wavelength filter. Namely, a part of the write beam introduced from the micro-scale waveguide is reflected by the wavelength filter, so that the reflected write beam from the wavelength filter overlaps with the incident write beam, inducing the "pulling water" effect.

Figure 4.19 shows propagation of 1.3 μm probe beams that are introduced from the input waveguide. For one-beam-writing SOLNET, before SOLNET formation, the probe beam expands with propagation due to diffraction, resulting in low coupling efficiency less than 0.1. After SOLNET formation, the probe beam is confined into SOLNET to be guided to the nano-scale waveguide with 2.5 times increase in the coupling efficiency. However, due to the misalignment between SOLNET and the output waveguide, considerable leakage is induced at the junction. Furthermore, light beams in higher order modes are excited in the output waveguide, so the light beam travels meandering.

For R-SOLNET, before SOLNET formation, coupling efficiency is slightly smaller comparing with the efficiency for the one-beam-writing SOLNET. This is attributed to loss at the wavelength filter. After SOLNET formation, the probe beam is guided to the wave-length filter to achieve a self-aligned optical coupling with 4 times increase in the coupling efficiency. Comparing with the case of the one-beam-writing SOLNET, larger coupling efficiency is obtained. Furthermore, the probe beam leakage at the junction is reduced, and

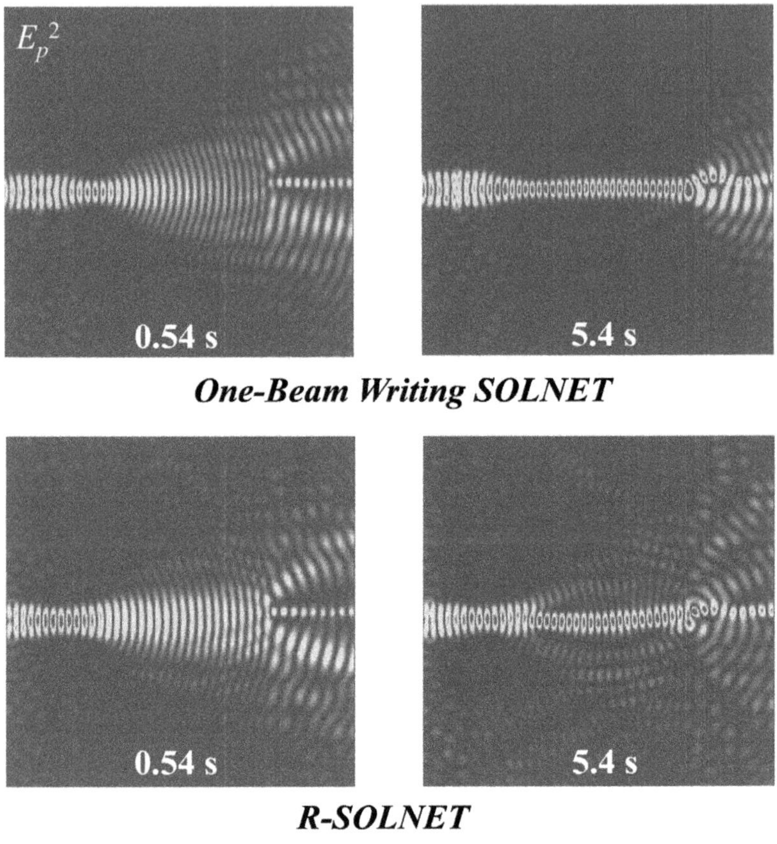

Figure 4.19. Propagation of probe beams in SOLNET. Reprinted from *Opt. Commun.*, **281**, H. Kaburagi and T. Yoshimura, Simulation of Micro/Nano-Scale Self-Organized Lightwave Network (SOLNET) Using the Finite Difference Time Domain Method, 4019–4022, Copyright (2008), with permission from Elsevier.

the degree of meandering in the output waveguide is suppressed, implying less mode disturbance. These results indicate that, by putting reflective elements, self-aligned coupling waveguides can be led to the element locations, and self-aligned coupling waveguides can be formed between a micro-scale waveguide and a nano-scale waveguide.

When Δn saturation limit, denoted by Δn_{sat}, is 0.3, R-SOLNET is formed at lateral misalignment of 0.5 μm. However, as Figs. 4.20(a)

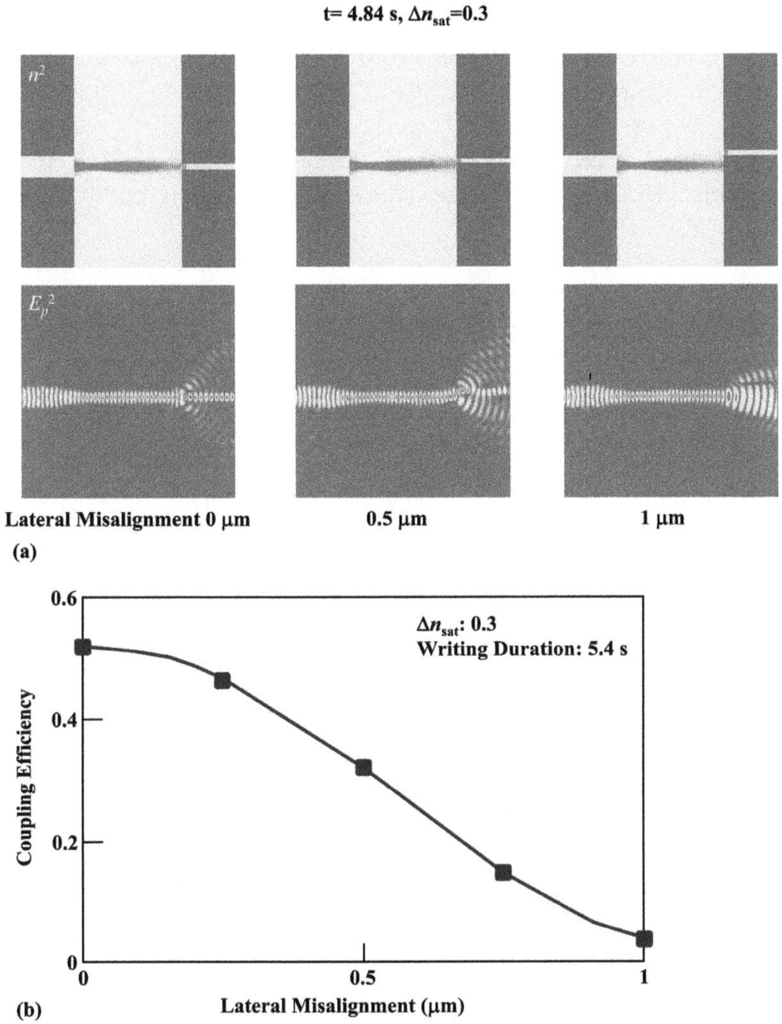

Figure 4.20. Influences of lateral misalignment on R-SOLNET formation: (a) Simulated refractive index distributions and electric field distributions of propagating probe beams. (b) Lateral misalignment dependence of coupling efficiency.

and (b) show, when the lateral misalignment increases up to 1 μm, R-SOLNET cannot be formed and leakage of guided light beams at the junction between the SOLNET and the output waveguide increases, resulting in decreases in coupling efficiency. Regarding Δn_{sat}, as can be seen from Fig. 4.21(a), light beam confinement becomes weak as Δn_{sat} decreases. It is found from Fig. 4.21(b) that considerable coupling efficiency is available for Δn_{sat} larger than 0.1.

4.4.4 *R-SOLNET for Y-Branching Self-Aligned Waveguides*

Figure 4.22 shows concept of Y-branching self-aligned waveguides of R-SOLNET [8]. This configuration is regarded as R-SOLNET for three-beam-writing SOLNET using two wavelength filters. By the "pulling water" effect, write beams from the input waveguide are led to the two wavelength filters to form a Y-branching waveguide. In the present simulation, as shown in Fig. 4.23(a), an input waveguide with core width of 2 μm and two output waveguides with core width of 0.5 μm are placed with PRI region in the gap. The output waveguides are misaligned by 0.5 μm from the input waveguide. An anti-reflection (AR) layer is inserted between the PRI region and air region to reduce write beam reflections from the boundary. It can be seen from Fig. 4.23(b) that Y-branching self-aligned coupling waveguides are formed. This suggests that a plurality of optical devices can be connected by putting wavelength filters on the core facets.

4.4.5 *R-SOLNET for Optical Z-Connections with Vertical Waveguides*

Figure 4.24 shows simulation model of an optical Z-connection with a vertical waveguide of R-SOLNET [9]. A 2 μm-thick core with a total internal reflection (TIR) 45° mirror is on a 0.5 μm-thick under clad layer to form an optical waveguide film. Two optical waveguide films are stacked with a gap filled with the PRI material. A wavelength filter is deposited at the TIR 45° mirror aperture in the upper optical waveguide film. Refractive index of the core (n_a) is 1.8 and that of the under clad (n_b) is 1.5. Refractive index of the PRI material (n_c) varies from 1.5 to 1.7 with write beam exposure. Refractive index of

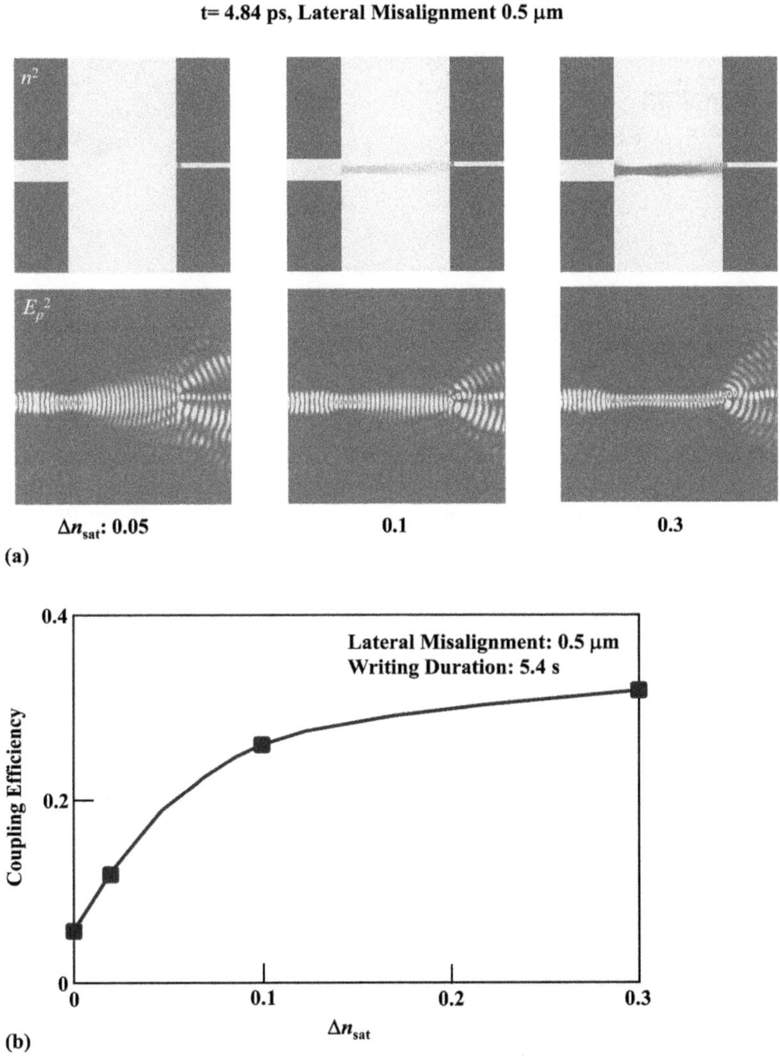

Figure 4.21. Influence of Δn_{sat} on R-SOLNET formation: (a) Simulated refractive index distributions and electric field distributions of propagating probe beams. (b) Δn_{sat} dependence of coupling efficiency.

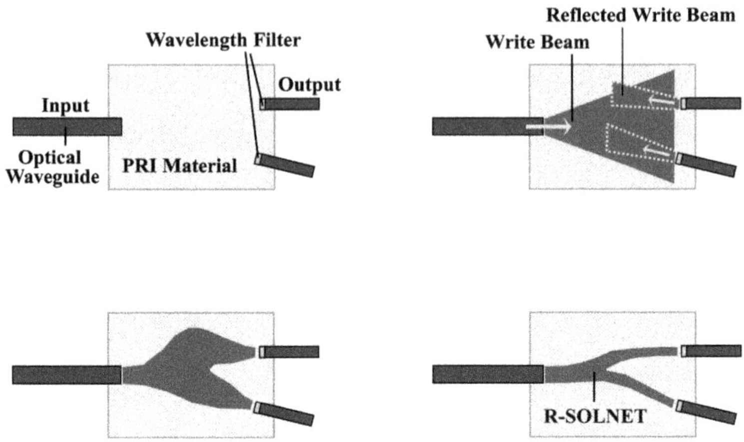

Figure 4.22. Concept of Y-branching self-aligned waveguides of R-SOLNET. [T. Yoshimura and H. Kaburagi, "Self-Organization of Optical Waveguides Between Misaligned Devices Induced by Write-Beam Reflection," *Appl. Phys. Express* **1**, 062007- 1-3 (2008).]

the surrounding region (n_d) is 1.0. Wavelengths of write beams and probe beams are 650 nm and 850 nm, respectively. Polarization is $E//z$.

Influences of lateral misalignment on R-SOLNET formation in the optical Z-connection are shown in Fig. 4.25. Negative lateral misalignment represents left-side dislocation of the upper optical waveguide film and positive lateral misalignment represents right-side dislocation from the just-aligned position. As can be seen from Fig. 4.25(a), in a lateral misalignment range from −0.12 μm to 0.75 μm, SOLNET is led to the wavelength filter location by the "pulling water" effect. In the case that considerable misalignment of −0.25 μm and 1.0 μm exists, SOLNET is led to the wavelength filter location partially, resulting in branching structures in SOLNET. Following the refractive index distributions shown in Fig. 4.25(a), relatively small leakage of probe beams is observed in the lateral misalignment range from −0.12 μm to 0.75 μm as shown in Fig. 4.25(b).

In Fig. 4.25, it should be noted that SOLNET exihibits left-side bending toward the TIR 45° mirror in the upper optical waveguide film even when the lateral misalignment is 0 μm. This is attributed to

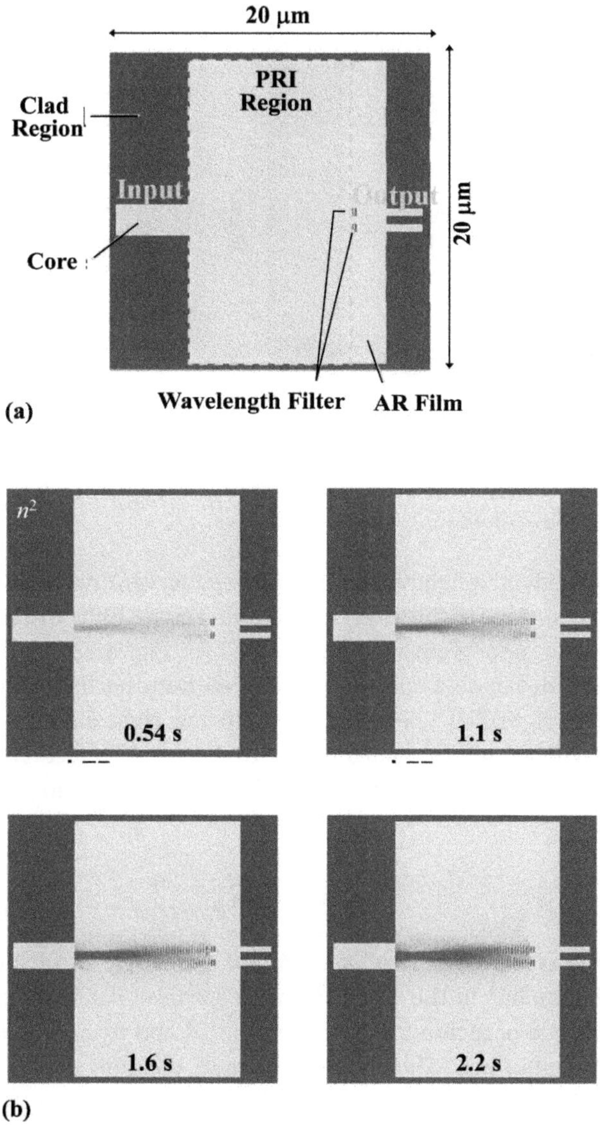

Figure 4.23. Simulation of Y-branching self-aligned waveguides of R-SOLNET: (a) Simulation model. (b) Simulated refractive index distributions. [T. Yoshimura and H. Kaburagi, "Self-Organization of Optical Waveguides Between Misaligned Devices Induced by Write-Beam Reflection," Appl. Phys. Express **1**, 062007- 1-3 (2008).]

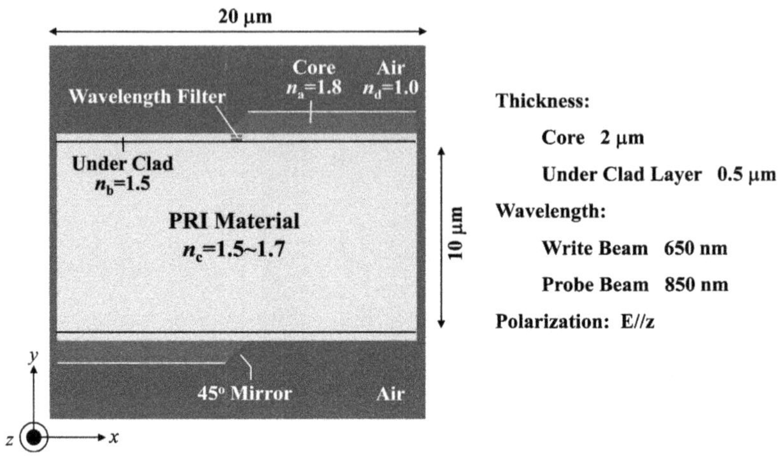

Figure 4.24. Simulation model of an optical Z-connection with a vertical waveguide of R-SOLNET.

the tunnel effect of light waves, namely, evanescent wave, at the TIR 45° mirror surface in the lower optical waveguide film. The situation is shown for one-beam-writing SOLNET in Fig. 4.26. In the case of TIR 45° mirror that corresponds to "without perfect conductor" (Fig. 4.26(a)), SOLNET location shifts to the right side by 0.2 μm from the TIR 45° mirror aperture due to the tunnel effect. The depth of the tunneling d_{tunnel} is given by the following expression for plane waves:

$$d_{\text{tunnel}} = \frac{1}{k_0\sqrt{n_a^2 \sin^2\theta - n_d^2}} \tag{4.5}$$

Here, k_0 is wavenumber in vacuum and θ is angle of incidence of the guided beams to the TIR 45° mirror surface. For wavelength of 650 nm, angle of incidence of 45°, $n_a = 1.8$ and $n_d = 1.0$, $d_{\text{tunnel}} =$ 0.13 μm is obtained. This value is close to the shift of 0.2 μm obtained by the simulation. When the mirror surface is covered by perfect conductor as shown in Fig. 4.26(b), the tunneling is suppressed to form SOLNET just above the 45° mirror aperture. Accordingly, as shown in Figs. 4.26(c) and (d), the optical coupling efficiency between the lower and the upper optical waveguides increases by the perfect conductor deposition. Coupling efficiency of the model

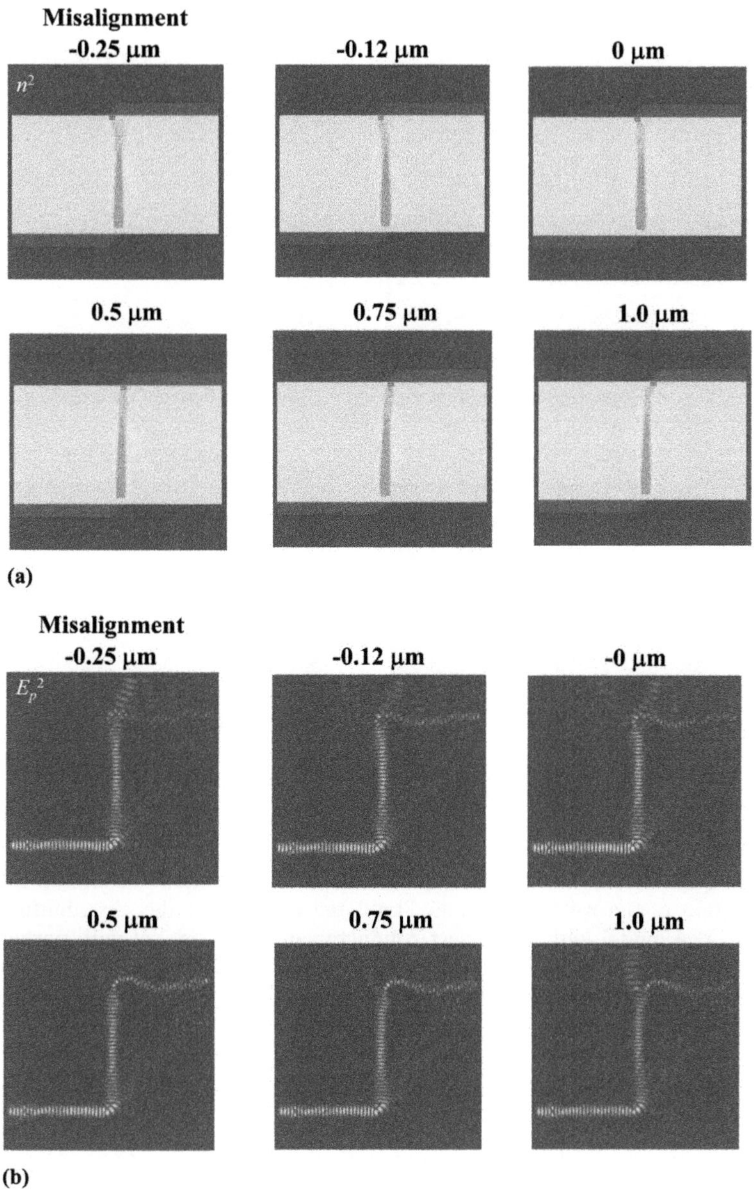

Figure 4.25. Influence of lateral misalignment on R-SOLNET formation in optical Z-connection: (a) Simulated refractive index distributions. (b) Intensity of propagated probe beams.

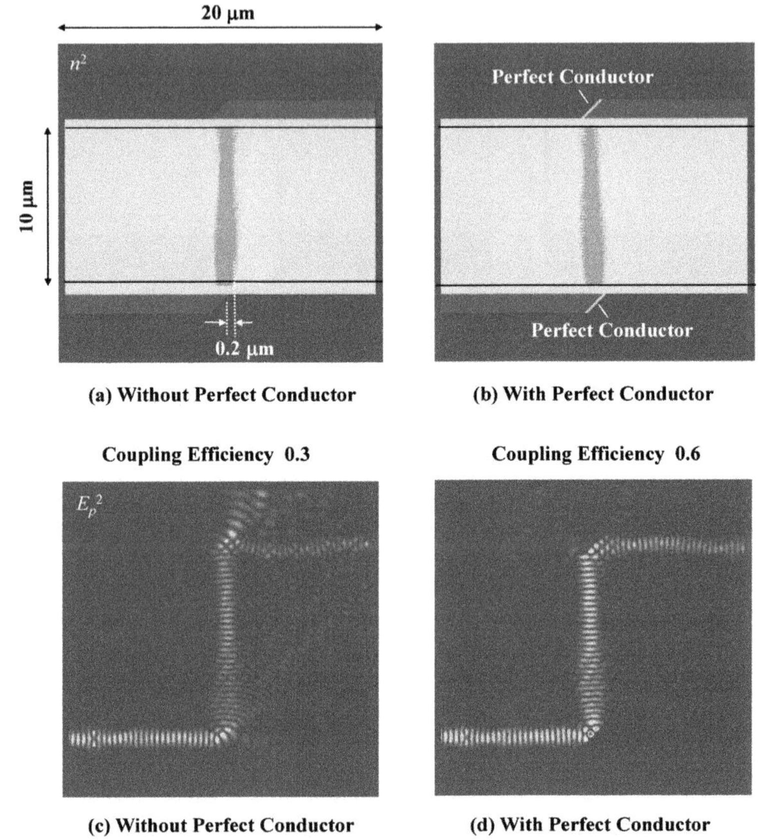

Figure 4.26. Influence of perfect conductor at the mirror surface on the position of the reflected beams. Simulated refractive index distributions for (a) a model without perfect conductor and (b) a model with perfect conductor. Electric field distributions of propagating probe beams for (c) a model without perfect conductor and (d) a model with perfect conductor.

without perfect conductor is 0.3 while that of the model with perfect conductor is 0.6.

It should also be noted in Fig. 4.25 that beam leakage at the 45° mirror is larger in the left-side-bending SOLNET than in the right-side-bending SOLNET. This result can be expalined by the model shown in Fig. 4.27. When the guided beams in SOLNET propagate along the left-side bending, the angle of incidence for the mirror

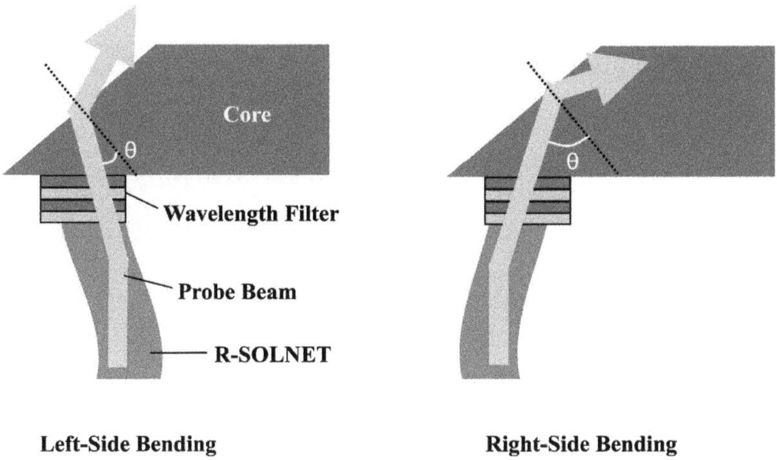

Figure 4.27. Model for light beam leakage at the vertical mirrors in R-SOLNET.

surface becomes small, reducing the beam reflection and increasing the beam leakage. When the guided beams in SOLNET propagate along the right-side bending, on the other hand, the angle of incidence for the mirror surface becomes large, enhancing the beam reflection and decreasing the beam leakage. By depositing perfect conductor on the mirror surface of the upper optical waveguide film, as shown in Fig. 4.28, the leakage at the mirror surface is drastically suppressed in the left-side-bending SOLNET.

Figure 4.29 summarizes influences of lateral misalignment on coupling efficiency of optical Z-connection with R-SOLNET. By putting the wavelength filter on the TIR 45° mirror aperture in the upper optical waveguide film, namely, by implementing R-SOLNET, tolerance of lateral misalignment increases. By depositing perfect conductor on the 45° mirror surface in the upper optical waveguide film, coupling efficiency increases. By the simultaneous implementation of the wavelength filter and the perfect conductor, a drastic coupling efficiency increase, and at the same time, an increase in tolerance, are observed, especially, in the negative lateral misalignment region. If the slight loss at the wavelength filter is suppressed, further improvement of the coupling efficiency is expected.

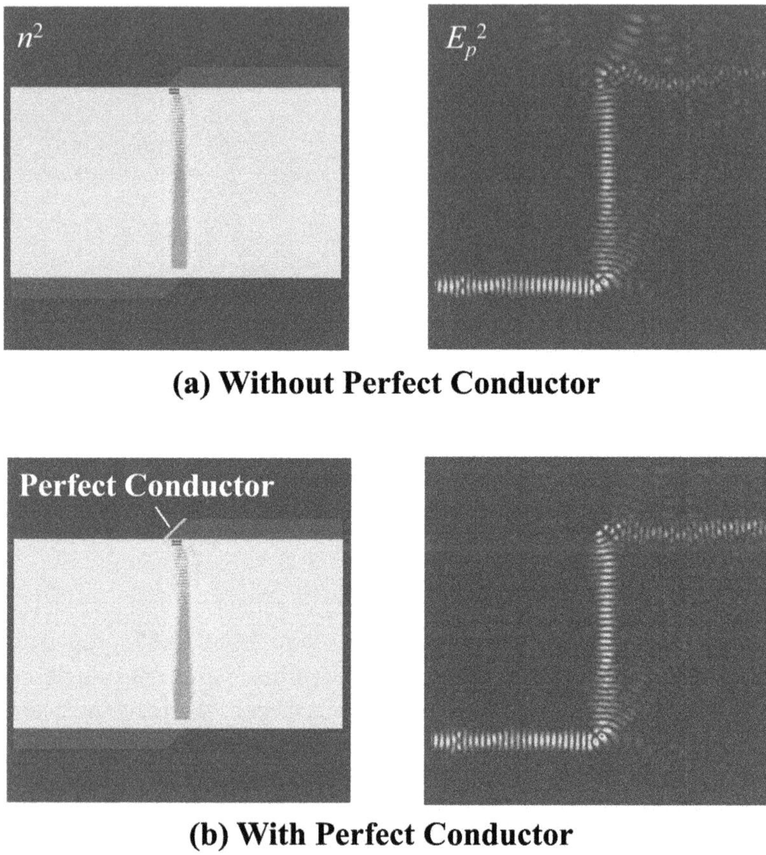

Figure 4.28. Influence of perfect conductor at the mirror surface of the upper optical waveguide film on the light beam leakage in R-SOLNET.

4.4.6 *R-SOLNET with Luminescent Materials*

We developed a SOLNET simulator that can treat photoluminescence generated from luminescent materials in the calculation region [10]. The electric field amplitude of the photoluminescence E_{PL} is defined as follows using time average of the write beam electric field $\overline{E_{\mathrm{Write}}}$ at the luminescent materials location and efficiency of light emission η_{PL}.

$$E_{\mathrm{PL}} = \sqrt{2}\eta_{\mathrm{PL}}\overline{E_{\mathrm{Write}}} \tag{4.6}$$

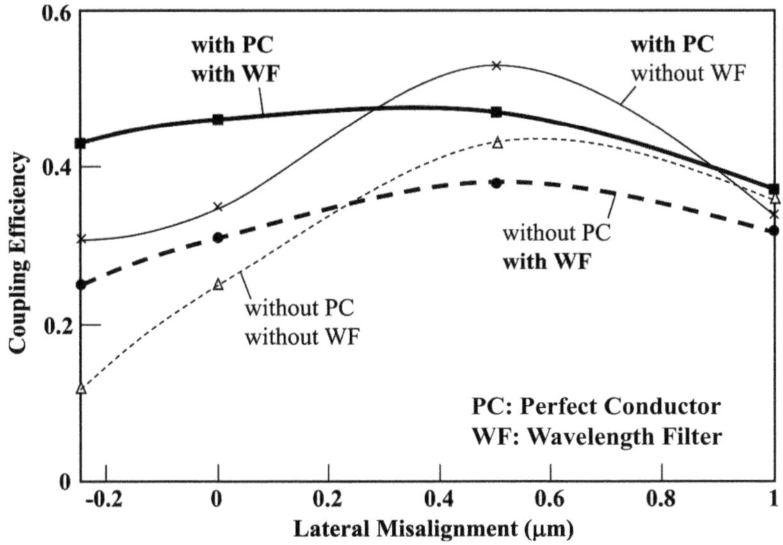

Figure 4.29. Influence of lateral misalignment on coupling efficiency of optical Z-connection with R-SOLNET.

Figure 4.30(a) depicts a reference model. An optical waveguide core of 1.2 μm wide and 3 μm long is put with a region of a PRI material. The refractive index of the core (n_{Core}) is 1.8 and that of the clad (n_{Clad}) is 1.0. The refractive index of the PRI material (n_{PRI}) increases from 1.5 to 1.7 with write beam exposure. Figure 4.30(b) depicts a model with a 600 nm-wide target of luminescent materials [10]. The target is located with a lateral offset of 600 nm from the axis of the optical waveguide core. Between the core and the target, there is a 6.4 μm-long gap filled with a PRI material. The wavelengths of the write beam, the photo-luminescence, and the probe beam are 650 nm, 700 nm, and 650 nm, respectively. η_{PL} is 0.36. The polarization is $E//z$.

The simulation result for the reference model is shown in Fig. 4.31. The upper and lower figures respectively represent dielectric constant ε, that is, square of refractive index n, and intensity of probe beams Ep^2. It is found that a straight SOLNET is grown from the waveguide core with writing time due to the self-focusing.

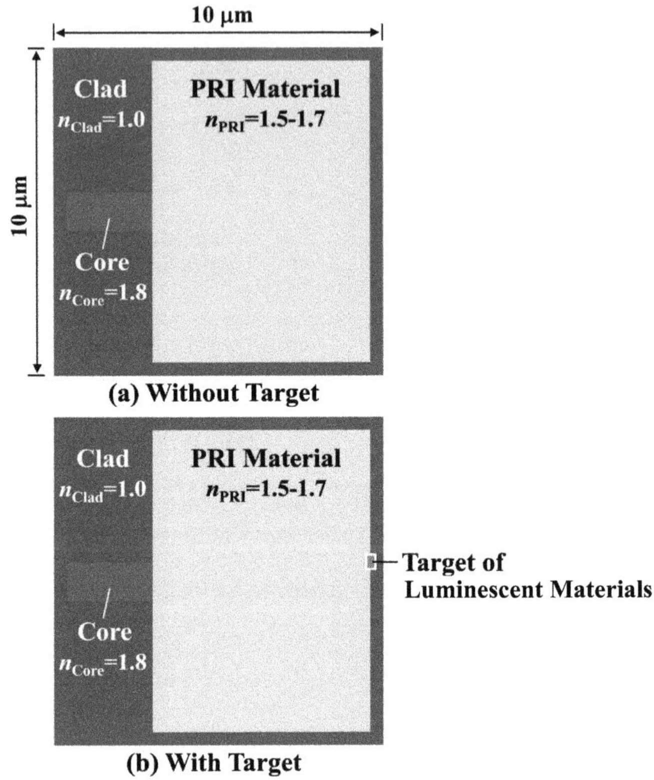

Figure 4.30. Simulation models for R-SOLNET with luminescent materials: (a) A reference model without targets of luminescent materials. (b) A model with a target of luminescent materials.

Accordingly, a probe beam is confined in the SOLNET to propagate straight.

Figure 4.32 shows a distribution of write beam and photo-luminescence intensity in a model with a target of luminescent materials. In addition to the write beam from the core, photo-luminescence from the target is observed. The photo-luminescence is generated by the write beam exposure on the target. These light beams construct R-SOLNET between the optical waveguide core and the target by the self-focusing.

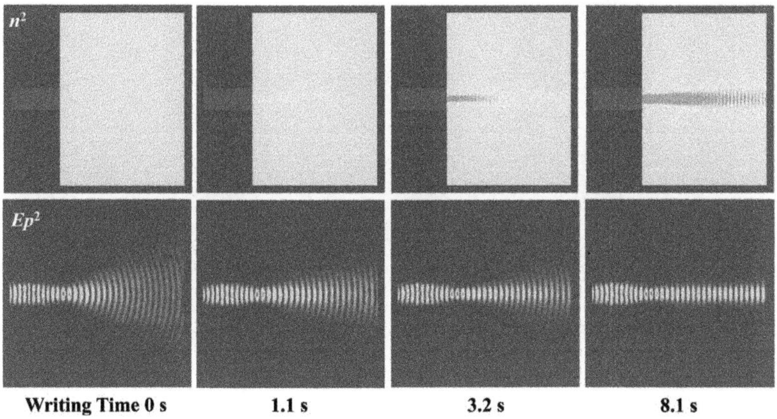

Figure 4.31. Distribution of refractive index (n^2) and probe beam intensity (Ep^2) in the reference model.

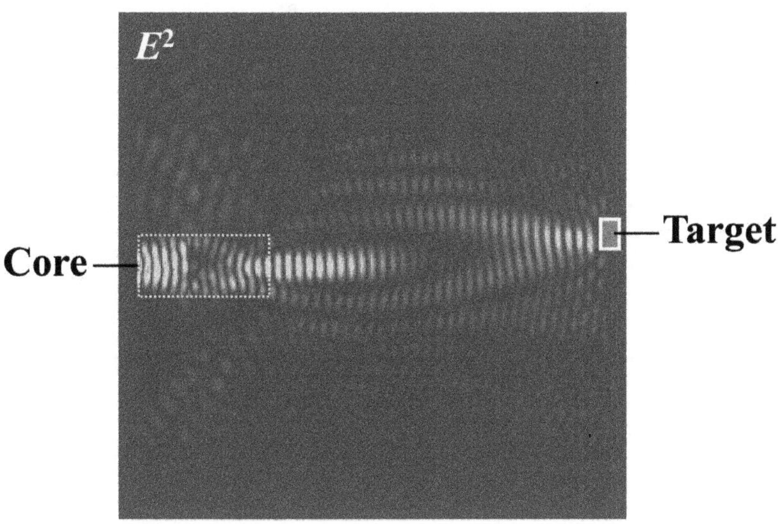

Figure 4.32. Distribution of write beam and photo-luminescence intensity (E^2) at writing time of 0.54 s in a model with a target of luminescent materials.

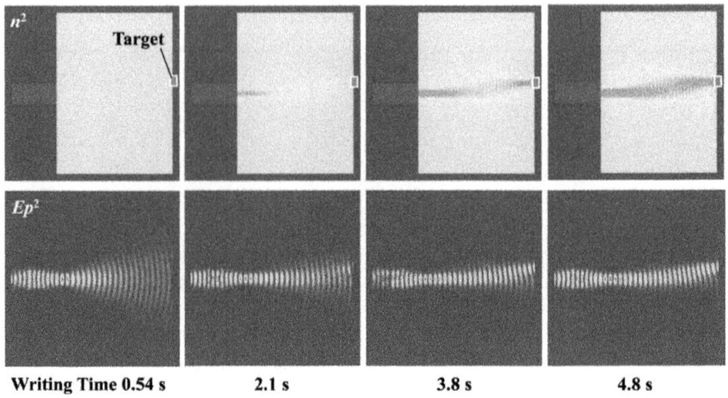

Figure 4.33. Distribution of refractive index (n^2) and probe beam intensity (Ep^2) in the model with a target of luminescent materials. See also Color Insert.

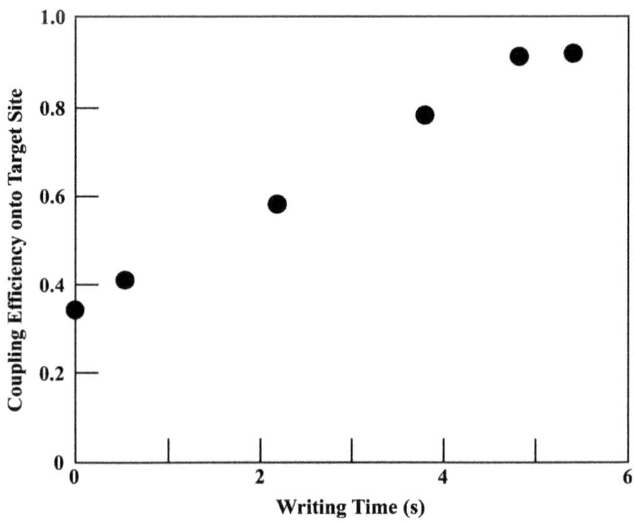

Figure 4.34. Coupling efficiency of the probe beam to the target site as a function of writing time.

The simulation result for the model with a target is shown in Fig. 4.33. It is found that with writing time, R-SOLNET is gradually constructed by the "pulling water" effect between the optical waveguide core and the target. As a result, the probe beam is guided to the target site. Figure 4.34 shows coupling efficiency of the probe beam to the target site as a function of writing time. Efficiency over 90% is expected after 4.8 s writing. These results reveal the prospect of the R-SOLNET constructed by luminescent materials.

Figure 4.35 shows distributions of write beam and photo-luminescence intensity, refractive index, and probe beam intensity

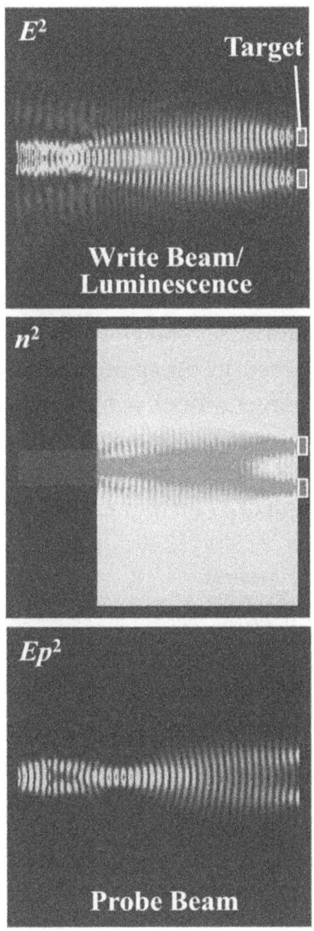

Figure 4.35. Distributions of write beam and photo-luminescence intensity (E^2), refractive index (n^2), and probe beam intensity (Ep^2) at writing time of 2.1 s in a model with two targets of luminescent materials.

in a model containing two targets [10]. Here, the wavelengths of the write beam, the photo-luminescence, and the probe beam are 650 nm, 700 nm, and 550 nm, respectively. It is found that R-SOLNET is constructed toward the two targets. This result suggests that the R-SOLNET has a capability to catch multiple targets.

4.5 SOLNET Using Two-Wavelength Write Beams

Some molecules are known to exhibit two-photon absorption characteristics [11, 12]. Using the two-photon absorption, SOLNET is expected to be formed by two write beams with different wavelengths. The two-photon absorption arises from the following processes shown in Fig. 4.36. In the three-level model, first, an electron in the molecule is excited to excited state 1 by absorbing a photon with a wavelength of λ_1. Then, the excited electron in the excited state 1 is excited to excited state 2 by absorbing a photon with another wavelength of λ_2. The electron or energy in the excited state 2 can be transferred to other molecules to initiate chemical reactions. In the four-level model, a photon with a wavelength of λ_1 excites an electron from S_0 state to S_n state in a molecule. The excited electron in the S_n state transfers to T_0 state. Then, a photon with another wavelength of λ_2 further excites the electron to T_n state

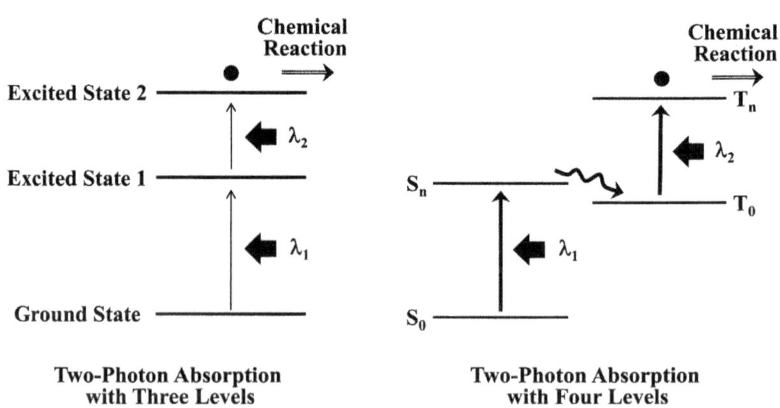

Figure 4.36. Two-photon absorption processes with two wavelengths.

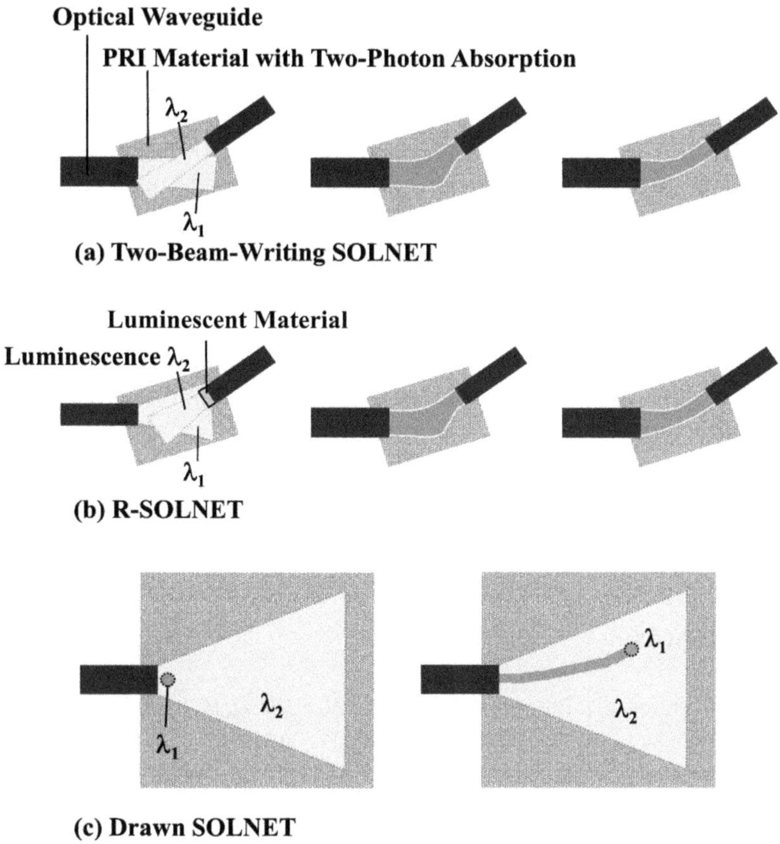

Figure 4.37. SOLNET construction processes utilizing two-wavelength write beams.

to induce chemical reaction. Due to these mechanisms, the chemical reactions start only when both λ_1 photons and λ_2 photons come into the material systems.

By using the two-photon absorption, unique SOLNET will be constructed [3]. In two-beam-writing SOLNET, as shown in Fig. 4.37(a), smooth self-focusing is expected by using two wavelengths write beams since a write beam with λ_2 is absorbed only in regions where a write beam with λ_1 exists, widening the overlap region of the write beams. R-SOLNET will be constructed by two-wavelength write beams using luminescent materials as shown Fig. 4.37(b). R-SOLNET grows by a write beam with λ_1 from an optical waveguide and photo-luminescence with λ_2 from the luminescent material.

In addition, the SOLNET using two-wavelength write beams will enable the drawn SOLNET shown in Fig. 4.37(c). In PRI material, a write beam with λ_2 propagates to provide sheet-like region where a write beam with λ_2 exists. The spot of a write beam with λ_1 is scanned in the region. Since refractive index increase is induced only at the cross points of the sheet-like region of λ_2 and the spot of the write beam with λ_1, channel-like SOLNET can be obtained by scanning the spot. By moving the sheet-like region with λ_2 up and down, three-dimensional drawing may become possible.

Although the SOLNET using two-wavelength write beams was not demonstrated experimentally, it might be one of the promising directions for future SOLNET improvement.

References

1. T. Yoshimura, J. Roman, Y. Takahashi, W. V. Wang, M. Inao, T. Ishitsuka, K. Tsukamoto, K. Motoyoshi, and W. Sotoyama, "Self-Organizing Waveguide Coupling Method 'SOLNET' and Its Application to Film Optical Circuit Substrates," *Proc. 50th Electron. Components Technol. Conf. (ECTC)*, 962–969 (2000).
2. T. Yoshimura, J. Roman, Y. Takahashi, W. V. Wang, M. Inao, T. Ishituka, K. Tsukamoto, K. Motoyoshi, and W. Sotoyama, "Self-Organizing Lightwave Network (SOLNET) and Its Application to Film Optical Circuit Substrates," *IEEE Trans. Comp., Packag. Technol.* **24**, 500–509 (2001).
3. T. Yoshimura, W. Sotoyama, K. Motoyoshi, T. Ishitsuka, K. Tsukamoto, S. Tatsuura, H. Soda, T. Yamamoto, "Method of Producing Optical Waveguide System, Optical Device and Optical Coupler Employing the Same, Optical Network and Optical Circuit Board," U.S. Patent 6,081,632 (2000).
4. T. Yoshimura, M. Ojima, Y. Arai, and K. Asama, "Three-Dimensional Self-Organized Micro Optoelectronic Systems for Board-Level Reconfigurable Optical Interconnects — Performance Modeling and Simulation," *IEEE J. Select. Top. Quantum Electron.* **9**, 492–511 (2003).

5. T. Yoshimura, T. Inoguchi, T. Yamamoto, S. Moriya, Y. Teramoto, Y. Arai, T. Namiki, and K. Asama, "Self-Organized Lightwave Network Based on Waveguide Films for Three-Dimensional Optical Wiring Within Boxes," *J. Lightwave Technol.* **22**, 2091–2100 (2004).
6. T. Yoshimura, A. Hori, Y. Yoshida, Y. Arai, H. Kurokawa, T. Namiki, and K. Asama, "Coupling Efficiencies in Reflective Self-Organized Lightwave Network (R-SOLNET) Simulated by the Beam Propagation Method," *IEEE Photon. Technol. Lett.* **17**, 1653–1655 (2005).
7. H. Kaburagi and T. Yoshimura, "Simulation of Micro/Nano-Scale Self-Organized Lightwave Network (SOLNET) Using the Finite Difference Time Domain Method," *Opt. Commun.* **281**, 4019–4022 (2008).
8. T. Yoshimura and H. Kaburagi, "Self-Organization of Optical Waveguides Between Misaligned Devices Induced by Write-Beam Reflection," *Appl. Phys. Express* **1**, 062007-1-062007-3 (2008).
9. T. Yoshimura, K. Wakabayashi, and S. Ono, "Analysis of Reflective Self-Organized Lightwave Network (R-SOLNET) for Z-Connections in Three-Dimensional Optical Circuits by the Finite Difference Time Domain Method," *IEEE J. Select. Top. Quantum Electron.* **17**, 566–570 (2011).
10. T. Yoshimura, C. Yoshino, K. Sasaki, T. Sato, and M. Seki, "Cancer Therapy Utilizing Molecular Layer Deposition and Self-Organized Lightwave Network -Proposal and Theoretical Prediction-," *IEEE J. Select. Topics in Quantum Electron.* **18**, Biophotonics 1, May/June 2012 [to be published].
11. C. Brauchle, U. P. Wild, D. M. Burland, G. C. Bjorkund, and D. C. Alvares, "Two-Photon Holographic Recording with Continuous-Wave Lasers in the 750–1100 nm Range," *Opt. Lett.* **7**, 177–179 (1982).
12. V. D. McGinniss and R. E. Schwerzel, "Photopolymerizable Composition Containing a Photosensitive Donor and Photoinitiating Acceptor," US Patent 4,571,377 (1986).

Chapter 5

Self-Organized Optical Waveguides: Experimental Demonstrations

As described in Chapter 4, proof of concept of SOLNET has been established by BPM and FDTD method theoretically. In this chapter, experimental demonstrations of one-beam-writing SOLNET, two-beam-writing SOLNET, and R-SOLNET are presented.

5.1 One-Beam-Writing SOLNET

5.1.1 *In Monomer/Binder-Type Photo-Polymers*

In order to observe one-beam-writing SOLNET construction in a free space filled with PRI materials [1, 2], the following setup was made. A single-mode (SM) optical fiber for 1.3 μm in wavelength with core diameter of 9.5 μm was put in a monomer/binder-type photo-polymer, which consists of carbazole monomers for high-refractive-index components and acrylic (or epoxy) polymer for low-refractive-index binder with addition of sensitizing agents and polymerization initiator. Maximum refractive index changes induced by write beams in the photo-polymer are typically ~0.02. A write beam of 80 nW from a high-pressure mercury lamp was introduced

Optical Electronics: Self-Organized Integration and Applications
Tetsuzo Yoshimura

ISBN 978-981-4310-90-1 (Hardcover), 978-981-4364-08-9 (eBook)
www.panstanford.com

from the optical fiber to the photo-polymer. A straight waveguide core was grown from the optical fiber core along the light beam propagation axis.

An interference microscopic photograph of a waveguide core of one-beam-writing SOLNET is shown in Fig. 5.1(a). The line grown from the core of the optical fiber by 2 min writing is the waveguide of SOLNET, where refractive index is higher than the surrounding area. The stretched optical waveguide has a width of almost the same as the fiber core diameter, showing an experimental proof of concept for one-beam-writing SOLNET. It can be seen from Fig. 5.1(b) that the light beam spreads due to the diffraction before SOLNET formation, while after SOLNET formation the light beam is confined in the SOLNET.

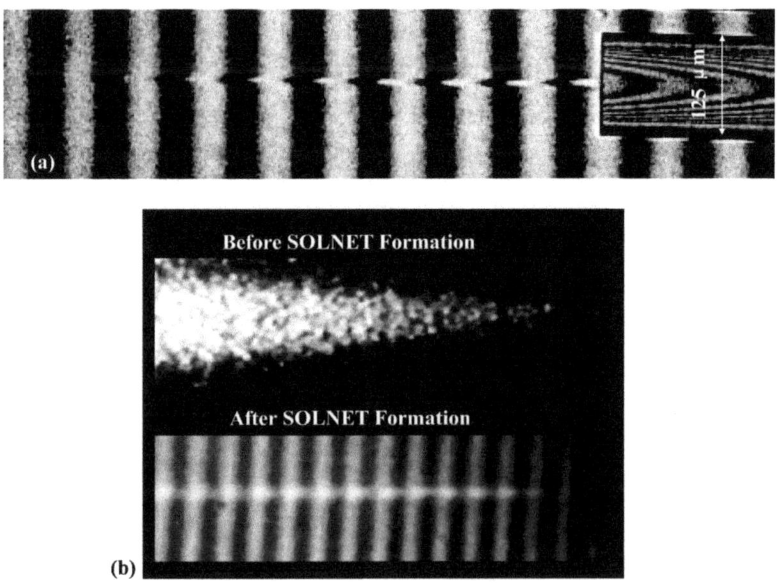

Figure 5.1. (a) Waveguide core of one-beam-writing SOLNET grown from an optical fiber with core diameter of 9.5 μm in monomer/binder type photo-polymer. (b) Light beam propagation before and after SOLNET formation. Reprinted with permission from T. Yoshimura, J. Roman, Y. Takahashi, W. V. Wang, M. Inao, T. Ishitsuka, K. Tsukamoto, K. Motoyoshi, and W. Sotoyama, "Self-Organizing Waveguide Coupling Method "SOLNET" and Its Application to Film Optical Circuit Substrates," *Proc. 50th Electron. Components Technol. Conf.* (ECTC), 962–969. © 2000 IEEE.

5.1.2 *In Monomer/Monomer-Type Photo-Polymers*

Figure 5.2 shows growth of one-beam-writing SOLNET from an optical fiber in monomer/monomer-type photo-polymer for writing time of 3 s and 60 s. Figures 5.3(a) and (b), respectively show one-beam-writing SOLNET grown from a multi-mode (MM) optical fiber with core diameter of 50 μm and a guided light beam in the SOLNET.

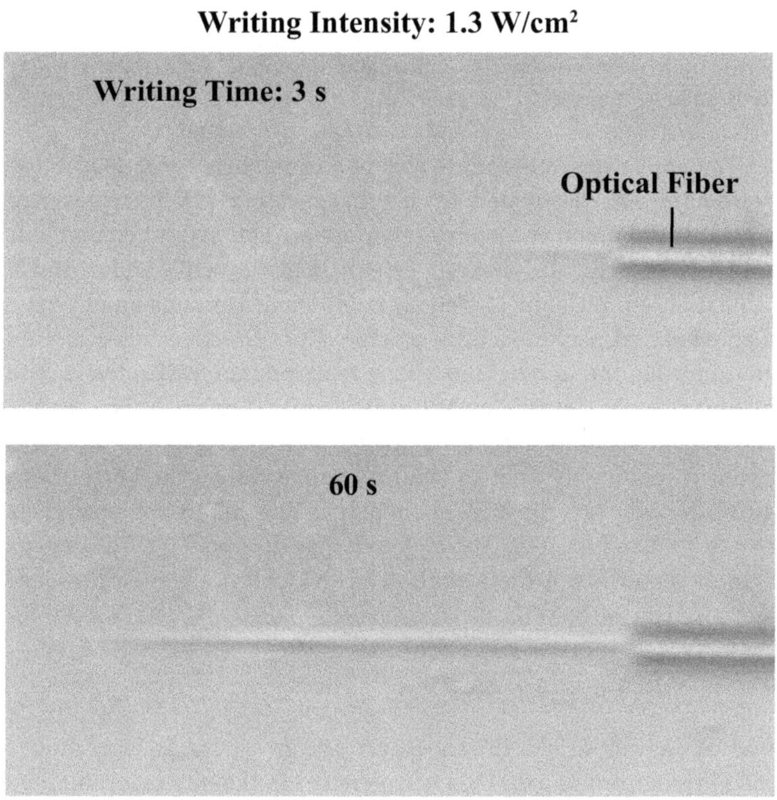

Figure 5.2. Waveguide core of one-beam-writing SOLNET grown from a MM optical fiber with core diameter of 50 μm in Monomer/Monomer type photo-polymer. T. Yoshimura, M. Miyazaki, Y. Miyamoto, N. Shimoda, A. Hori, and K. Asama, "Three-Dimensional Optical Circuits Consisting of Waveguide Films and Optical Z-Connections," *J. Lightwave Technol.* **24**, 4345–4352. © 2006 IEEE.

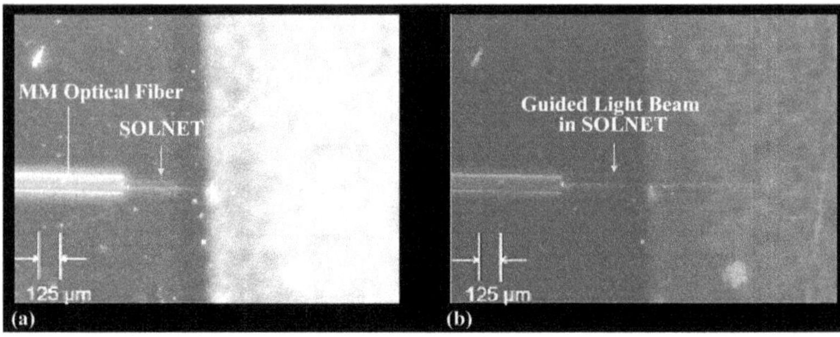

Figure 5.3. (a) One-beam-writing SOLNET grown from a MM optical fiber with core diameter of 50 μm in monomer/monomer type photo-polymer. (b) Guided light beams in the SOLNET.

The monomer/monomer-type photo-polymer used in the experiments was developed by JSR Corporation [3]. It consists of acrylic component and epoxy component. The acrylic component contains an aromatic diacrylate with high refractive index and a radical photo-initiator. The epoxy component contains an aliphatic diepoxide and a photoacid generator. The refractive index is 1.53 and 1.51 for the acrylic and epoxy components, respectively. The mixing ratio of the two components is 50:50 in wt%. The acrylic component exhibits larger reactivity than the epoxy component upon irradiation of light in the range of wavelength shorter than 400 nm, inducing the PRI phenomena. The photo-polymer gives rise to a maximum refractive index change around 0.02. The energy density required for the saturation is ~6 J/cm^2 of the high-pressure mercury lamp.

5.1.3 *Direct Growth from LD*

One-beam writing SOLNET can directly be grown from a laser diodes (LD) [4]. In front of a 680 nm wavelength LD, the monomer/binder-type photo-polymer was put. In this experiment the photo-polymer was sensitized by doping dye molecules for red region. As shown in Fig. 5.4(a), by emitting a light beam from the LD, an optical waveguide stretching from the LD appeared in the photo-polymer in ~10 s. The write beam from the LD was spontaneous emission

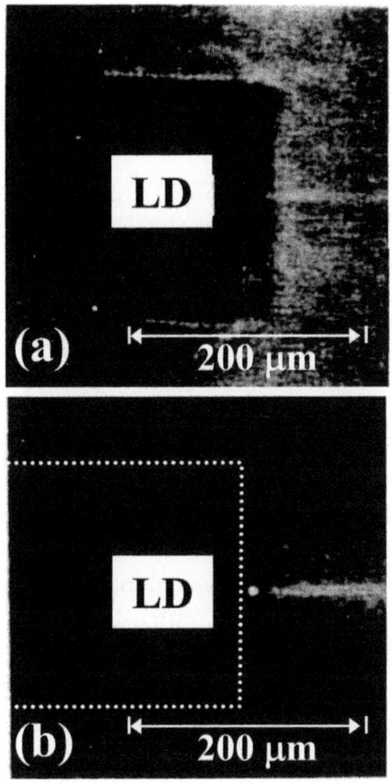

Figure 5.4. One-beam-writing SOLNET growth from 680 nm wavelength LD in photo-polymer. (a) A self-organized waveguide stretching from a waveguide edge of the LD. (b) LD light beams confined into the SOLNET. Reprinted with permission from T. Yoshimura, T. Inoguchi, T. Yamamoto, S. Moriya, Y. Teramoto, Y. Arai, T. Namiki, and K. Asama, "Self-Organized Lightwave Network Based on Waveguide Films for Three-Dimensional Optical Wiring Within Boxes," *J. Lightwave Technol.* **22**, 2091–2100. © 2004 IEEE.

with power of 20 μW, and was modulated with a duty ratio of 1/10, implying that the effective write beam power is 2 μW. The LD light beam was confined into the optical waveguide as shown in Fig. 5.4(b). Coupling loss was measured between the LD and an optical fiber with core diameter of 9.5 μm placed in a counter configuration with a distance of 100 μm. Initial loss was 14 dB. After

SOLNET formation, the loss was reduced to 7 dB. The result indicates that SOLNET is expected applicable to couplings between LD and optical waveguides, and the mode size conversion of LD beams from wide-diffraction-angle state of 10–20° to appropriate beam shapes.

In the SOLNET grown from a LD directly, the wavelength of write beams and signal beams are the same. This causes a concern of an insertion loss increase due to light absorption by photo-polymers. However, the dyes doped in photo-polymers for spectral sensitization are bleached by absorbing write beams because chemical reactions are initiated by dissociation of the dye molecules. Furthermore, since the length of SOLNET is in the order of several tens to several hundreds μm, light absorption of the material does not affect largely on insertion losses.

One of the challenges in direct SOLNET growth from a LD is post curing of the photo-polymer. After forming SOLNET by a write beam, signal laser beams of the same wavelength as the write beam are transmitted in the SOLNET with much higher intensity, causing a concern that further reactions occur in the photo-polymer, which destroys the refractive index distribution in the SOLNET. Post curing conditions should be optimized for blanket UV light exposure or baking.

5.2 Two-Beam-Writing SOLNET

5.2.1 *In Monomer/Binder-Type Photo-Polymers*

In order to demonstrate two-beam-writing SOLNET [1, 2], two optical fibers with core diameter of 9.5 μm were put in a glass cylinder with counter directions. The gap of 400 μm between the optical fibers was filled with the monomer/binder-type photo-polymer. Write beams of 40 nW from a high-pressure mercury lamp were introduced from the two optical fibers into the photo-polymer. An optical waveguide connecting the two optical fibers was gradually formed with time.

An optical waveguide formed by 5 min writing is shown in Fig. 5.5(a). The coupling efficiency between the two optical

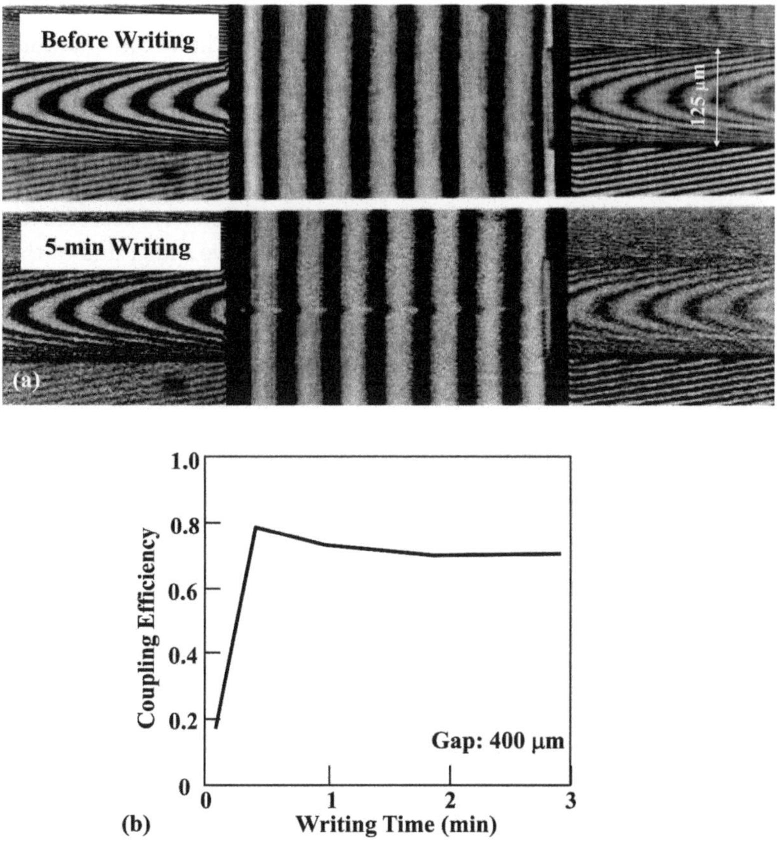

Figure 5.5. (a) Waveguide core of two-beam-writing SOLNET grown between two optical fibers with core diameter of 9.5 μm in monomer/binder type photo-polymer. (b)Coupling efficiency versus writing time characteristics in the two-beam-writing SOLNET. Reprinted with permission from T. Yoshimura, J. Roman, Y. Takahashi, W. V. Wang, M. Inao, T. Ishitsuka, K. Tsukamoto, K. Motoyoshi, and W. Sotoyama, "Self-Organizing Waveguide Coupling Method "SOLNET" and Its Application to Film Optical Circuit Substrates," *Proc. 50th Electron. Components Technol. Conf. (ECTC)*, 962–969. © 2000 IEEE.

fibers was measured using a probe light beam of 1.3 μm in wavelength, to which the photo-polymer is insensitive. As shown in Fig. 5.5(b), with writing time, the coupling efficiency increases up to 0.8, then decreases slightly toward a saturated value. The writing time dependence of coupling efficiency is similar to that predicted by the BPM simulation (see Fig. 4.9). The results demonstrate an experimental proof of concept for two-beam-writing SOLNET.

The decrease in the coupling efficiency beyond the peak, which appears in Fig. 5.5(b), might be attributed to over-exposure. The over-exposure may cause the coupling efficiency decrease in following two ways. Excessive change in refractive index due to the over-exposure of write beams (1) increases the beam confinement ability of SOLNET to make the mode size in SOLNET smaller than that in input/output waveguides or (2) widens the waveguide core of SOLNET to make the mode size larger than that of input/output waveguides. These mode size mismatches cause loss at the interface between SOLNET and input/output waveguides, giving rise to the coupling efficiency reduction. After sufficient amount of write beam exposure, the write beams are completely confined in the SOLNET core, and at the same time, the refractive index reaches its saturation limit. Then, no more waveguide shape changes are induced, resulting in the coupling efficiency saturation.

5.2.2 *In Monomer/Monomer-Type Photo-Polymers*

Figure 5.6 shows two-beam-writing SOLNET connecting two optical fibers with a core diameter of 9.5 μm, where, the monomer/monomer-type photo-polymer is used [5]. For lateral misalignments of 1 μm to 9 μm, self-aligned coupling waveguides of SOLNET are constructed. For lateral misalignment of 20 μm, a coupling waveguide is not constructed. Optical waveguides grown from the two optical fibers do not merge, but, remain as two separated optical waveguides. This is attributed to the fact that overlap of the write beams emitted from the two optical fibers becomes small when the lateral misalignment increases to 20 μm.

Dependence of coupling efficiency on lateral misalignment for the two-beam-writing SOLNET and for butt joint is compared in Fig. 5.7(a). Two-beam-writing SOLNET exhibits wider tolerance for misalignment than the butt joint. These results confirm the

Figure 5.6. Waveguide core of two-beam-writing SOLNET grown between two optical fibers with lateral misalignment of 1, 3, 9, and 20 μm.

superiority of two-beam-writing SOLNET as a self-aligned optical coupling method.

In Fig. 5.7(b), coupling efficiency change during two-beam-writing SOLNET formation is shown for lateral misalignment of 1 μm. With writing time, the coupling efficiency increases up to 0.85, then decreases slightly to reach a saturated point. The maximum coupling efficiency of 0.85 corresponds to coupling loss of ~0.7 dB.

The writing time dependence of the coupling efficiency is similar to that in the experiment using the monomer/binder-type photopolymer (see Fig. 5.5(b)), and also similar to that predicted by BPM simulation (see Fig. 4.9). The results suggest that the prediction by the simulations using BPM provides general characteristics of SOLNET independently of PRI materials.

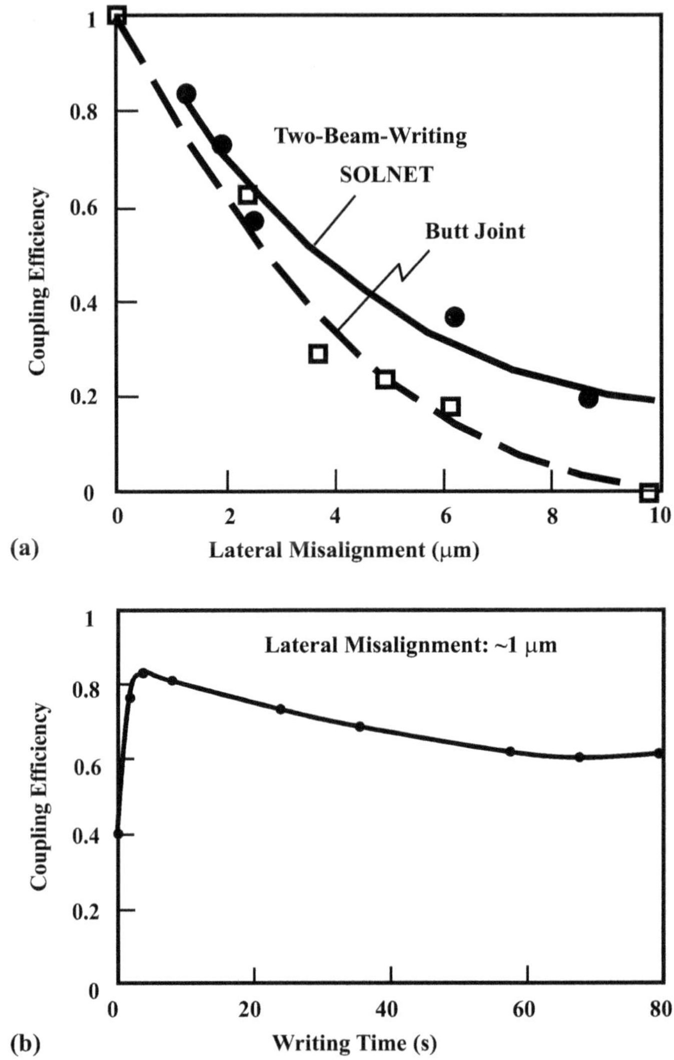

Figure 5.7. (a) Dependence of coupling efficiency on lateral misalignment for the two-beam-writing SOLNET and for butt joint. (b)Coupling efficiency change observed during two-beam-writing SOLNET formation for lateral misalignment of 1 μm. Reprinted with permission from T. Yoshimura, A. Hori, Y. Yoshida, Y. Arai, H. Kurokawa, T. Namiki, and K. Asama, "Coupling Efficiencies in Reflective Self-Organized Lightwave Network (R-SOLNET) Simulated by the Beam Propagation Method," *IEEE Photon. Technol. Lett.* **17**, 1653–1655. © 2005 IEEE.

5.3 R-SOLNET

5.3.1 *Between a Window and a Mirror*

To demonstrate the proof of concept for R-SOLNET, a setup shown in Fig. 5.8(a) was prepared [4]. The monomer/monomer-type photo-polymer is inserted between two glass substrates placed in parallel with a gap of ~200 μm. On the upper substrate, Al mask with a circle window is deposited. The diameter of the window is 100 μm. On the lower substrate, Al mirrors with various kinds of patterns are formed. By introducing an UV light beam from a high-pressure

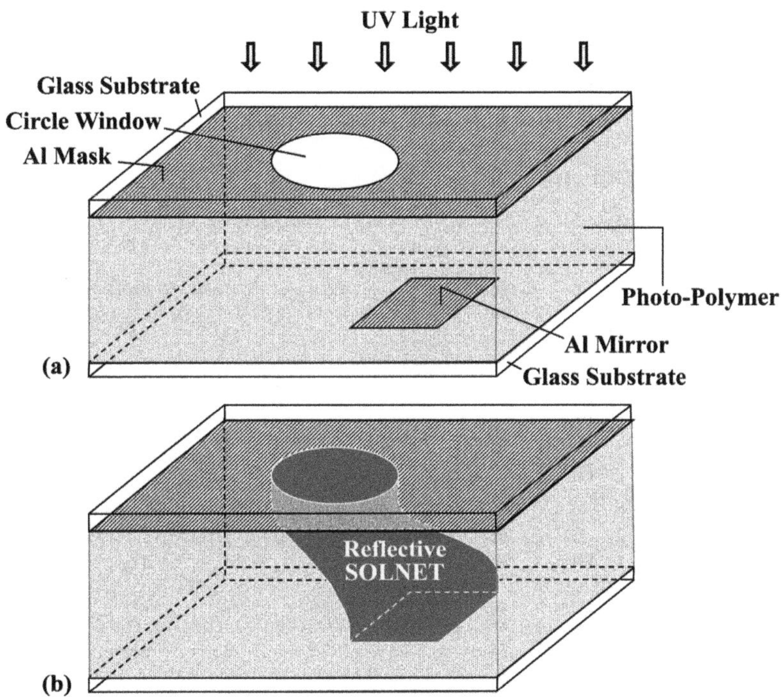

Figure 5.8. (a) Setup for experimental demonstration of R-SOLNET and (b) a schematic illustration of expected structure of R-SOLNET. Reprinted with permission from T. Yoshimura, T. Inoguchi, T. Yamamoto, S. Moriya, Y. Teramoto, Y. Arai, T. Namiki, and K. Asama, "Self-Organized Lightwave Network Based on Waveguide Films for Three-Dimensional Optical Wiring Within Boxes," *J. Lightwave Technol.* **22**, 2091–2100. © 2004 IEEE.

mercury lamp into the setup through the circle window on the upper substrate, R-SOLNET is expected to be formed as shown in Fig. 5.8(b). A write beam comes from the circle window and is reflected from the mirror. It should be noted that in actual cases the mirrors are replaced with wavelength filters.

Figure 5.9 shows photographs of SOLNET. They were taken from the lower-substrate side by detaching the substrate after SOLNET formation.

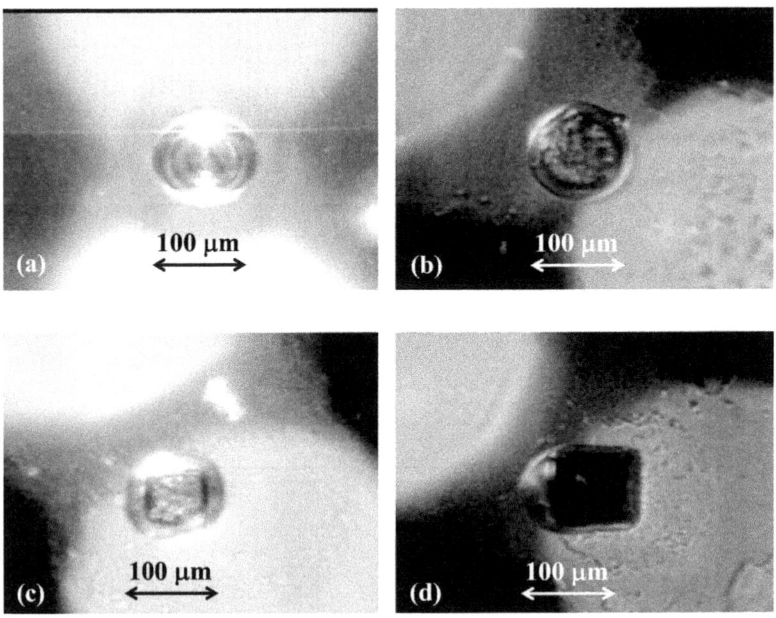

Figure 5.9. Photographs of one-beam-writing SOLNET and R-SOLNET. (a) One-beam-writing SOLNET stretching from a circle window on the upper substrate. (b) R-SOLNET between a circle window and a circle mirror with no misalignment. (c) R-SOLNET between a circle window and a square mirror, which is smaller than the circle window, with no misalignment. (d) R-SOLNET between a circle window and a square mirror with misalignment. Reprinted with permission from T. Yoshimura, T. Inoguchi, T. Yamamoto, S. Moriya, Y. Teramoto, Y. Arai, T. Namiki, and K. Asama, "Self-Organized Lightwave Network Based on Waveguide Films for Three-Dimensional Optical Wiring Within Boxes," textitJ. Lightwave Technol. **22**, 2091–2100. © 2004 IEEE.

Figure 5.9(a) is a photograph of one-beam-writing SOLNET stretching from the window on the upper substrate. In this case, the lower substrate was not placed during write beam exposure. A focused down-tapered optical waveguide is observed. Figure 5.9(b) shows R-SOLNET for a case that a 100 μm-diameter circle mirror is formed on the lower substrate with no misalignment against the circle window. Figure 5.9(c) shows R-SOLNET between the circle window and a square mirror, which is smaller than the circle window and is formed with no misalignment. An optical waveguide with a circle bottom and a square top is observed. Figure 5.9(d) shows R-SOLNET for a square mirror on the lower substrate with 30 μm misalignment. An optical waveguide with a circle bottom and a square top can be seen, where a displacement exists between the top and the bottom according to the misalignment between the circle window and the square mirror. These results demonstrate R-SOLNET construction.

5.3.2 *Between an Optical Fiber and a Mirror with Angular Misalignment*

Figures 5.10 and 5.11 show experimental demonstration of R-SOLNET using optical fibers [6]. R-SOLNET formation is attempted between a MM optical fiber with a core diameter of 50 μm and a mirror deposited on the facet of an optical fiber. Figure 5.10(a) shows a setup for experiments of R-SOLNET formation with angular misalignment. The MM optical fiber and the mirror are placed with a gap of ~800 μm in the monomer/monomer-type photo-polymer. An angular misalignment of 3° exists between the fiber direction and surface-normal direction of the mirror. Figure 5.10(b) shows a photograph of the MM optical fiber and the optical fiber with the Al mirror on the facet in the photo-polymer before SOLNET formation. When a write beam from a high-pressure mercury lamp was introduced from the MM optical fiber into the photo-polymer, R-SOLNET was formed. It is found from Fig. 5.10(c) that a probe beam of 650 nm in wavelength, which is introduced from the MM optical fiber, propagates in the optical waveguide of SOLNET. The core width of the R-SOLNET is about 50 μm, which is close to that of the MM optical fiber.

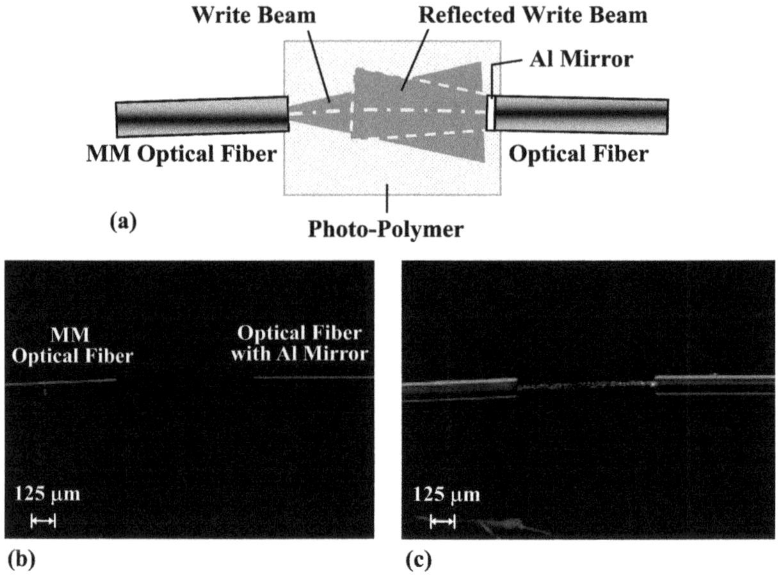

Figure 5.10. Experiments of R-SOLNET formation. Angular misalignment of 3° exists between MM optical fiber direction and surface-normal direction of the mirror. (a) Experimental setup. (b) A photograph of the MM optical fiber and the optical fiber with the Al mirror on the facet in the photo-polymer before SOLNET formation. (c) Probe beam propagation in the bow-shaped self-aligned waveguide of R-SOLNET. Reprinted with permission from T. Yoshimura and H. Kaburagi, "Self-Organization of Optical Waveguides between Misaligned Devices Induced by Write-Beam Reflection," *Appl. Phys. Express* **1**, 062007- 1-3 (2008). See also Color Insert.

If the "pulling water" effect does not exist, two optical waveguides of SOLNET with directional difference of 6° should be formed from the mirror by the incident write beam and the reflected write beam. However, in the photograph shown in Fig. 5.10(c), only one bow-shaped optical waveguide of R-SOLNET is observed. This indicates that a self-aligned optical waveguide of R-SOLNET is formed due to the "pulling water" effect induced by the reflected write beam from the mirror.

As can be seen from Fig. 5.11, R-SOLNET is constructed for angular misalignment of 6° or less. When the angular misalignment becomes larger than 6°, two separated optical waveguides of SOLNET grow from the mirror.

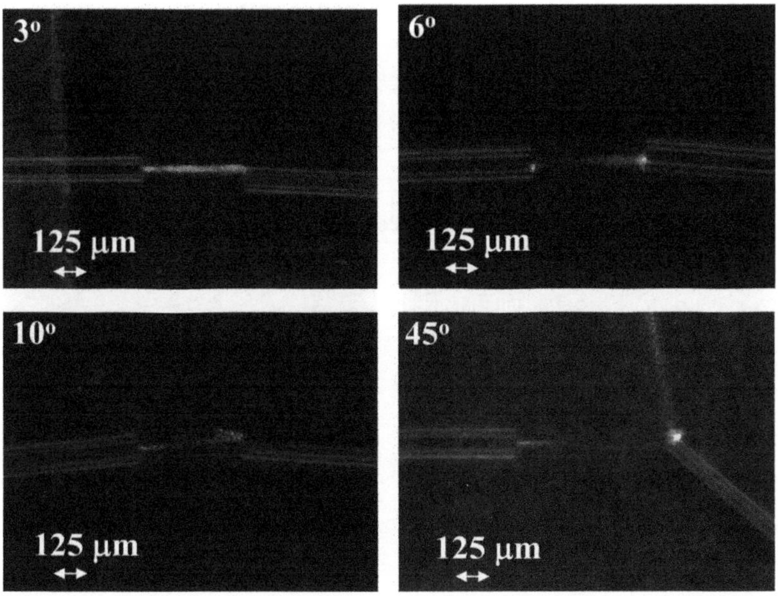

Figure 5.11. Experiments of R-SOLNET formation for angular misalignment of 3°, 6°, 10°, and 40°.

5.3.3 *Between an Optical Fiber and a Mirror with Lateral Misalignment*

Figures 5.12(a) and (b) show a setup for experiments of R-SOLNET formation and a photograph of the MM optical fiber and the optical fiber with the Al mirror in the photo-polymer before SOLNET formation [6]. The lateral misalignment between the optical fiber and the Al mirror is about 60 μm. When a write beam was introduced from the MM optical fiber into the photo-polymer, R-SOLNET was formed, connecting the MM optical fiber to the misaligned mirror. It is found from Fig. 5.12(c) that a probe beam of 650 nm propagates in the S-shaped self-aligned waveguide of R-SOLNET with core width of about 50 μm.

It should be noted that R-SOLNET is connected to the upper part of the fiber facet. As can be seen from Fig. 5.13, the mirror is not deposited uniformly on the fiber facet, but is deposited partially. This is the reason why R-SOLNET is pulled to the upper part. Actually,

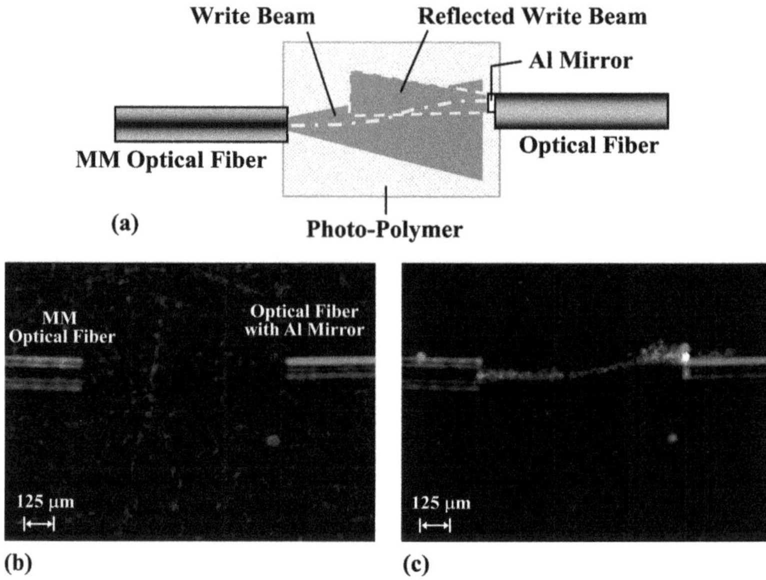

Figure 5.12. Experiments of R-SOLNET formation. Lateral misalignment of 60 μm exists between the MM optical fiber and the Al mirror. (a) Experimental setup. (b) A photograph of the MM optical fiber and the optical fiber with the Al mirror in the photo-polymer before SOLNET formation. (c) Probe beam propagation in the S-shaped self-aligned waveguide of R-SOLNET. Reprinted with permission from T. Yoshimura and H. Kaburagi, "Self-Organization of Optical Waveguides between Misaligned Devices Induced by Write-Beam Reflection," *Appl. Phys. Express* **1**, 062007- 1-3 (2008). See also Color Insert.

trace of SOLNET is found on the upper part of the fiber facet, where the mirror exists.

In order to measure the coupling efficiency variation during R-SOLNET formation, R-SOLNET was formed by a semitransparent Al mirror. The setup for the experiments is shown in Fig. 5.14(a). R-SOLNET can be formed by an incident write beam from the MM optical fiber and a reflected beam from the semitransparent Al mirror deposited on a glass substrate. The write beam on the semitransparent mirror surface was observed by a camera from the back side. The results are shown in Fig. 5.14(b). Initially, the write beam reaches left side of the semitransparent mirror. With writing time the write beam is pulled to the semitransparent mirror.

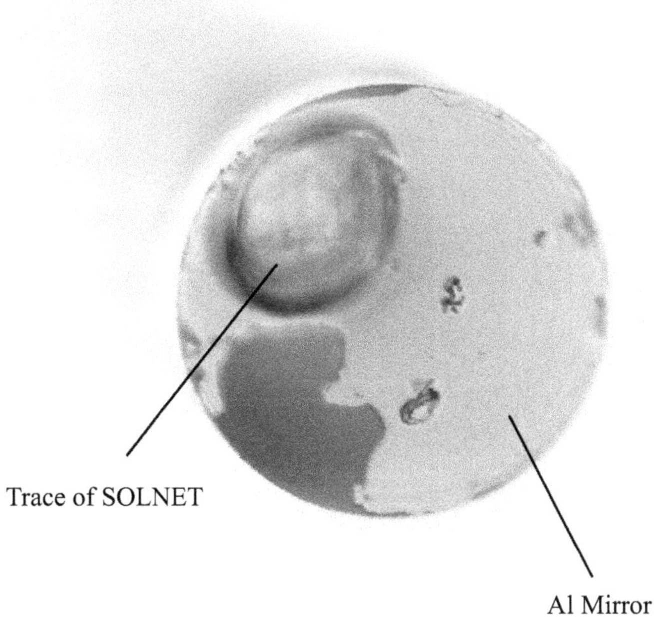

Figure 5.13. A photograph of the mirror on the fiber facet. Reprinted with permission from T. Yoshimura and H. Kaburagi, "Self-Organization of Optical Waveguides between Misaligned Devices Induced by Write-Beam Reflection," *Appl. Phys. Express* **1**, 062007- 1-3 (2008).

Figure 5.15 shows coupling efficiency variation with writing time. The efficiency was monitored using an optical fiber placed just behind the semitransparent Al mirror. With R-SOLNET formation, the coupling efficiency increases. After reaching the peak, the coupling efficiency decreases due to over-exposure.

5.3.4 *Between an Optical Fiber and a Luminescent Target of Phosphor*

In order to demonstrate R-SOLNET with luminescent materials described in Section 4.4.6, as depicted in Fig. 5.16(a), a luminescent

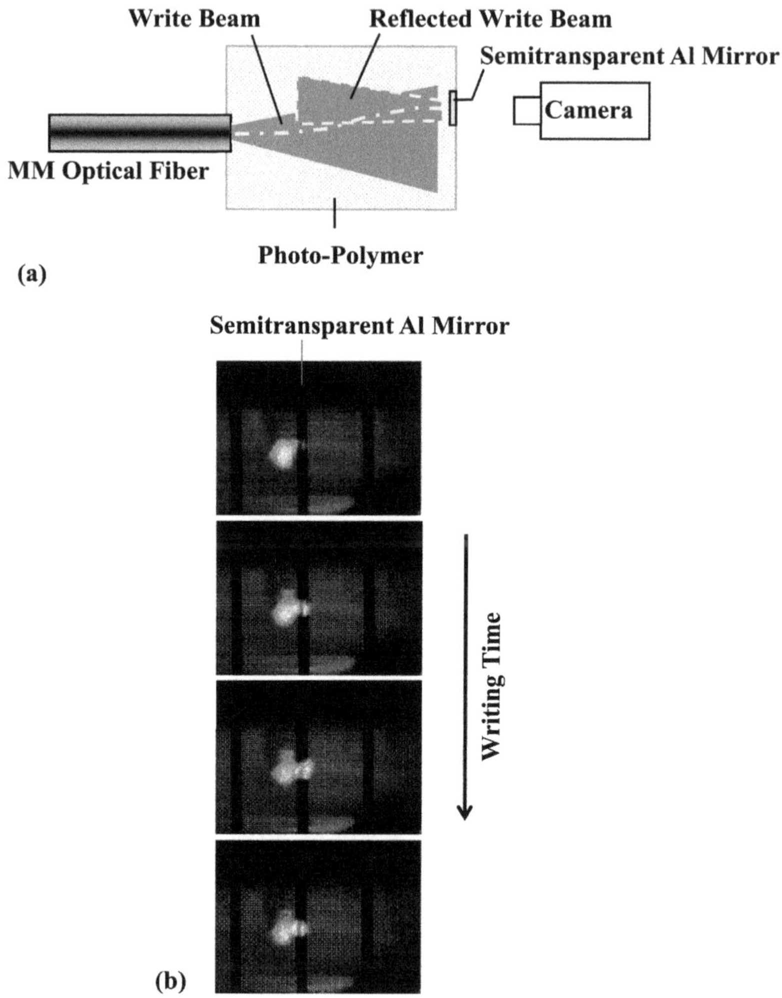

Figure 5.14. Experiments of R-SOLNET formation using semi-transparent Al mirrors. (a) Experimental setup. (b) Photographs taken by a camera from the backside of the semi-transparent mirror.

target was put in a PRI material [7]. In the present experiment, tris(8-hydroxyquinolinato)aluminum (Alq3) was used for the luminescent target. The PRI material was dye-sensitized to extend its spectral response to green-red light regions, enabling the PRI material to work for both the write beam and photo-luminescence.

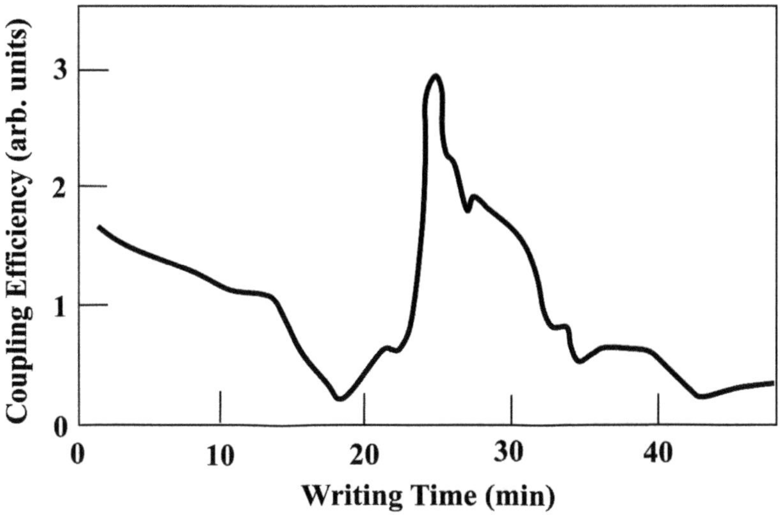

Figure 5.15. Coupling efficiency variation with writing time.

As shown in Fig. 5.16(b), when a write beam of blue LD (405 nm) was introduced into the PRI material from an optical fiber, photo-luminescence was emitted from the luminescent target of Alq3. It is found in the figure that the write beam is pulled to the Alq3 target. Accordingly, a probe beam of red LD (650 nm) introduced from the optical fiber is guided toward the Alq3 target. These results indicate that R-SOLNET with luminescent materials was successfully formed by the write beam and the photo-luminescence from the luminescent target.

5.4 High-Index-Contrast SOLNET

A PRI sol-gel material, which was developed by Nissan Chemical Industries, LTD., is a promising material since the material exhibits a large refractive index increase from 1.65 to 1.85 with UV/blue-light exposure and baking. This index contrast is sufficient to produce nano-scale waveguide structures [8]. Therefore, the PRI sol-gel material enables us to fabricate SOLNET with strong light beam confinement [9]. The mechanism of the refractive index change in the material is described in Section 6.4.1.

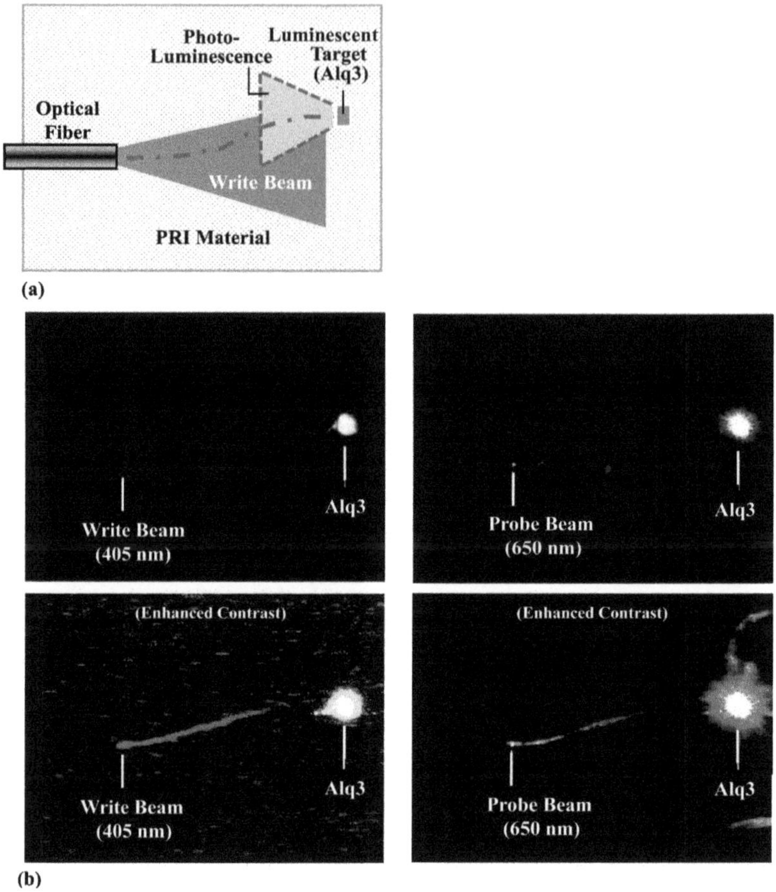

Figure 5.16. Experiments of R-SOLNET formation between a MM optical fiber and a luminescent target of Alq3 with lateral misalignment. (a) Experimental setup. (b) Photographs of write beam propagation and probe beam propagation in R-SOLNET. The contrast of the images at the bottom is enhanced. (This work was performed by Mitsutoshi Shirasaki, a temporal student researcher from Purdue University, in the summer 2011 in T. Yoshimura Lab). See also Color Insert.

5.4.1 *SOLNET in PRI Sol-Gel Materials*

The concept of SOLNET formation using the PRI sol-gel materials is shown in Fig. 5.17. When a write beam is introduced from an optical device into a film of the PRI sol-gel material under baking,

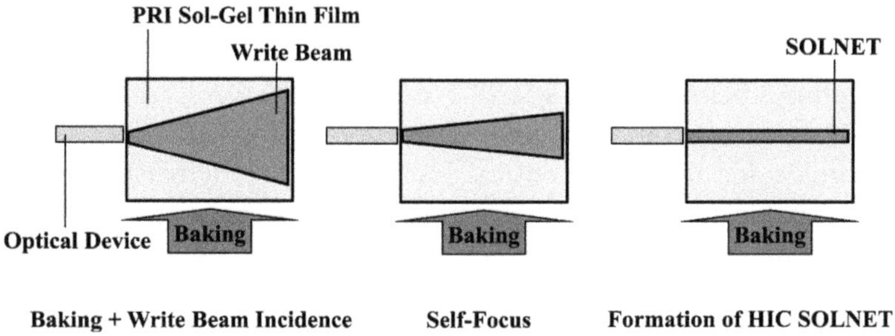

Figure 5.17. The concept of SOLNET formation using PRI sol-gel materials. Reprinted with permission from S. Ono, T. Yoshimura, T. Sato, J. Oshima, "Fabrication of Self-Organized Optical Waveguides in Photo-Induced Refractive Index Variation Sol-Gel Materials with Large Index Contrast," *J. Lightwave Technol.* **27**, 5308–5313. © 2009 IEEE.

self-focus is induced, resulting in a high-index-contrast (HIC) SOLNET. The details of SOLNET formation process is shown in Fig. 5.18. On a Si substrate with a 2000 nm-thick SiO_2 film for an under-clad layer, the PRI sol-gel material is coated and prebaked to form the PRI sol-gel thin film of approximately 230 nm thick. A write beam with a wavelength of 405 nm is introduced from a SM mode optical fiber with a core diameter of 3 μm into the PRI sol-gel thin film that is baked during SOLNET formation. After SOLNET is formed, the PRI sol-gel thin film is cooled to room temperature, then,

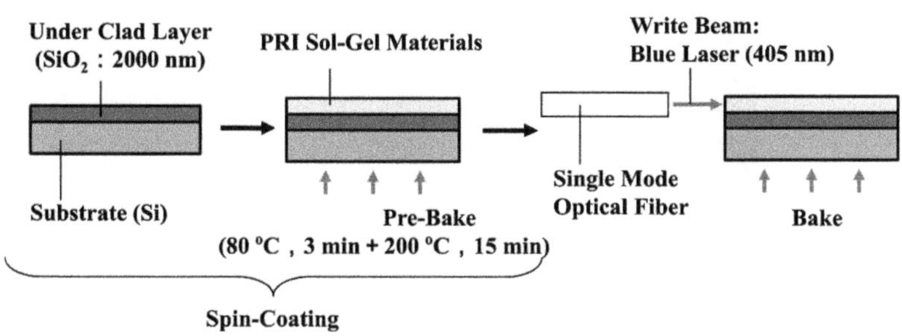

Figure 5.18. The details of the SOLNET formation process using PRI sol-gel materials. Reprinted with permission from S. Ono, T. Yoshimura, T. Sato, J. Oshima, "Fabrication of Self-Organized Optical Waveguides in Photo-Induced Refractive Index Variation Sol-Gel Materials with Large Index Contrast," *J. Lightwave Technol.* **27**, 5308–5313. © 2009 IEEE.

probe beams with wavelengths of 405 nm or 633 nm are introduced into the SOLNET to evaluate the light beam confinement effect and coupling efficiency of the SOLNET.

It has been reported that short-wavelength light beams are favorable for intra-chip optical interconnects [10] because they are suitable for the miniaturization of optical circuits and high-sensitivity photo-detection. For these reasons, we focused our attention on the short-wavelength region and used probe beam wavelengths of 405 nm and 633 nm.

Figure 5.19 shows SOLNET formed at various baking temperatures. The writing time was 1 min. An output beam of 0.5 mW from the SM optical fiber was introduced into the PRI sol-gel thin film. Although the actual light power guided into the PRI sol-gel thin film

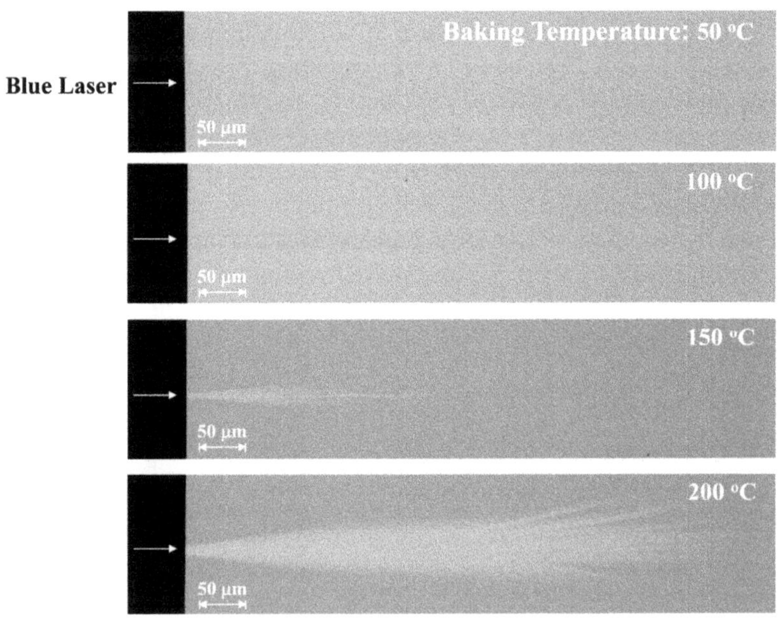

Figure 5.19. Baking temperature dependence of SOLNET formation. The write beam power is 0.5 mW and the writing time is 1 min. Reprinted with permission from S. Ono, T. Yoshimura, T. Sato, J. Oshima, "Fabrication of Self-Organized Optical Waveguides in Photo-Induced Refractive Index Variation Sol-Gel Materials with Large Index Contrast," *J. Lightwave Technol.* **27**, 5308–5313. © 2009 IEEE.

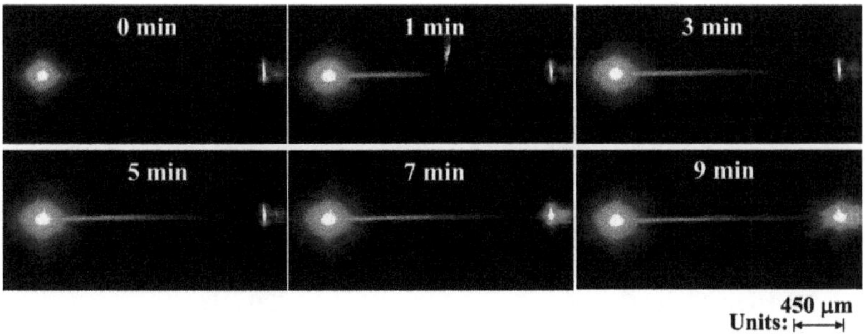

Figure 5.20. SOLNET formation with write beam power of 1.0 mW at a baking temperature of 200°C. Reprinted with permission from S. Ono, T. Yoshimura, T. Sato, J. Oshima, "Fabrication of Self-Organized Optical Waveguides in Photo-Induced Refractive Index Variation Sol-Gel Materials with Large Index Contrast," *J. Lightwave Technol.* **27**, 5308–5313. © 2009 IEEE.

was less than 0.5 mW due to coupling loss, we define the output beam power from the optical fiber as the writing power.

As the baking temperature was increased to 100°C, 150°C, and 200°C, SOLNET appeared, the refractive index contrast was enhanced, and the length of SOLNET increased. These results indicate that SOLNET formation reactivity of the PRI sol-gel materials increases with baking temperatures.

Figure 5.20 shows SOLNET formed with a writing power of 1.0 mW at a baking temperature of 200°C for writing time of 0, 1, 3, 5, 7, and 9 min. It can be seen that SOLNET grows with time toward the thin film right-side edge. Photographs of write beam propagation before and after SOLNET formation are shown in Fig. 5.21 for writing power of 1.0, 0.5, and 0.1 mW. In each case, as time advances, SOLNET is stretched and reaches the right-side edge, where, confined beam spots are observed.

5.4.2 *Light Beam Confinement and Coupling Efficiency*

The light beam confinement effect along the core width direction was investigated using near field patterns (NFPs). The NFPs of the output beams are shown in Fig. 5.22. As a measure of the light beam confinement, the half-width of the output beam spot was used.

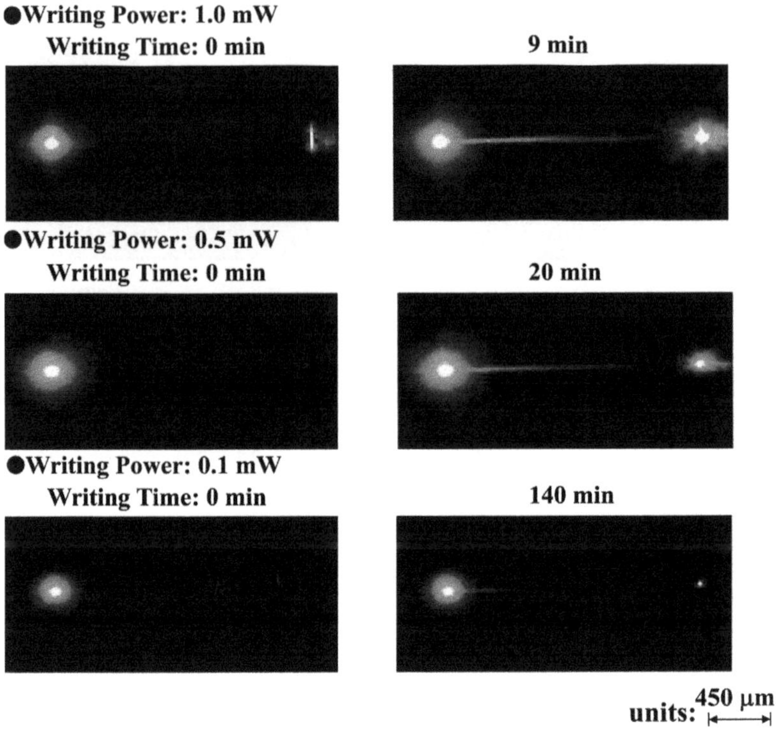

Figure 5.21. Photographs of write beam propagation before and after SOLNET formation for write beam power of 1 mW, 0.5 mW, and 0.1 mW. Reprinted with permission from S. Ono, T. Yoshimura, T. Sato, J. Oshima, "Fabrication of Self-Organized Optical Waveguides in Photo-Induced Refractive Index Variation Sol-Gel Materials with Large Index Contrast," *J. Lightwave Technol.* **27**, 5308–5313. © 2009 IEEE.

The relationship between the writing power and the light beam confinement is shown in Fig. 5.23 for the probe beam wavelength of 405 nm. The half-width of the output beam spot after SOLNET formation decreases from 23.8 μm to 11.8 μm as writing power decreases from 1.0 mW to 0.1 mW. The result indicates that SOLNET widths become narrow as writing power is reduced, suggesting that SOLNET formed with low writing power exhibits high-index-contrast characteristics.

Light beam confinement in SOLNET along the film thickness direction is determined by the PRI sol-gel film thickness. The

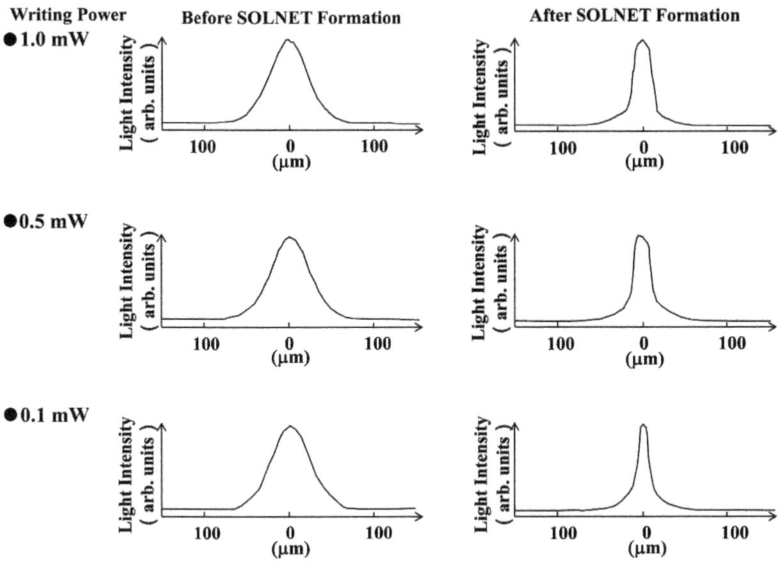

Figure 5.22. NFPs of output beams from the core edge of the fabricated SOLNET. Reprinted with permission from S. Ono, T. Yoshimura, T. Sato, J. Oshima, "Fabrication of Self-Organized Optical Waveguides in Photo-Induced Refractive Index Variation Sol-Gel Materials with Large Index Contrast," *J. Lightwave Technol.* **27**, 5308–5313. © 2009 IEEE.

half-width of the output beam spot along the film thickness direction is estimated to be 300 nm in the fabricated SOLNET.

As described in Section 4.4, we demonstrated by simulations using the FDTD method that light beam confinement of 0.5 μm are theoretically achievable in SOLNET [6,11] when the PRI material exhibits a large Δn, say, 0.2~0.3. However, experimentally available light beam confinement is not as strong as the predicted one. This might be attributed to the power and beam sizes of the input write beam. In simulations, writing power can be as small as desired. In the experiment, however, there are limitations in reducing the writing power because SOLNET formation with low writing power requires long writing time. In practice, it is difficult to keep the optical alignment of the measurement system within submicron accuracy for a long time due to vibration disturbance. Furthermore, in the simulation, a 0.5 μm-diameter input write beam was used, but

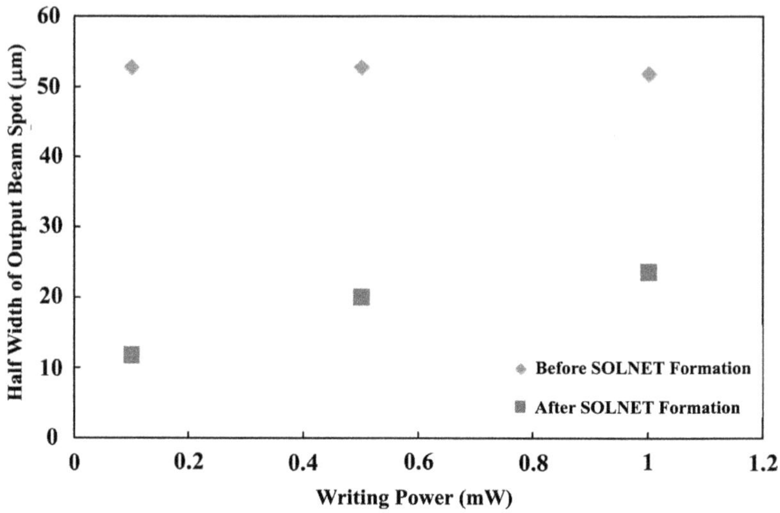

Figure 5.23. The relationship between write beam power and light beam confinement for a probe beam wavelength of 405 nm. Reprinted with permission from S. Ono, T. Yoshimura, T. Sato, J. Oshima, "Fabrication of Self-Organized Optical Waveguides in Photo-Induced Refractive Index Variation Sol-Gel Materials with Large Index Contrast," *J. Lightwave Technol.* **27**, 5308–5313. © 2009 IEEE.

in the experiment, a 3 μm-diameter input write beam was used. This is not favorable for obtaining narrow SOLNETs. In order to produce stronger light beam confinement, the optimization of writing power and diameter with improved measurement systems is necessary.

The system for measuring the coupling efficiency is shown in Fig. 5.24. From the SM optical fiber, a write beam was introduced into a PRI sol-gel thin film under baking. The light power received by an optical fiber with core diameter of 9.5 μm at the output edge of the PRI sol-gel thin film was measured before and after SOLNET formation. The light power before and after SOLNET formation is denoted by A and B, respectively. The relative coupling efficiency between the two optical fibers before and after SOLNET formation can be estimated by the ratio of B/A. The influence of writing power on the relative coupling efficiency is shown in Fig. 5.25 for wavelengths of 405 nm and 633 nm. In each case, the coupling efficiency increases as writing power is reduced. The result confirms

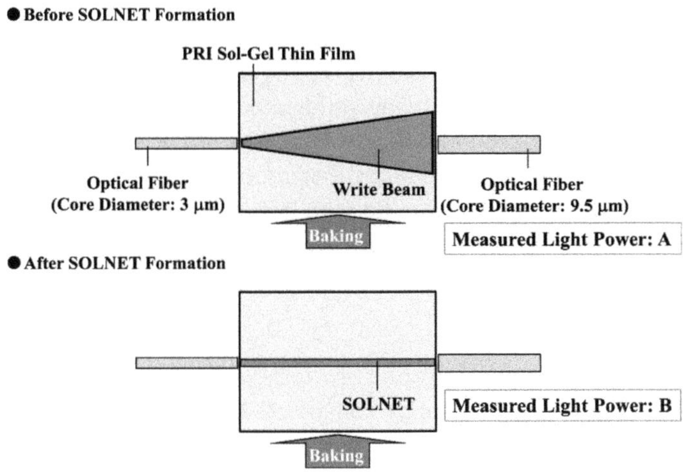

Figure 5.24. Coupling efficiency measurement system. Reprinted with permission from S. Ono, T. Yoshimura, T. Sato, J. Oshima, "Fabrication of Self-Organized Optical Waveguides in Photo-Induced Refractive Index Variation Sol-Gel Materials with Large Index Contrast," *J. Lightwave Technol.* **27**, 5308–5313. © 2009 IEEE.

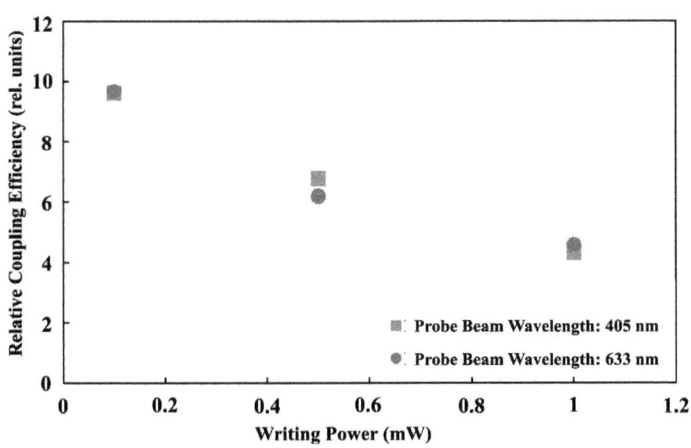

Figure 5.25. Influence of write beam power on the relative coupling efficiency. Reprinted with permission from S. Ono, T. Yoshimura, T. Sato, J. Oshima, "Fabrication of Self-Organized Optical Waveguides in Photo-Induced Refractive Index Variation Sol-Gel Materials with Large Index Contrast," *J. Lightwave Technol.* **27**, 5308–5313. © 2009 IEEE.

that SOLNET formed with low writing power provides strong light beam confinement effects.

Propagation loss of fabricated SOLNET can be estimated from the propagation loss in optical waveguides made of the same PRI sol-gel material to be around 1.9 dB/cm at 633 nm. It is expected that there is a graded refractive index distribution in SOLNET along the core width direction. This can be seen in simulations using BPM and FDTD method as described in Chapter 4.

5.4.3 *R-SOLNET*

By putting a silver paste droplet as the reflective site in the PRI sol-gel thin film, R-SOLNET can be constructed. A setup and the results for experimental observation of R-SOLNET construction are shown in Figs. 5.26 and 5.27, respectively. The writing power was 0.5 mW

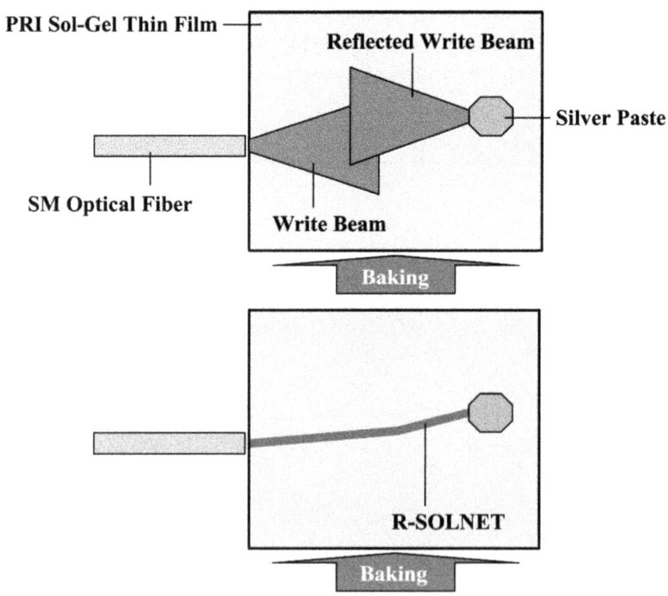

Figure 5.26. The R-SOLNET construction system. Reprinted with permission from S. Ono, T. Yoshimura, T. Sato, J. Oshima, "Fabrication of Self-Organized Optical Waveguides in Photo-Induced Refractive Index Variation Sol-Gel Materials with Large Index Contrast," *J. Lightwave Technol.* **27**, 5308–5313. © 2009 IEEE.

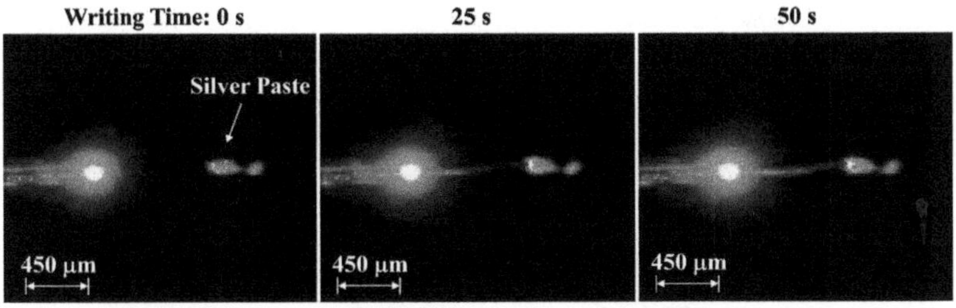

Figure 5.27. Experimental observation of R-SOLNET construction. Reprinted with permission from S. Ono, T. Yoshimura, T. Sato, J. Oshima, "Fabrication of Self-Organized Optical Waveguides in Photo-Induced Refractive Index Variation Sol-Gel Materials with Large Index Contrast," *J. Lightwave Technol.* **27**, 5308–5313. © 2009 IEEE.

and the baking temperature was 200°C. At writing time of 25 s, SOLNET is slightly drawn to the silver paste droplet. At writing time of 50 s, the SOLNET is completely drawn to the droplet. Thus, it is concluded that PRI sol-gel materials has R-SOLNET construction capability.

5.5 Influence of Write Beam Absorption in PRI Materials on SOLNET Growth Dynamics

As shown in Fig. 5.28, SOLNET shapes are influenced by write beam intensity and write beam absorption in PRI materials. When write beam intensity is too high, molecular reactions reach a saturated condition before self-focusing occurs. So, write beam intensity should be low enough to occur self-focusing.

When write beam absorption in PRI materials is too large, write beams propagate only a short distance. Therefore, short SOLNET is formed. Once SOLNET is formed, the absorption in the SOLNET region decreases due to decomposition of sensitizers, then next-step SOLNET formation starts. By repeating this process, SOLNET is constructed step-by-step, just like "a mole is digging a tunnel to go forward." This limits the overlap of write beams in multi-beam-writing SOLNET and R-SOLNET. When write beam absorption is low enough, on the other hand, write beams can propagate over a long distance to induce uniform self-focusing all over the

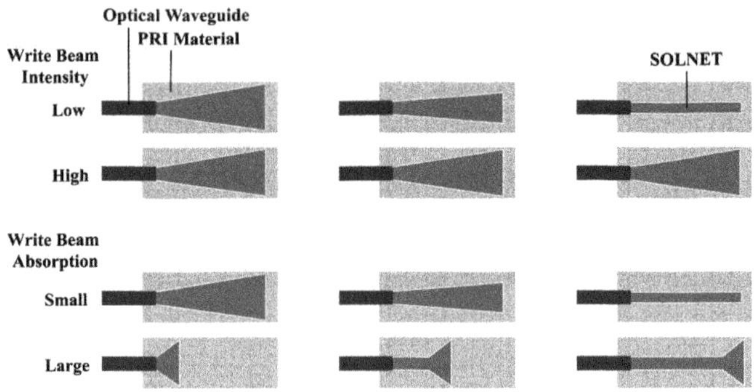

Figure 5.28. Influence of write beam intensity and write beam absorption in PRI materials on SOLNET construction.

photo-polymer. This increases the overlap of write beams from optical waveguides and reflected write beams, which is favorable for multi-beam-writing SOLNET and R-SOLNET construction.

In order to examine the influence of the write beam absorption on the SOLNET growth dynamics, SOLNET was formed in dye-sensitized photo-polymers. Dye sensitization is effective to control not only the spectral responses but also the amount of the write beam absorption. The photo-polymer was prepared by mixing acrylic materials (NORLAND) of NOA81 with refractive index of 1.56 and NOA65 with refractive index of 1.52. The mixing ratio was NOA81:NOA65=2:1. Crystal violet (CV) was added into the mixture as a sensitizer. SOLNET was grown by emitting write beams of 532 nm in wavelength from a MM optical fiber with a core diameter of 50 μm.

As shown in Fig. 5.29, in a photo-polymer containing 3wt% CV, the short-distance growth of SOLNET is observed. In a photo-polymer containing 0.15wt% CV, the long-distance uniform growth is observed. Here, the bending of the SOLNET may occur by slight flow of the photo-polymer. From these results, it is confirmed that low write beam absorption condition is suitable to grow uniformly-self-focused SOLNET.

5.6 Emissive SOLNET

It should be noted in Fig. 5.29 that although the SOLNET formation was carried out by green LD of 532 nm in wavelength, SOLNET

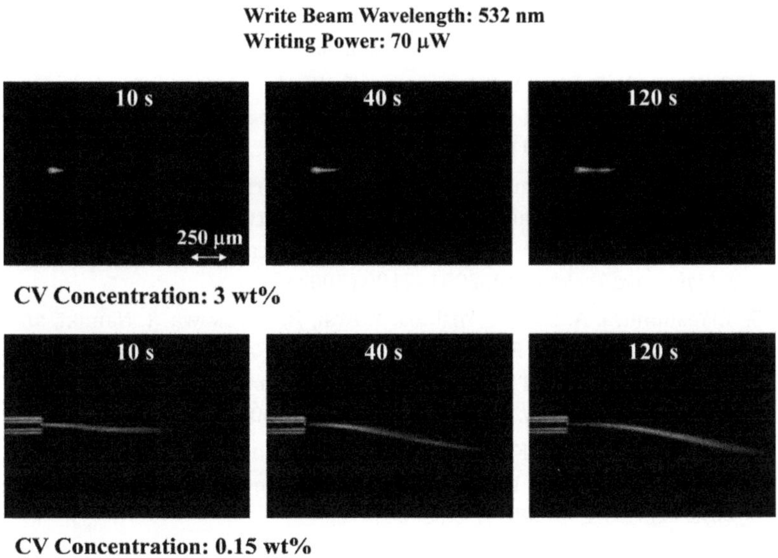

Figure 5.29. Influence of CV concentration in PRI materials on SOLNET growth dynamics. The red images of SOLNET are observed by photo-luminescence from CV excided by 532 nm write beams.

was observed as red images. This phenomenon is attributed to red photo-luminescence from CV doped in the photo-polymer. The 532 nm write beams that act as excitation light beams for CV are confined in SOLNET, increasing the excitation light intensity in SOLNET. Thus, the red images corresponding to the SOLNET shape appear. This unique property is useful for observation of SOLNET growth dynamics and will be used for phosphor SOLNET and other novel SOLNET applications.

References

1. T. Yoshimura, J. Roman, Y. Takahashi, W. V. Wang, M. Inao, T. Ishitsuka, K. Tsukamoto, K. Motoyoshi, and W. Sotoyama, "Self-Organizing Waveguide Coupling Method 'SOLNET' and Its Application to Film Optical Circuit Substrates," *Proc. 50th Electron. Components Technol. Conf. (ECTC)*, 962–969 (2000).
2. T. Yoshimura, J. Roman, Y. Takahashi, W. V. Wang, M. Inao, T. Ishituka, K. Tsukamoto, K. Motoyoshi, and W. Sotoyama, "Self-Organizing

Lightwave Network (SOLNET) and Its Application to Film Optical Circuit Substrates," *IEEE Trans. Comp. Packag. Technol.* **24**, 500–509 (2001).

3. F. Huang, H. Takase, Y. Eriyama, and T. Ukachi, "Optical Fiber Interconnection by Using Self-written Waveguides," *Proc. 9th Int. Conf. Rad. Tech. Asia*, Japan, 637–639 (2003).
4. T. Yoshimura, T. Inoguchi, T. Yamamoto, S. Moriya, Y. Teramoto, Y. Arai, T. Namiki, and K. Asama, "Self-Organized Lightwave Network Based on Waveguide Films for Three-Dimensional Optical Wiring Within Boxes," *J. Lightwave Technol.* **22**, 2091–2100 (2004).
5. T. Yoshimura, A. Hori, Y. Yoshida, Y. Arai, H. Kurokawa, T. Namiki, and K. Asama, "Coupling Efficiencies in Reflective Self-Organized Lightwave Network (R-SOLNET) Simulated by the Beam Propagation Method," *IEEE Photon. Technol. Lett.* **17**, 1653–1655 (2005).
6. T. Yoshimura and H. Kaburagi, "Self-Organization of Optical Waveguides between Misaligned Devices Induced by Write-Beam Reflection," *Appl. Phys. Express* **1**, 062007-1-3 (2008).
7. T. Yoshimura, C. Yoshino, K. Sasaki, T. Sato, and M. Seki, "Cancer Therapy Utilizing Molecular Layer Deposition and Self-Organized Lightwave Network -Proposal and Theoretical Prediction-," *IEEE J. Select. Topics in Quantum Electron.* **18**, Biophotonics 1, May/June 2012 [to be published].
8. S. Ono, T. Yoshimura, T. Sato, and J. Oshima, "Fabrication and Evaluation of Nano-Scale Optical Circuits Using Sol-Gel Materials with Photo-Induced Refractive Index Variation Characteristics," *J. Lightwave Technol.* **27**, 1229–1235 (2009).
9. S. Ono, T. Yoshimura, T. Sato, J. Oshima, "Fabrication of Self-Organized Optical Waveguides in Photo-Induced Refractive Index Variation Sol-Gel Materials with Large Index Contrast," *J. Lightwave Technol.* **27**, 5308–5313 (2009).
10. A. Bhatnagar, S. Latif, C. Debaes, D. A. B. Miller, "Pump-Probe Measurements of CMOS Detector Rise Time in the Blue," *J. Lightwave Technol.* **22**, 2213–2217 (2004).
11. H. Kaburagi and T. Yoshimura, "Simulation of Micro/Nano-Scale Self-Organized Lightwave Network (SOLNET) Using the Finite Difference Time Domain Method," *Opt. Commun.* **281**, 4019–4022 (2008).

Chapter 6

Optical Waveguide Films with Vertical Mirrors

Optical waveguide films with vertical mirrors are backbone for the 3-D OE platform and one of the core technologies in optical electronics. Vertical mirrors in the optical waveguide films are necessary for couplings between the optical waveguides and optical devices such as VCSELs and PDs in the optical waveguide films. The vertical mirrors are also used for optical Z-connections in 3-D optical circuits consisting of stacked waveguide films.

In this chapter, fabrication processes and characterization of the optical waveguide films with vertical mirrors are reviewed.

6.1 Duplication Process of Optical Waveguide Films of Photo-Definable Materials

Figures 6.1 and 6.2, respectively, show a duplication process and the process flow for the duplication of optical waveguide films with vertical mirrors [1, 2]. The process is called the built-in mask method. On a glass substrate, a reflective film is deposited with window patterns of waveguide cores to make a built-in mask. After a removable layer and an under-clad layer are successively formed

Optical Electronics: Self-Organized Integration and Applications
Tetsuzo Yoshimura

ISBN 978-981-4310-90-1 (Hardcover), 978-981-4364-08-9 (eBook)
www.panstanford.com

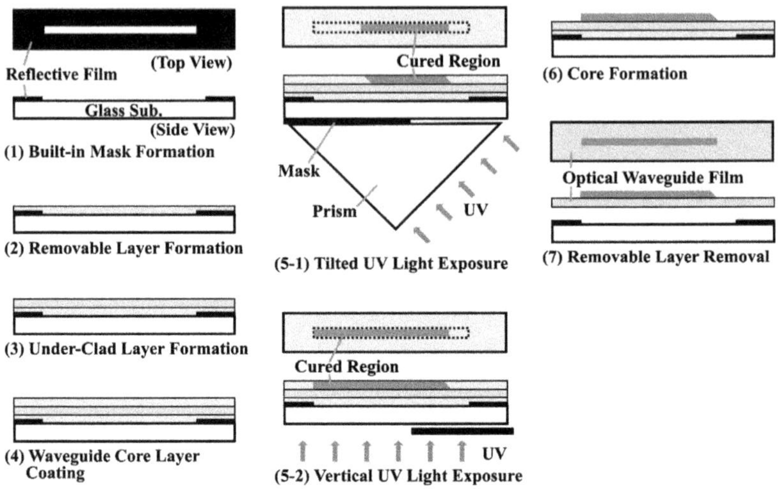

Figure 6.1. Duplication process of optical waveguide films with vertical mirrors: Built-in mask method.

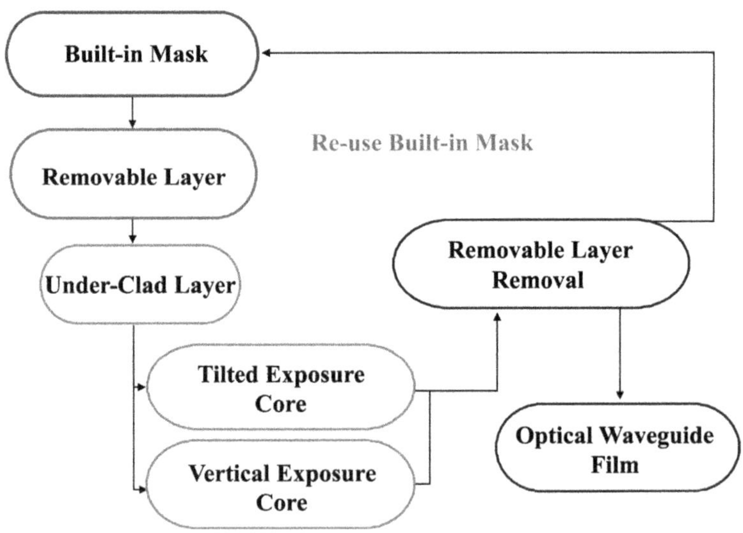

Figure 6.2. Process flow of the built-in mask method. Reprinted with permission from T. Yoshimura, M. Miyazaki, Y. Miyamoto, N. Shimoda, A. Hori, and K. Asama, "Three-Dimensional Optical Circuits Consisting of Waveguide Films and Optical Z-Connections," *J. Lightwave Technol.* **24**, 4345–4352. © 2006 IEEE.

on the mask, a core layer of a photo-definable material is coated. The core layer is selectively exposed to tilted UV light through a prism put on the back of the built-in mask. By this process, cured structures of beveled walls with designated angle slopes are constructed at the waveguide core ends, providing waveguide core end facets for vertical mirrors. For 45° mirror formation, tilted UV light with 45° incident angle is used. Vertical UV light exposure is carried out with a mask that covers the beveled wall regions to construct planar waveguide cores. By developing, the core formation is completed. Finally, the removable layer is removed to obtain an optical waveguide film with vertical mirrors. The built-in masks are reused for the duplication of the optical waveguide films.

6.2 Polymer Optical Waveguide Films Fabricated by the Built-In Mask Method

In the present study [1, 2], Al film was used for the reflective film of the built-in mask. The removable layer was 10 μm-thick poly-vinyl-alcohol (PVA). For the under clad layer and waveguide cores, acrylic photo-definable materials (OPSTAR, JSR Corporation) were used [3]. For curing them, UV light from a high-pressure mercury lamp was used. The required energy density was ~1 J/cm^2. Refractive index of the under-clad layer was 1.56 and that of the core 1.58. The removable layer was solved in water.

Figure 6.3 shows CAD pattern and the whole view of the built-in mask. The mask size is 5 × 5 cm^2. Various types of waveguide patterns are included. Core width ranges from 4 to 64 μm. The built-in mask was set on the apparatus for core and vertical mirror construction shown in Fig. 6.4. The core layer was exposed to UV light under an aligner.

6.2.1 *Waveguide Core Fabrication*

Figure 6.5 shows a top view and a cross section of cores in an optical waveguide film obtained by the vertical UV light exposure. Nominal core width and height are 64 μm and 32 μm, respectively. Sharp core patterns are formed, indicating that patterns of built-in mask are

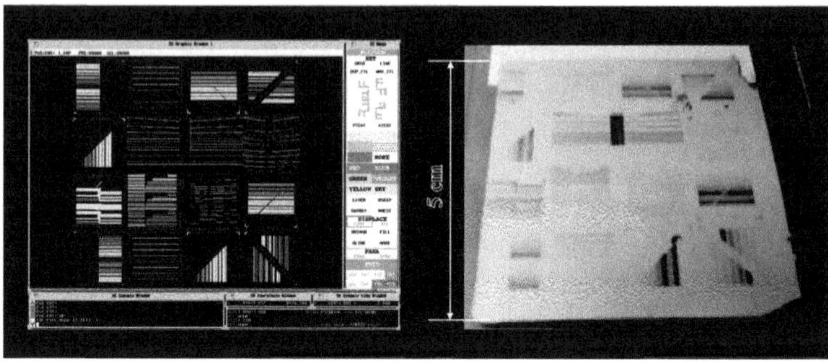

Figure 6.3. CAD patterns and a whole view of a built-in mask. Reprinted with permission from T. Yoshimura, T. Inoguchi, T. Yamamoto, S. Moriya, Y. Teramoto, Y. Arai, T. Namiki, and K. Asama, "Self-Organized Lightwave Network Based on Waveguide Films for Three-Dimensional Optical Wiring Within Boxes," *J. Lightwave Technol.* **22**, 2091–2100. © 2004 IEEE. See also Color Insert.

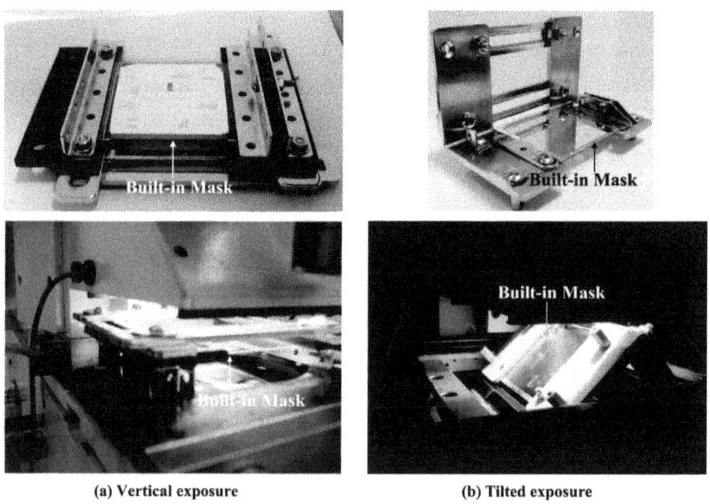

Figure 6.4. Exposure apparatus for core and vertical mirror construction. See also Color Insert.

appropriately duplicated into the core layer. Figure 6.6 shows NFPs of guided light beams of 1.3 μm wavelength in a 64 μm × 32 μm core and a 32 μm × 32 μm core in optical waveguide films. The confinement of guided beams into the core regions is observed. The

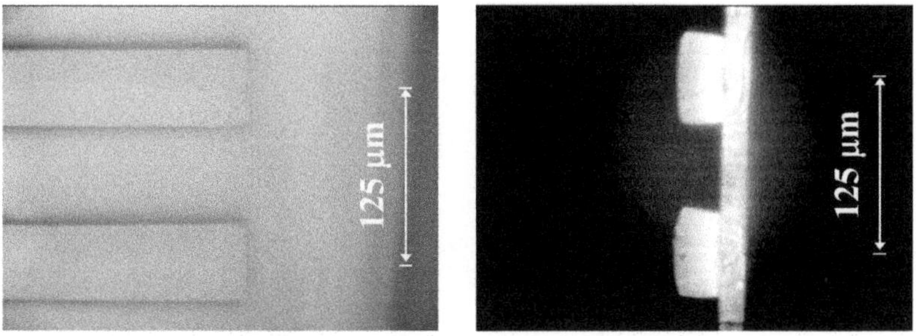

(a) Top View (b) Cross-Section

Figure 6.5. Top view and cross-section of cores of an optical waveguide film with nominal core width of 64 μm fabricated by the vertical UV light exposure. Reprinted with permission from T. Yoshimura, T. Inoguchi, T. Yamamoto, S. Moriya, Y. Teramoto, Y. Arai, T. Namiki, and K. Asama, "Self-Organized Lightwave Network Based on Waveguide Films for Three-Dimensional Optical Wiring Within Boxes," *J. Lightwave Technol.* **22**, 2091–2100. © 2004 IEEE.

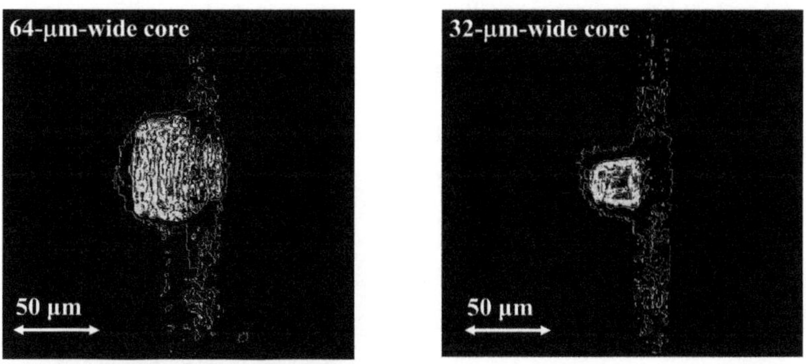

Figure 6.6. Near field patterns of optical waveguide films with nominal core width of 64 μm and 32 μm.

propagation loss measured by the cutback method was 0.8 dB/cm at 1.3 μm.

Figure 6.7 shows optical waveguide films with nominal core width of 4 μm. As can be seen from Fig. 6.7(a), optical circuits with core width of 4 μm were duplicated by the built-in mask method. Top view and cross section of a 4 μm-wide core are shown in Fig. 6.7(b).

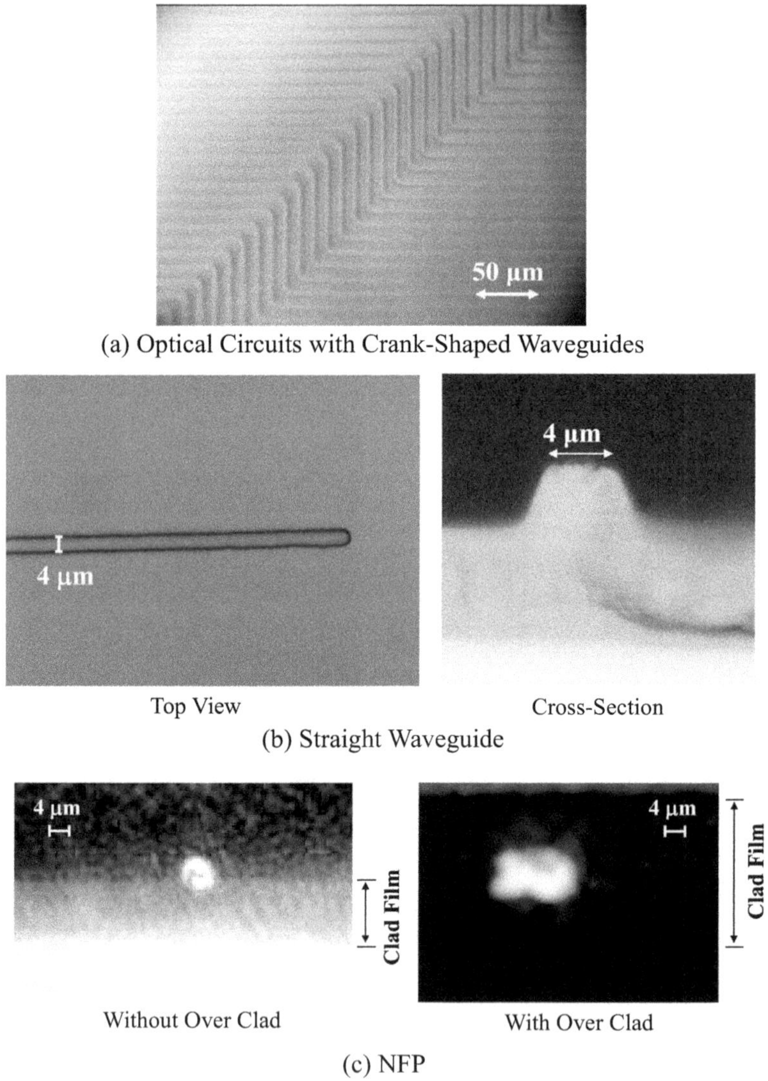

Figure 6.7. Optical waveguide films with nominal core width of 4 μm. (a) Optical circuits with crank-shaped waveguides. (b) Straight waveguide. (c) Near field patterns for optical waveguides without over clad and with over clad. Reprinted with permission from T. Yoshimura, K. Ogushi, Y. Kitabayashi, K Naito, Y. Miyamoto, and M. Miyazaki, "Optical Waveguide Films with Two-Layer Skirt-Type Core End Facets for Beam Leakage Reduction at 45° Mirrors," *Proc. SPIE* **6899**, 689913-1-11 (2008).

Although slopes are found at core sides, a core of 4 μm-width is obtained. As shown in Fig. 6.7(c), the width of the NFP is about 4 μm in the optical waveguide without over clad, indicating a strong confinement of the guided beams. In the optical waveguide with over clad, guided beam size increases due to evanescent waves arising from the tunnel effect. The result suggests that air-clad waveguides, which mean optical waveguides without over clad, are preferable for intra-chip optical interconnects from the viewpoint of guided beam confinement. In the optical waveguide shown in Fig. 6.7(b), however, the bottom of the core contacts the clad film. This causes reduction of beam confinement strength for vertical directions. Ideal conditions for strong beam confinement can be achieved in the all-air-clad waveguide that is described in Sections 6.4.3 and 8.5.2.

6.2.2 *Vertical Mirror Fabrication*

Figure 6.8 shows a side view of a duplicated optical waveguide film with a vertical mirror, and NFP of a reflected beam of 1.3 μm in wavelength from the vertical mirror. Smooth mirror surface and a clear reflected beam are obtained. Although the UV light exposure was carried out with an incident angle of 45°, the mirror angle is deviated from the 45°, making the mirror wall steeper. This is caused by the fact that we did not use a prism on the backside of the glass substrate during the tilted exposure: namely, the refraction at the boundary between the air and the glass substrate changes the UV light direction.

Figure 6.9(a) shows exposure apparatus for vertical mirror construction with a prism on the backside of the glass substrate [4]. Top view, side view, and far field pattern (FFP) of an optical waveguide film with 45° mirrors fabricated by the apparatus are shown in Figs. 6.9(b), (c), and (d), respectively. Nominal core width and height are 32 μm for lower three cores shown in the top view. Core end facets can be observed at the right edge of the cores. Beveled sidewalls are also found in the cross marker. It can be seen in the side view that the core end facet has a flat surface with exact 45° slope. The FFP for the reflected beams from the 45° mirror indicates that the beams are reflected in the surface-normal direction from the optical waveguide film, confirming that the mirror angle is exactly

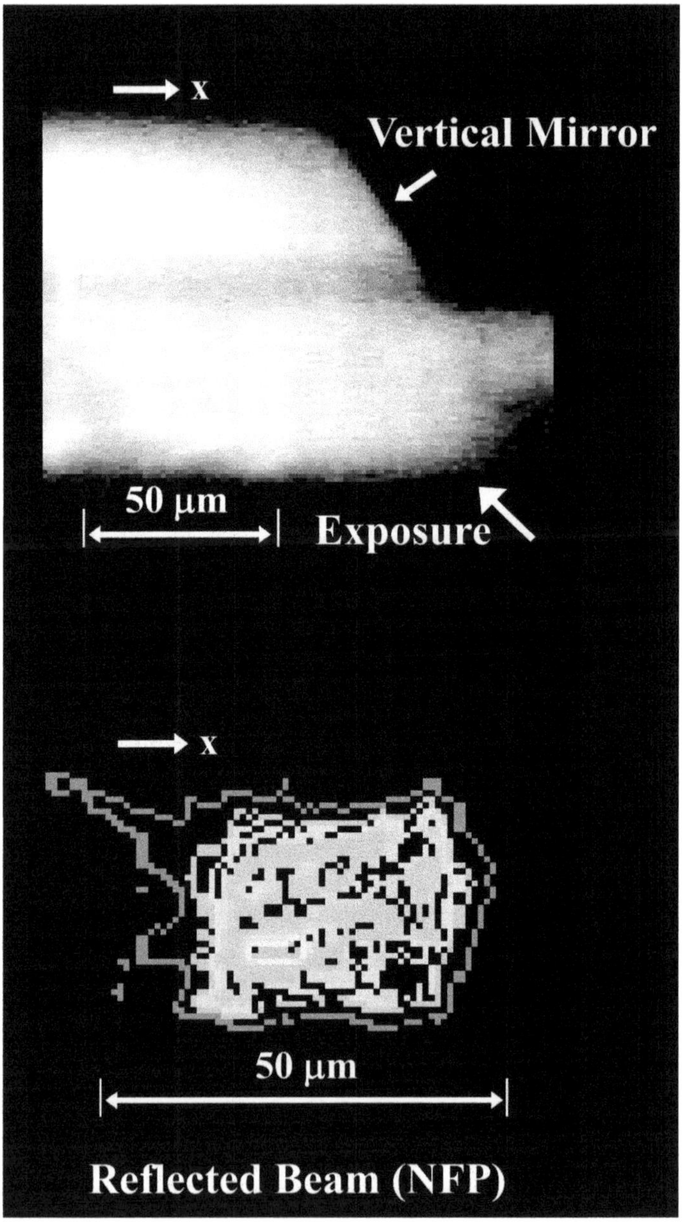

Figure 6.8. A duplicated optical waveguide film with a vertical mirror formed by the tilted exposure with incident angle of 45° without the prism on the backside of the glass substrate and NFP of a reflected beam.

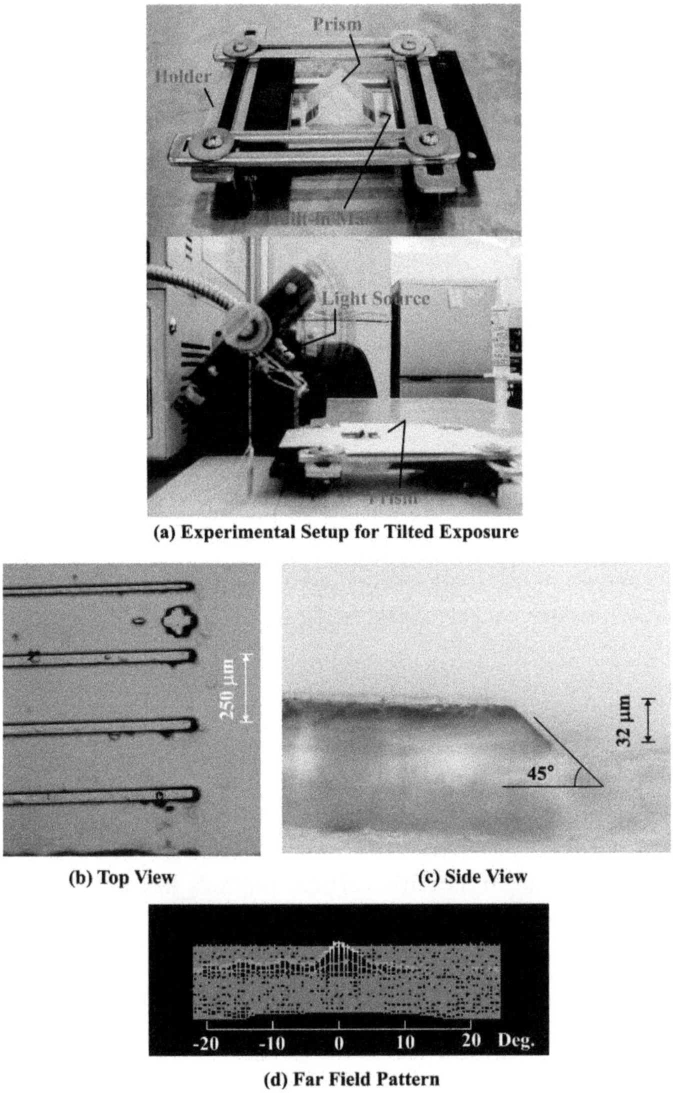

Figure 6.9. (a) Exposure apparatus for vertical mirror construction with a prism on the backside of the glass substrate. (b) Top view, (c) side view and (d) far field pattern of an optical waveguide film with 45° mirrors fabricated by the built-in mask method. Reprinted with permission from T. Yoshimura, M. Miyazaki, Y. Miyamoto, N. Shimoda, A. Hori, and K. Asama, "Three-Dimensional Optical Circuits Consisting of Waveguide Films and Optical Z-Connections," *J. Lightwave Technol.* **24**, 4345–4352. © 2006 IEEE.

Figure 6.10. A reflected beam from the vertical mirror of a 4 μm-wide core. Reprinted with permission from T. Yoshimura, K. Ogushi, Y. Kitabayashi, K Naito, Y. Miyamoto, and M. Miyazaki, "Optical Waveguide Films with Two-Layer Skirt-Type Core End Facets for Beam Leakage Reduction at 45° Mirrors," *Proc. SPIE* **6899**, 689913-1-11 (2008).

45°. As shown in Fig. 6.10, a reflected beam from a vertical mirror of a 4 μm-wide core was also observed.

6.3 Three-Layer Skirt-Type Core Structures

To increase the coupling efficiency between optical waveguides or between optical waveguides and optical devices, quality improvement of the vertical mirrors is important. To do this, we proposed optical waveguide films with vertical mirrors having "three-layer skirt-type core structure" [5, 6].

6.3.1 *Observation of Leakage and Scattering at Vertical Mirrors by SOLNET*

Figure 6.11 shows an example of observation of guided beam leakage and scattering at a vertical mirror in a conventional optical waveguide film using SOLNET. Photo-polymer layer was coated on the backside of an optical waveguide film as shown in Fig. 6.11(a). When a blue laser beam of 405 nm wavelength was introduced into the core, four lines grown from the vertical mirror were observed in the photo-polymer (Fig. 6.11(b)). One is in surface-normal direction,

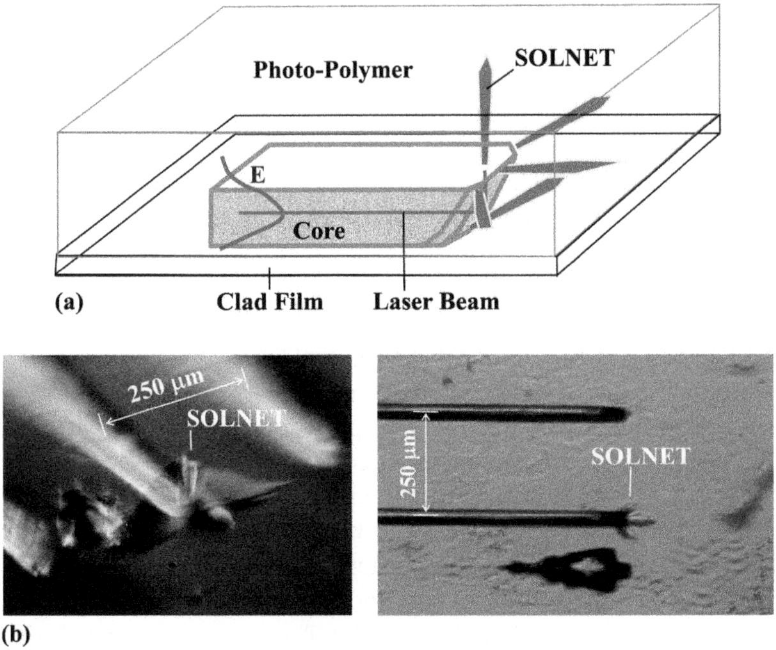

Figure 6.11. Observation of guided beam leakage and scattering at a vertical mirror in an optical waveguide film by SOLNET. (a) Schematic illustration of core structures and SOLNET at a vertical mirror in a conventional optical waveguide film. (b) Photographs of SOLNET grown from a vertical mirror. Reprinted with permission from T. Yoshimura, K. Ogushi, Y. Kitabayashi, Y. Miyamoto, M. Miyazaki, "Optical Waveguide Films with Two-Layer Skirt-Type Core Structures for Beam Leakage/Scattering Reduction at Tapered Mirrors," *IEEE Photon. Technol. Lett.* **19**, 1427–1429. © 2007 IEEE.

and the other three are in front, left-side, and right-side directions, respectively. The surface-normal line is the expected one that is the SOLNET formed by the beam reflected from the mirror. The front line and the left/right-side lines are undesirable ones.

As can be seen in Fig. 6.11(a), the front SOLNET line may be induced by leaked beams in the clad film. Namely, a portion of the guided beam penetrates into the clad film by the tunnel effect. The beam that exists in the clad film cannot be reflected by the vertical mirror, passing through the mirror to form SOLNET in front direction. When corner patterns are made by etching, sometimes round-shaped corners appear due to photolithographic imperfection, resulting in guided beam reflection and refraction at the corners. The reflected beams from the round-shaped core end top, in addition to the tunnel effect, may contribute to the front SOLNET line. The left/right-side SOLNET lines may be induced by refracted beams at round-shaped core end side corners.

6.3.2 *Three-Layer Skirt-Type Core Structures*

In Fig. 6.12, a proposed "three-layer skirt-type core structure," where a core with a layered structure consisting of low-index (LI) part, high-index (HI) part, and LI part is formed on a clad film, is illustrated. In the conventional core structure, as described in Section 6.3.1, the tunneling and the reflection/refraction at round-shaped core end corners give rise to beam leakage/scattering. In the three-layer skirt-type core structure, on the other hand, the LI part located between the HI part and the clad film prevents guided beams from tunneling into the clad film. The skirt-type shape of core end facets puts the round-shaped core end side corners away from the guided beam path. Similarly, the upper LI part puts the round-shaped core end top away from the guided beam path. Thus, the beam leakage/scattering can be reduced.

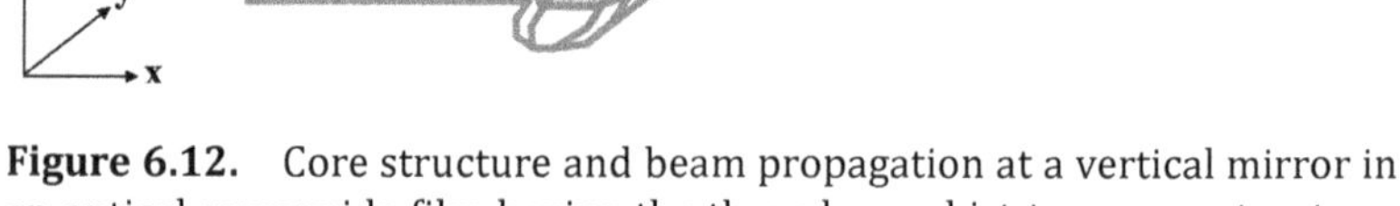

Figure 6.12. Core structure and beam propagation at a vertical mirror in an optical waveguide film having the three-layer skirt-type core structure. Reprinted with permission from T. Yoshimura, K. Ogushi, Y. Kitabayashi, Y. Miyamoto, M. Miyazaki, "Optical Waveguide Films with Two-Layer Skirt-Type Core Structures for Beam Leakage/Scattering Reduction at Tapered Mirrors," *IEEE Photon. Technol. Lett.* **19**, 1427–1429. © 2007 IEEE.

6.3.3 *Simulation by BPM and FDTD Method*

Figure 6.13 shows results of the beam leakage and scattering effect calculated by a coupled simulation using the BPM and the FDTD method for 1.3 μm. FDTD is applied to calculations for the region, where a mirror exists. Figure 6.13(a) shows the results for beam leakage due to the tunneling. In the conventional core structure, refractive index of the clad film and that of the core are assumed to be 1.5 and 1.505, respectively. Core thickness is 4 μm. In the two-layer skirt-type core structure, the refractive index of the clad film and that of the LI part are 1.5, and the refractive index of the HI part is 1.505. LI part and HI part thicknesses are 4 μm. It is found that beam leakage is drastically reduced in the two-layer core structure. The efficiency of the reflection from the vertical mirror increases to ~0.85 in the new structure from ~0.4 in the conventional structure.

Figure 6.13(b) shows results for the beam scattering effect at round-shaped core end side corners. Refractive index of the clad region and that of the core are assumed 1 and 1.505, respectively. Core width and skirt width are 4 μm and 10 μm, respectively. Core corners are cut by 1 μm along directions parallel and perpendicular to the core direction. In the conventional core structure, beam propagation in the left-side/right-side directions appears, which is consistent with the observation shown in Fig. 6.11. In the skirt-type core structure, on the other hand, the beam propagation in the left-side/right-side directions disappears.

Figure 6.13(c) shows results for the beam scattering effect at the round-shaped core end top. In the conventional core structure, disturbed reflected beams can be seen while in the three-layer core structure, smooth reflection of the propagating beams occurs at the vertical mirror part. These results indicate that the three-layer skirt-type core structure is effective to suppress undesired beam leakage/scattering at vertical mirrors.

6.3.4 *Fabrication by the Built-In Mask method*

Figure 6.14 shows a fabrication process for optical waveguide films with vertical mirrors having the three-layer skirt-type core structure. The process is basically the same as the built-in mask

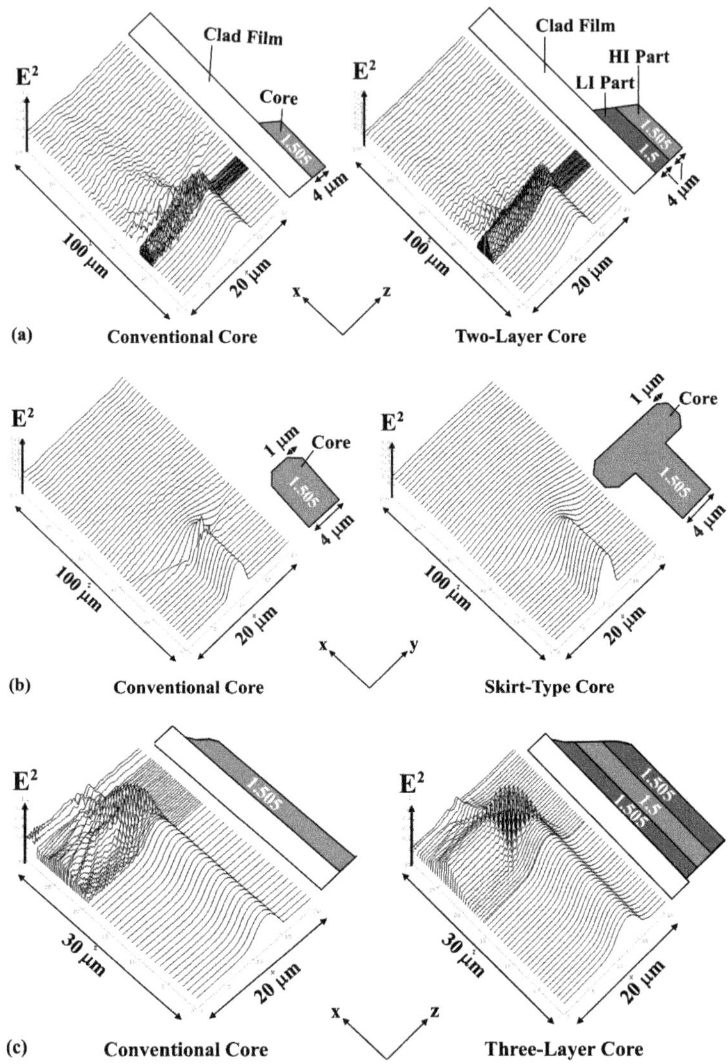

Figure 6.13. BPM/FDTD simulation results of (a) beam leakage due to the tunneling, (b) beam scattering due to the refraction by the round-shaped core end side corners, and (c) beam scattering due to the reflection by the round-shaped core end top. Reprinted with permission from T. Yoshimura, K. Ogushi, Y. Kitabayashi, Y. Miyamoto, M. Miyazaki, "Optical Waveguide Films with Two-Layer Skirt-Type Core Structures for Beam Leakage/Scattering Reduction at Tapered Mirrors," *IEEE Photon. Technol. Lett.* **19**, 1427–1429. © 2007 IEEE.

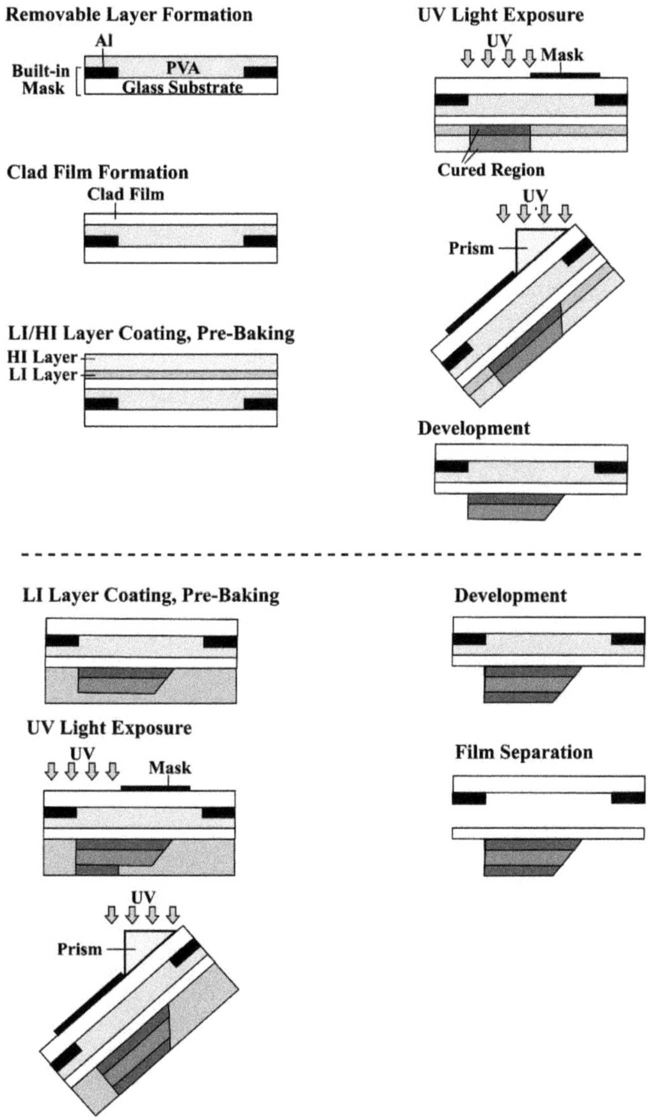

Figure 6.14. Fabrication process of optical waveguide films having the three-layer skirt-type core structure. Reprinted with permission from T. Yoshimura, K. Ogushi, Y. Kitabayashi, Y. Miyamoto, M. Miyazaki, "Optical Waveguide Films with Two-Layer Skirt-Type Core Structures for Beam Leakage/Scattering Reduction at Tapered Mirrors," *IEEE Photon. Technol. Lett.* **19**, 1427–1429. © 2007 IEEE.

method shown in Fig. 6.1. After PVA removable layer of 10 μm thickness and a clad film of 20 μm thickness were formed on a built-in mask, LI layer of 10 μm thickness and HI layer of 32 μm thickness were successively coated/pre-baked. The LI/HI layer was selectively exposed to UV light vertically. By this process, cured structures for planar waveguide cores are formed. To construct core end facets, tilted UV light exposure was carried out through a prism with a mask that covers the planar waveguide core regions. By development, two-layer core formation was completed. Next, LI layer was coated/pre-baked again. By the vertical and tilted UV light exposure followed by development, three-layer core formation was completed. Finally, the removable layer was removed in water to obtain an optical waveguide film with vertical mirrors.

Figure 6.15 shows photo-mask patterns for skirt-type core end facets. Skirt width is four times of the core width. Skirt length L varies as 1/2, 1/1, and 2 of the core width. Acrylic photo-definable polymers with refractive index of 1.50 were used for the clad film and the LI part, and those with refractive index of 1.52 for the HI part.

To examine influences of L on reflected beams, an optical waveguide film with one-layer skirt-type core structures was

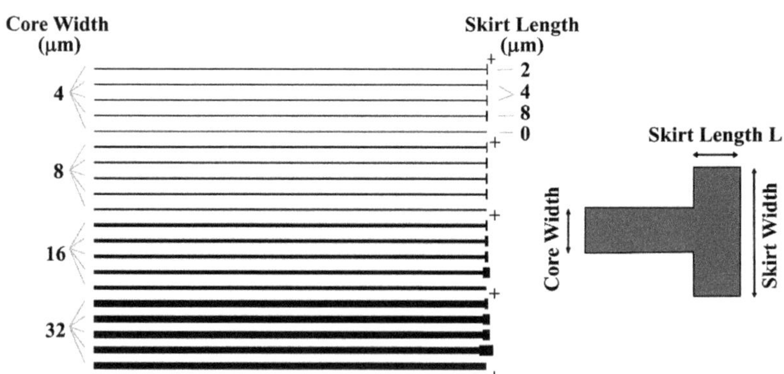

Figure 6.15. Mask patterns for the built-in mask. Reprinted with permission from T. Yoshimura, K. Ogushi, Y. Kitabayashi, Y. Miyamoto, M. Miyazaki, "Optical Waveguide Films with Two-Layer Skirt-Type Core Structures for Beam Leakage/Scattering Reduction at Tapered Mirrors," *IEEE Photon. Technol. Lett.* **19**, 1427–1429. © 2007 IEEE .

(a)

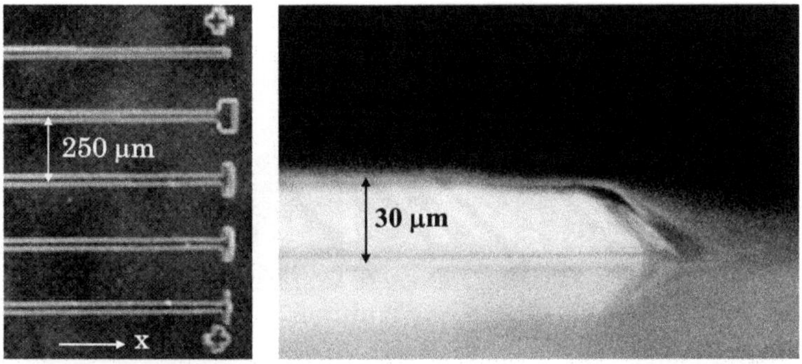

(b)

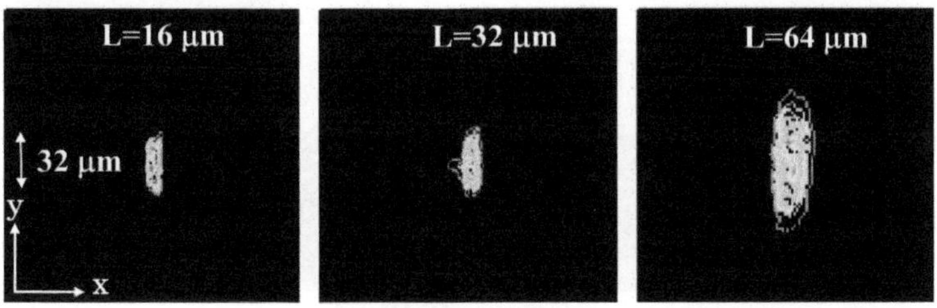

Figure 6.16. (a) Top and side views of an optical waveguide film having the skirt-type core structure. (b) NFPs of reflected beams from vertical mirrors of the skirt-type core end facets. Reprinted with permission from T. Yoshimura, K. Ogushi, Y. Kitabayashi, Y. Miyamoto, M. Miyazaki, "Optical Waveguide Films with Two-Layer Skirt-Type Core Structures for Beam Leakage/Scattering Reduction at Tapered Mirrors," *IEEE Photon. Technol. Lett.* **19**, 1427–1429. © 2007 IEEE.

fabricated. Fig. 6.16(a) shows top and side view of the optical waveguide film. Nominal core width and thickness are 32 μm. It is found that vertical mirrors are constructed. NFPs of reflected beams of 1.3 μm in wavelength are shown in Fig. 6.16(b) for $L = 16$, 32, and 64 μm. Although, with increasing L, the NFP size increases due to beam diameter expansion during propagation in a free space of the skirt region, drastic enlargement does not occur for $L \leq 32$ μm. Therefore, it is suggested that appropriate skirt length is around the core width.

Figure 6.17(a) shows cross sections and NFPs at the cross sections of an optical waveguide film with two-layer cores of 32 μm and 4 μm widths. The wavelength of guided beams is 650 nm. It is found that two-layer core structures are clearly formed, and

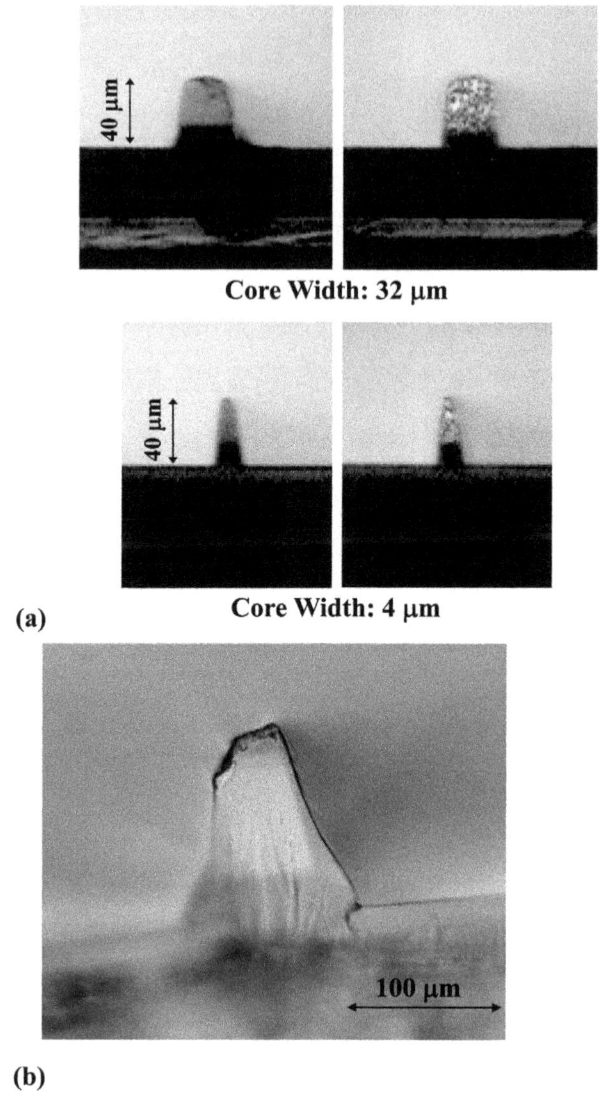

Figure 6.17. (a) Cross sections and NFPs at the cross section of an optical waveguide film having the two-layer core structure. (b) Cross sections of an optical waveguide film having the three-layer core structure. Reprinted with permission from T. Yoshimura, K. Ogushi, Y. Kitabayashi, Y. Miyamoto, M. Miyazaki, "Optical Waveguide Films with Two-Layer Skirt-Type Core Structures for Beam Leakage/Scattering Reduction at Tapered Mirrors," *IEEE Photon. Technol. Lett.* **19**, 1427–1429. © 2007 IEEE.

guided beams are confined in the HI part. These results confirm that the two-layer core structure can be constructed by the built-in mask method. Propagation loss of the optical waveguide was 0.9 dB/cm at 1.3 μm. The relatively large propagation loss might be caused by air clad, namely, be caused from scattering due to multi-mode characteristics of the optical waveguide with a high core/clad index contrast. Reflected beams from the core end facet in the two-layer skirt-type core structure were observed. Reflection loss of the vertical mirror was estimated to be ~2 dB. Although the structure is not perfect, an optical waveguide film having the three-layer core structure was also constructed as shown in Fig. 6.17(b).

To investigate the influence of round-shaped corners on light beam propagation, FFP measurements were carried out for the reflected light beams at vertical mirrors. The results are shown in Fig. 6.18. In a vertical mirror of three-layer core without skirt, the FFP is spread along y-direction, that is, a direction perpendicular to the guided beam propagation in the film plane (see Fig. 6.12). This suggests that the guided beams are refracted to the left-side/

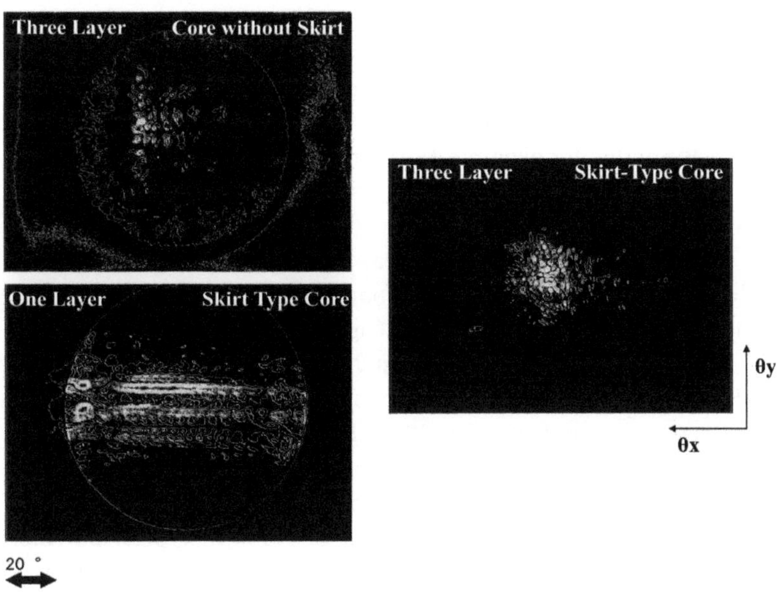

Figure 6.18. FFPs of reflected beams from vertical mirrors.

right-side directions by the round-shaped core end corners as can be seen in Fig. 6.11(a). In a vertical mirror of one-layer skirt-type core, the FFP is spread along x-direction, that is, a direction parallel to the guided beam propagation. This suggests that the guided beams pass through the vertical mirror due to the tunnel effect and/or are reflected to the front direction by the round-shaped core end top. In a vertical mirror of three-layer skirt-type core, the spread of the FFP is suppressed both along the x- and y-directions. This suggests that the influence of the tunneling and round-shaped core end corners is removed successfully.

In addition, the three-layer skirt-type core structure is effective to widen the tolerance of alignment between core end facets and metal/dielectric films for mirrors/filters, avoiding direct deposition of the metal/dielectric films on the HI part.

6.4 Nano-Scale Waveguides of PRI Sol-Gel Materials

Nano-scale optical circuits are required in intra-LSI optical interconnects. Si waveguides [7] are usually used for the nano-scale optical circuits because they have large refractive index contrast between core and clad. The fabrication processes of Si waveguides are well established in the LSI industry. However, there are several problems in Si waveguides, for example, the fabrication process is complicated, and short-wavelength light cannot transmit owing to a narrow band gap of Si.

The PRI sol-gel material that was developed by Nissan Chemical Industries, LTD is suitable for the nano-scale optical circuits due to its large-refractive-index-increase capability. In spin-coated films of the PRI sol-gel material, the refractive index increases from 1.65 to 1.85 at $\lambda = 633$ nm by UV light exposure and baking, giving rise to a refractive index contrast of $\sim$13%. Although the contrast is smaller than $\sim$60% in Si waveguides, it is sufficient to provide nano-scale waveguide structures as described in the below part of the current section. Note that the index contrast of $\sim$1% in typical polymer waveguides is insufficient for nano-scale waveguide structures.

The large refractive index contrast in the PRI sol-gel material implies that the materials enable us to fabricate high-index-contrast

(HIC) waveguides without etching/developing and vacuum processes, simplifying the fabrication processes of the nano-scale waveguides. Another advantage of the PRI sol-gel material is that it has a wide band gap. This enables us to use short-wavelength light that is favorable for miniaturizing the optical circuits since their dimension can be reduced in proportion to $1/\lambda$, where λ is wavelength, achieving volume reduction by $1/\lambda^3$. Furthermore, by using the sol-gel material, optical circuits consisting of HIC waveguides are available in structures of ultra-thin planar films with thickness of 100 nm order. This is a favorable property for stacking optical circuits on LSIs, or for constructing three-dimensional optical circuits with stacked films [2, 4]. Therefore, the PRI sol-gel material is expected promising for nano-scale optical circuits of 3-D OE platforms in LSIs and MCMs.

In the following section, fabrication and evaluation of nano-scale waveguides made of the PRI sol-gel material are reviewed. [8]

6.4.1 *Linear, Bending, and Branching Waveguides*

6.4.1.1 Fabrication process

The PRI sol-gel material is a silicon-oxide-based material incorporating titanium oxide. The fabrication process of nano-scale waveguides using the sol-gel material is shown in Fig. 6.19. On

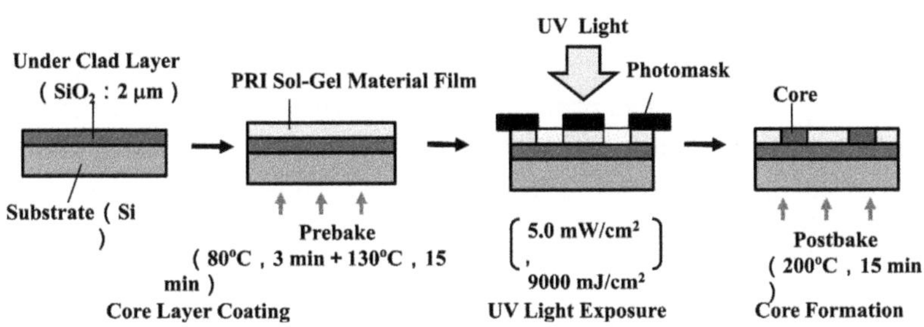

Figure 6.19. Fabrication process of optical waveguides using the PRI sol-gel material. Reprinted with permission from S. Ono, T. Yoshimura, T. Sato, and J. Oshima, "Fabrication and Evaluation of Nano-Scale Optical Circuits Using Sol-Gel Materials with Photo-Induced Refractive Index Variation Characteristics," *J. Lightwave Technol.* **27**, 1229–1235. © 2009 IEEE.

a Si substrate with SiO_2 of 2 μm thickness, the sol-gel material was spin-coated and then prebaked for solvent removal to form a film. The film was selectively exposed to UV light from a high-pressure mercury lamp with light intensity of ~5 mW/cm^2 through a photomask with energy of 9000 mJ/cm^2. Nominal core pattern width is 1 μm in the photomask. The UV light exposure has a role to activate the hydrolytic condensation of the titanium oxide. The film was postbaked at 200°C for 15 min. During the postbaking, the hydrolytic condensation proceeds more rapidly in UV light exposed areas than in unexposed areas, yielding higher refractive index in the exposed areas than in the unexposed surrounding areas. Thus, the UV light exposed areas become cores.

The refractive index difference between core and clad (Δn), i.e., the amount of activation of hydrolytic condensation, is controllable with UV light dose. Initial refractive index with no UV light dose, which corresponds to the refractive index of the clad, is 1.65. Δn increases with UV light dose, and is saturated to be ~0.2 with UV light dose of 9000 mJ/cm^2. Δn is basically not affected by the prebake/postbake temperature below 200°C. However, a high-temperature baking (ex. over 300°C) induces the hydrolytic condensation in unexposed areas, and hence the Δn tends to be reduced with increasing the baking temperature. Photographs of the fabricated nano-scale optical circuits are shown in Fig. 6.20. The core thickness is ~230 nm and the core width is ~1 μm.

In intra-chip optical interconnects, 850 nm is the most popular as signal light wavelength. However, from view points of detector sensitivity, short-wavelength light is preferable [9]. The short-wavelength light is also preferable to miniaturize optical circuits for increasing wiring density. For these reasons, all the measurements of optical circuits reported in the present chapter were carried out using a He-Ne laser of 633 nm.

6.4.1.2 Linear waveguide

Figure 6.21 shows a guided beam and an output beam from a core edge in a linear waveguide. Propagation loss was found from the cutback method to be 1.86 dB/cm for the TE mode (polarization//x-axis) and 1.89 dB/cm for the TM mode (polarization//y-axis),

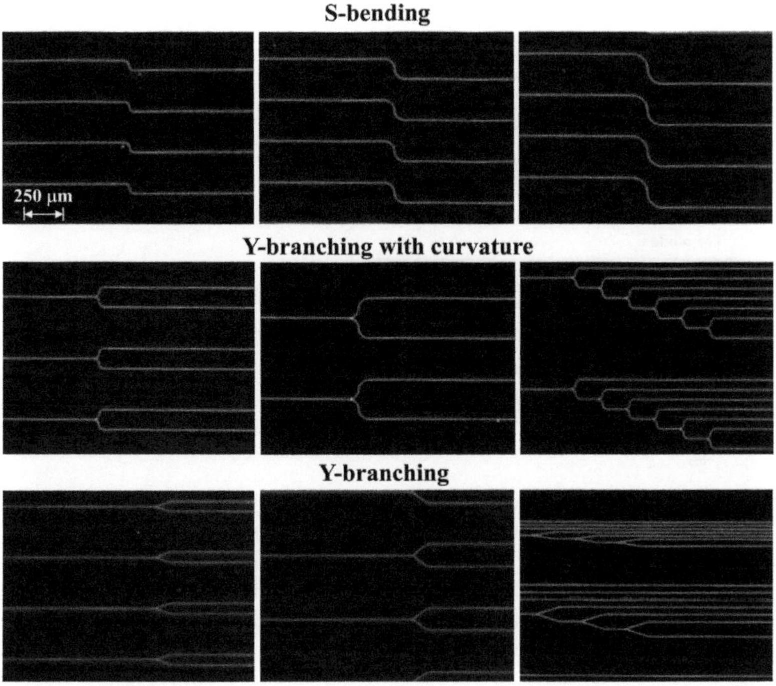

Figure 6.20. Optical waveguides fabricated by the process shown in Fig. 6.19. Reprinted with permission from S. Ono, T. Yoshimura, T. Sato, and J. Oshima, "Fabrication and Evaluation of Nano-Scale Optical Circuits Using Sol-Gel Materials with Photo-Induced Refractive Index Variation Characteristics," *J. Lightwave Technol.* **27**, 1229–1235. © 2009 IEEE.

indicating small polarization dependence. For wavelengths of 1300 nm and 1550 nm that are widely used in optical communications, light beams were hard to propagate in the cores. This might be due to very thin core thickness with asymmetric refractive index of over/under clad.

Guided beam confinement was investigated by FFPs of output beams from the core edge. The results are shown in Fig. 6.22. Spread angles of the output beams are determined by width of the FFPs at light intensity of $1/e$ of the maximum height. Spread angles along core width direction ($//x$-axis) and along core thickness direction ($// y$-axis) are 20° and 40°, respectively, for both the TE mode and the TM mode.

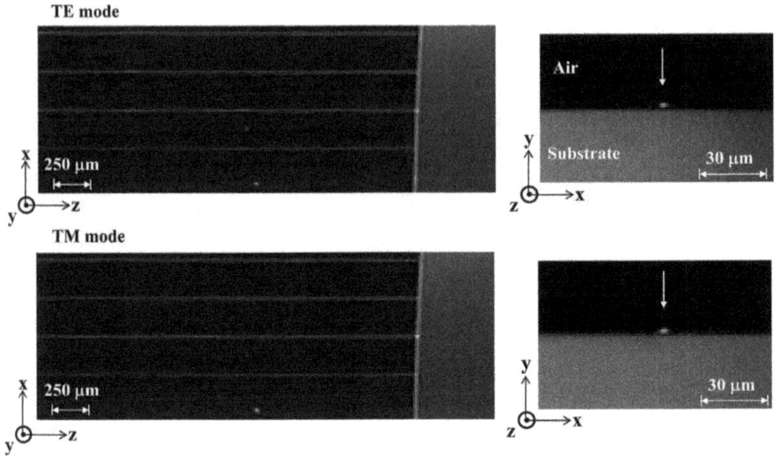

Figure 6.21. Linear waveguides, a guided beam, and an output beam from the core edge. Reprinted with permission from S. Ono, T. Yoshimura, T. Sato, and J. Oshima, "Fabrication and Evaluation of Nano-Scale Optical Circuits Using Sol-Gel Materials with Photo-Induced Refractive Index Variation Characteristics," *J. Lightwave Technol.* **27**, 1229–1235. © 2009 IEEE.

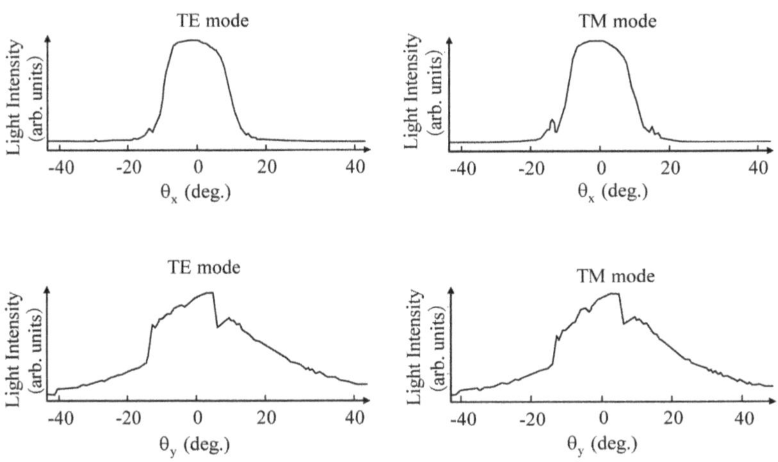

Figure 6.22. FFPs of output beams from the core edge of the linear waveguide. Reprinted with permission from S. Ono, T. Yoshimura, T. Sato, and J. Oshima, "Fabrication and Evaluation of Nano-Scale Optical Circuits Using Sol-Gel Materials with Photo-Induced Refractive Index Variation Characteristics," *J. Lightwave Technol.* **27**, 1229–1235. © 2009 IEEE.

Spot radius w of guided beams can be calculated from the spread angles of output beams using the following equation:

$$w = \frac{180 \times \lambda}{2\theta \times \pi^2 n} \tag{6.1}$$

Here, 2θ[°] is spread angle, λ [μm] is guided beam wavelength, and n is refractive index of the free space, where emitted beams propagate. In the present case, n is 1. Spot radii of guided beams along x-axis and along y-axis were calculated to be 0.6 μm and 0.3 μm, respectively, for both the TE mode and the TM mode. From these results, it is concluded that the sol-gel material can realize guided beam confinement within a cross section area less than 1.0 μm^2 with small polarization dependence.

Light intensity profiles in the optical waveguides were simulated by BPM. The model is shown in Fig. 6.23. Core thickness is 230 nm, core width is 1 μm, refractive index of core is 1.85, refractive index of clad is 1.65, and SiO_2 (under clad) thickness is 2 μm and the refractive index is 1.46. The wavelength of guided beams is 633 nm. The results are shown in Fig. 6.24. Spot radii of guided beams along x-axis and along y-axis are 0.5 μm and 0.2 μm, respectively, for both the TE mode and the TM mode. The spot radii obtained by experiments and the simulations fairly agree.

It should be noted that the guided beam confinement is slightly stronger in the simulations than in the experiments. The difference might be attributed to the fact that in the fabricated optical waveguides Δn is slightly smaller than Δn in the simulation model, that is, less than ~0.2, resulting in an increase in penetration of evanescent waves into the clad to increase in spot radii of guided beams.

Si waveguides enable strong guided beam confinement in cores with a width of ~0.5 μm due to its high core refractive index of ~3.5. However, the high refractive index causes a decrease in the propagation speed of optical signals in the Si waveguides. The sol-gel material waveguides, on the other hand, exhibit faster propagation speed of optical signals by factor of ~2 compared with Si waveguides since the core refractive index is ~1.85. Considering the trade-off relationship between the guided beam confinement and the propagation speed, it is expected that the sol-gel material

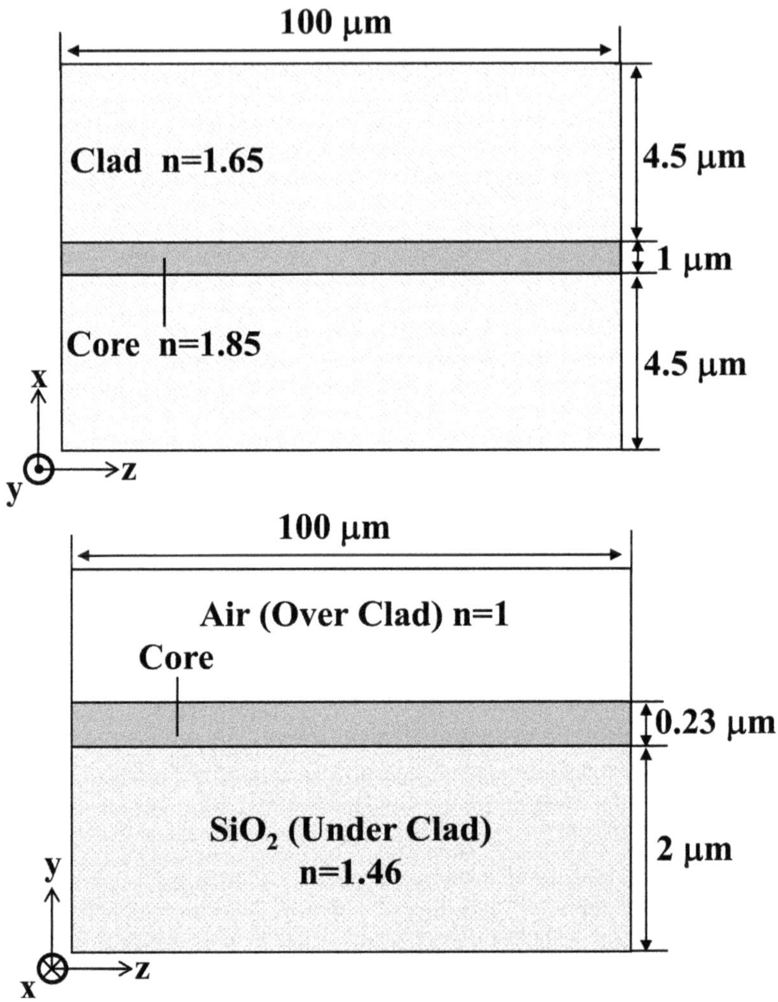

Figure 6.23. A model for simulation by the BPM for an optical waveguide using the PRI sol-gel material. Reprinted with permission from S. Ono, T. Yoshimura, T. Sato, and J. Oshima, "Fabrication and Evaluation of Nano-Scale Optical Circuits Using Sol-Gel Materials with Photo-Induced Refractive Index Variation Characteristics," *J. Lightwave Technol.* **27**, 1229–1235. © 2009 IEEE.

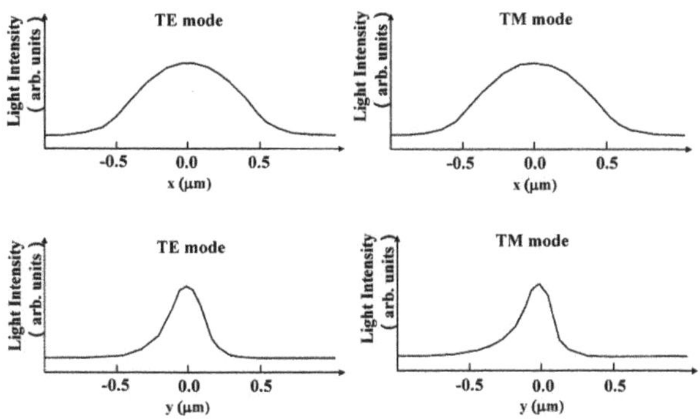

Figure 6.24. Results of the BPM simulation for light intensity profiles in optical waveguides. Reprinted with permission from S. Ono, T. Yoshimura, T. Sato, and J. Oshima, "Fabrication and Evaluation of Nano-Scale Optical Circuits Using Sol-Gel Materials with Photo-Induced Refractive Index Variation Characteristics," *J. Lightwave Technol.* **27**, 1229–1235. © 2009 IEEE.

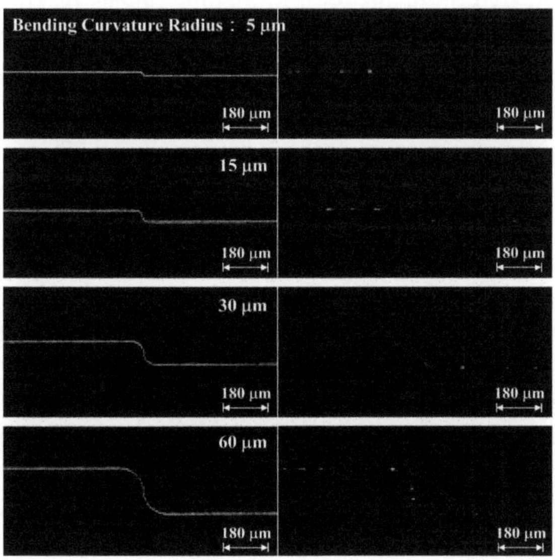

Figure 6.25. S-bending waveguides and guided beams in them for TE mode. Reprinted with permission from S. Ono, T. Yoshimura, T. Sato, and J. Oshima, "Fabrication and Evaluation of Nano-Scale Optical Circuits Using Sol-Gel Materials with Photo-Induced Refractive Index Variation Characteristics," *J. Lightwave Technol.* **27**, 1229–1235. © 2009 IEEE. See also Color Insert.

waveguides can preferably be applied to global wiring in chips and MCMs.

It is expected that further reduction of core width is achieved in the PRI sol-gel material by optimizing the fabrication processes to increase Δn, which enables core width less than 1 μm.

6.4.1.3 Bending and branching waveguides

S-bending waveguides and 633 nm guided beams in them are shown in Fig. 6.25 for the TE mode. Core thickness and width of the optical waveguides are respectively ~230 nm and ~1 μm. The beams propagate in the S-bending waveguides with bending curvature radius of 5, 15, 30, and 60 μm. We similarly observed TM mode beam propagation in the S-bending waveguides.

Output beams from core edges of the Y-branching waveguides are shown in Fig. 6.26 for the TE mode. It is found that input beams are divided into two branches with branching angles of 20°, 40°, and 80°. For the TM mode, same kinds of divided beams were observed.

The S-bending loss was evaluated with the following process (Fig. 6.27):

(1) Output power P_1 at waveguide length L_1 is measured.
(2) After cutting the waveguide, output power P_2 at waveguide length L_2 is measured.
(3) After the optical waveguide is cut beyond the S-bend, output power P_3 at waveguide length L_3 is measured.
(4) The averaged output power in unit of dBm at each measurement point is plotted against waveguide length.
(5) (L_1, P_1) and (L_2, P_2) are connected with a straight line of $y = ax + b$. Since slope "a" gives propagation loss of the linear waveguide, another straight line of slope "a", namely, $y = ax + c$, is traced on (L_3, P_3) to represent power versus waveguide length characteristics before reaching the S-bend part.
(6) Bending loss is obtained from the gap at the S-bend part, '$c - b$."

Y-branching loss was also measured by the same method. Bending and Y-branching losses are expressed as follows. Bending

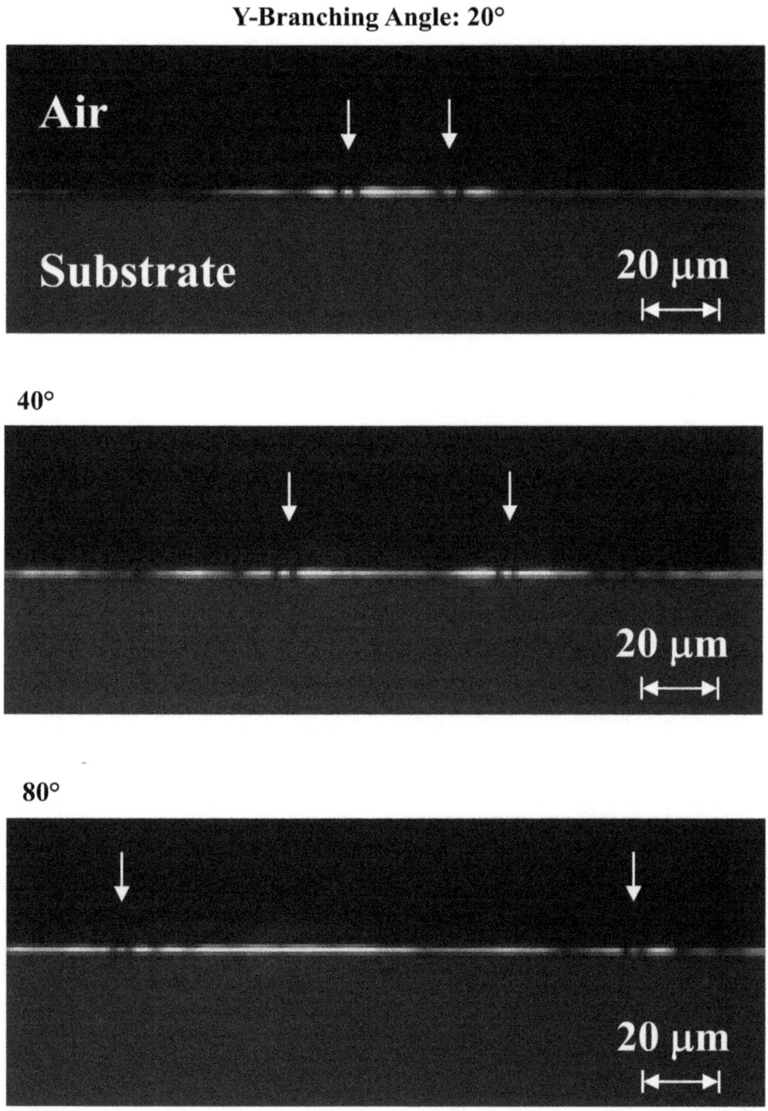

Figure 6.26. Output beams from core edges of Y-branching waveguides for TE mode. Reprinted with permission from S. Ono, T. Yoshimura, T. Sato, and J. Oshima, "Fabrication and Evaluation of Nano-Scale Optical Circuits Using Sol-Gel Materials with Photo-Induced Refractive Index Variation Characteristics," *J. Lightwave Technol.* **27**, 1229–1235. © 2009 IEEE.

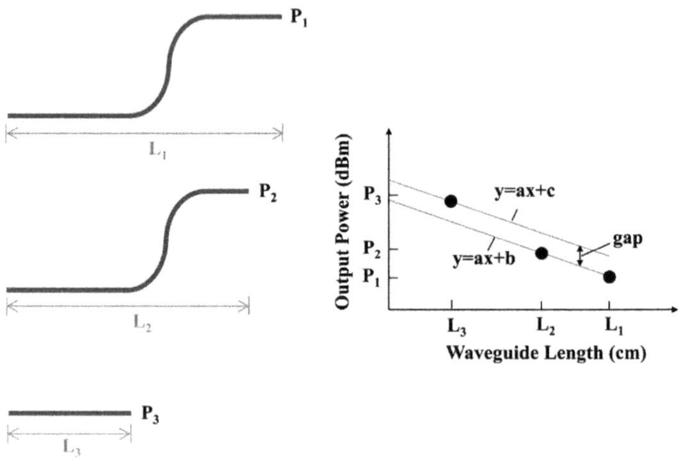

Figure 6.27. Evaluation procedure for bending loss. Reprinted with permission from S. Ono, T. Yoshimura, T. Sato, and J. Oshima, "Fabrication and Evaluation of Nano-Scale Optical Circuits Using Sol-Gel Materials with Photo-Induced Refractive Index Variation Characteristics," *J. Lightwave Technol.* **27**, 1229–1235. © 2009 IEEE.

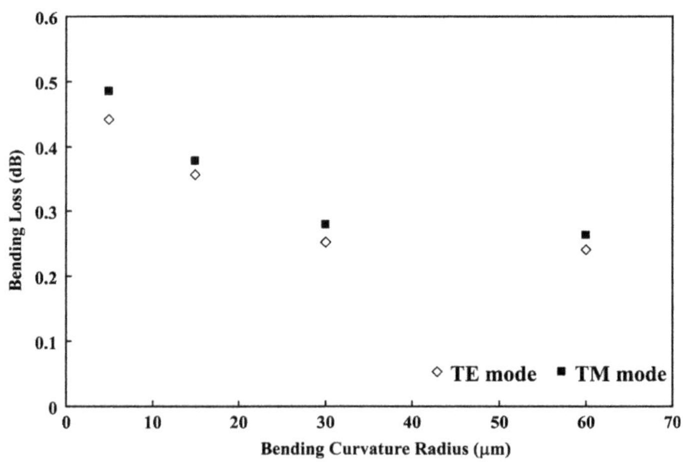

Figure 6.28. Relationship between bending loss and bending curvature radius. Reprinted with permission from S. Ono, T. Yoshimura, T. Sato, and J. Oshima, "Fabrication and Evaluation of Nano-Scale Optical Circuits Using Sol-Gel Materials with Photo-Induced Refractive Index Variation Characteristics," *J. Lightwave Technol.* **27**, 1229–1235. © 2009 IEEE.

loss for one bend is a half of the S-bending loss.

$$\begin{aligned} \text{S-Bending loss} &= c - b[\text{dB}] \\ \text{Bending loss} &= (c - b)/2[\text{dB}] \\ \text{Y-Branching loss} &= c - b - 3[\text{dB}] \end{aligned} \tag{6.2}$$

The relationship between bending loss and bending curvature radius is shown in Fig. 6.28. With increasing the bending curvature radius from 5 μm to 60 μm, bending loss is reduced from ~0.43 dB to ~0.24 dB. The result indicates that the sol-gel material can provide miniaturized optical circuits with bending curvature radius of several tens μm. The relationship between Y-branching loss and branching angle is shown in Fig. 6.29. With decreasing the branching angle from 80° to 20°, Y-branching loss is reduced from ~1.35 dB to ~0.1 dB. If Y-branching loss of ~ 0.4 dB is permitted, the sol-gel material can be used to fabricate optical circuits with a branching angle of ~40°.

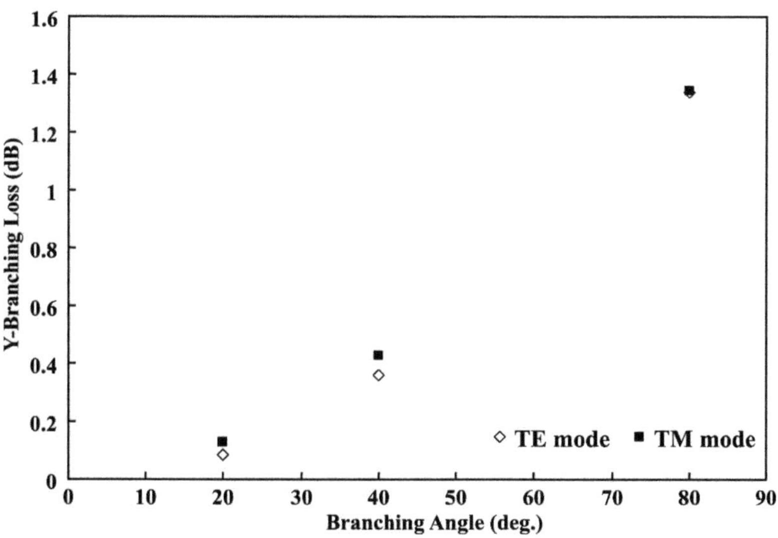

Figure 6.29. Relationship between Y-branching loss and branching angle. Reprinted with permission from S. Ono, T. Yoshimura, T. Sato, and J. Oshima, "Fabrication and Evaluation of Nano-Scale Optical Circuits Using Sol-Gel Materials with Photo-Induced Refractive Index Variation Characteristics," *J. Lightwave Technol.* **27**, 1229–1235. © 2009 IEEE.

Fabrication process optimization will reduce the bending and branching loss as well as the propagation loss. In fine patterns having sizes less than 1 μm, it is speculated that dimension distortions are caused by diffraction in UV light exposure using a photomask. The electron beam lithography might be required for accurate pattern formation.

6.4.2 *Vertical Mirrors*

Vertical mirrors in the nano-scale waveguide were constructed by the tilted exposure using UV light with an incident angle of 45°. In this preliminary experiment, the prism for the tilted exposure was not used. So, the slope of the beveled wall in the vertical mirror deviated from 45°. The expected slope was estimated to be 65° due to refraction at the air/Sol-Gel material boundary. FFP measurements revealed that the reflected beams from the vertical mirror propagate along an axis tilted from the surface-normal direction by 40°. This confirms that the fabricated vertical mirror has a slope of 65°.

6.4.3 *Fine 3-D Structures for All-Air-Clad Waveguides*

The PRI sol-gel material developed by Nissan Chemical Industries, LTD., becomes insolvable in Al etchant by exposing the material to UV/blue light followed by post-bake. By using this property, fine 3-D structures including all-air-clad waveguides can be fabricated.

Figure 6.30 shows an example of the process for fabrication of 3-D waveguide structures. After patterned Al films are formed on a substrate, a sol-gel film is spin-coated on it. Through a photomask, the sol-gel film is exposed to UV light with a pattern of waveguide cores and is post-baked. By put the sample in Al etchant, the unexposed regions of the sol-gel film and the Al film are solved. Thus fine 3-D patterns for a 3-D waveguide structure are completed. The example shown in Fig. 6.30 is a 3-D structure for all-air-clad waveguides with spacers. Figure 6.31 shows a SEM image of an experimentally obtained all-air-clad waveguide structure.

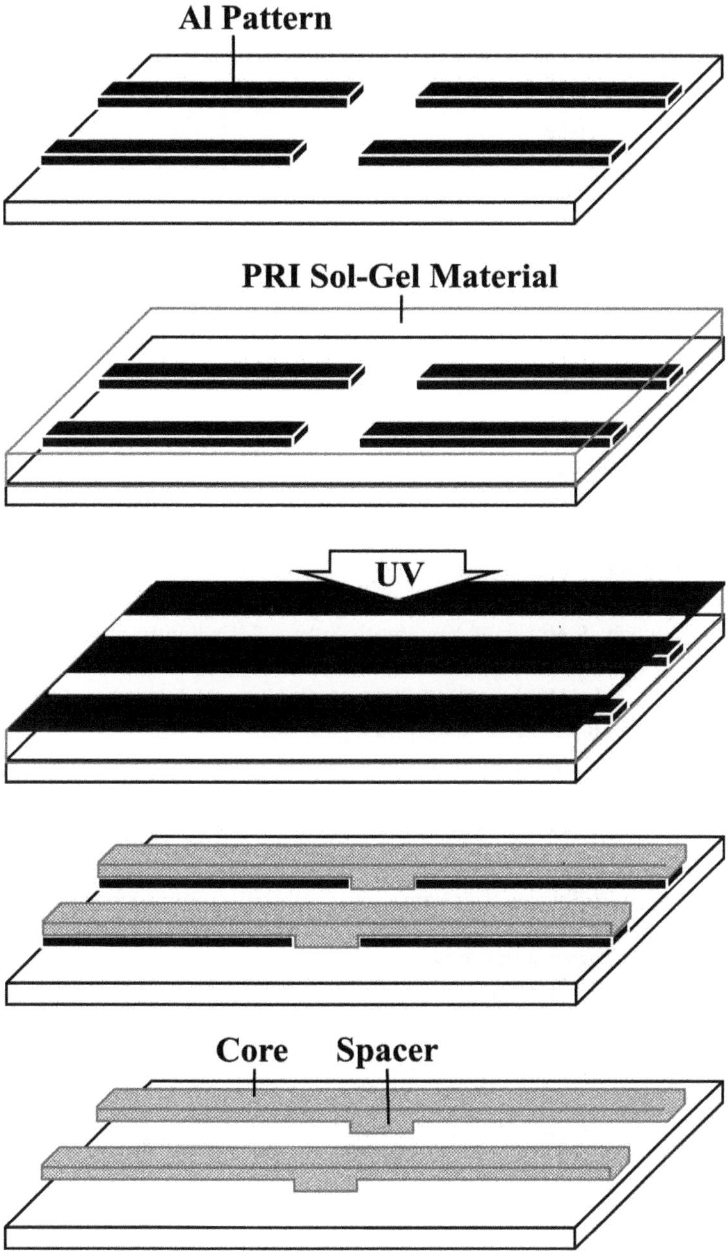

Figure 6.30. Process for 3-D waveguide structures.

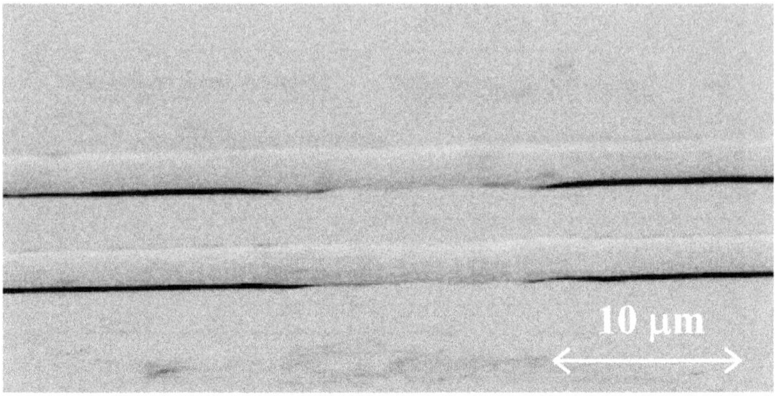

Figure 6.31. A SEM image of a 3-D waveguide structure for fabricating all-air-clad waveguides with spacers. [The SEM image was provided by Dr. Tetsuo Sato of Nissan Chemical Industries, LTD.]

References

1. T. Inoguchi, S. Moriya, T. Yamamoto, Y. Arai, T. Yoshimura, and K. Asama, "Film Waveguides with Micro Mirrors for Three-Dimensional Optical Circuits Fabricated by the Built-In Mask Method," *J. Jpn. Inst. Electron. Packaging* **8**, 237–242 (2005) [in Japanese].
2. T. Yoshimura, T. Inoguchi, T. Yamamoto, S. Moriya, Y. Teramoto, Y. Arai, T. Namiki, and K. Asama, "Self-Organized Lightwave Network Based on Waveguide Films for Three-Dimensional Optical Wiring Within Boxes," *J. Lightwave Technol.* **22**, 2091–2100 (2004).
3. F. Huang, H. Takase, Y. Eriyama, and T. Ukachi, "Optical Fiber Interconnection by Using Self-Written Waveguides," *Proc. 9th Int. Conf. Rad. Tech. Asia*, Japan, 637–639 (2003).
4. T. Yoshimura, M. Miyazaki, Y. Miyamoto, N. Shimoda, A. Hori, and K. Asama, "Three-Dimensional Optical Circuits Consisting of Waveguide Films and Optical Z-Connections," *J. Lightwave Technol.* **24**, 4345–4352 (2006).
5. T. Yoshimura, K. Ogushi, Y. Kitabayashi, Y. Miyamoto, M. Miyazaki, "Optical Waveguide Films with Two-Layer Skirt-Type Core Structures for Beam Leakage/Scattering Reduction at Tapered Mirrors," *IEEE Photon. Technol. Lett.* **19**, 1427–1429 (2007).
6. T. Yoshimura, K. Ogushi, Y. Kitabayashi, K Naito, Y. Miyamoto, and M. Miyazaki, "Optical Waveguide Films with Two-Layer Skirt-Type Core

End Facets for Beam Leakage Reduction at 45° Mirrors," *Proc. SPIE* **6899**, from Photonics West 2008, San Jose, California, 689913-1-11 (2008).

7. E. Cassan, S. Laval, S. Lardenois, and A. Koster, "On-Chip Optical Interconnects with Compact and Low-Loss Light Distribution in Silicon-on-Insulator Rib Waveguides," *IEEE J. Select. Topics Quantum Electron.* **9**, 460–464 (2003).
8. S. Ono, T. Yoshimura, T. Sato, and J. Oshima, "Fabrication and Evaluation of Nano-Scale Optical Circuits Using Sol-Gel Materials with Photo-Induced Refractive Index Variation Characteristics," *J. Lightwave Technol.* **27**, 1229–1235 (2009).
9. A. Bhatnagar, S. Latif, C. Debaes, D. A. B. Miller, "Pump-Probe Measurements of CMOS Detector Rise Time in the Blue," *J. Lightwave Technol.* **22**, 2213–2217 (2004).

Chapter 7

3-D Optical Circuits with Stacked Waveguide Films

In the 3-D optical circuit, which is one of the core technologies in optical electronics, optical waveguide films are stacked and optical links between stacked waveguide films are carried out by optical Z-connections using 45° mirrors.

In this chapter, 3-D optical circuits fabricated by stacking two optical waveguide films with vertical mirrors are demonstrated [1, 2]. As a challenge to reduce the loss at the optical Z-connections, we fabricate an optical waveguide film with a vertical waveguide grown above a vertical mirror aperture using SOLNET [1, 2].

7.1 Structures of 3-D Optical Circuits

Figure 7.1 shows two types of 3-D optical circuits investigated in the present experiments. Type 1 is "Stacked waveguide films with vertical mirrors." Two optical waveguide films, Film A and B, are stacked by contacting their under-clad layers. No over-clad layer is coated. A probe beam is introduced into Input in Film A. The beam is transferred into Film B through Optical Z-Connection to reach Output in Film B.

Optical Electronics: Self-Organized Integration and Applications
Tetsuzo Yoshimura

ISBN 978-981-4310-90-1 (Hardcover), 978-981-4364-08-9 (eBook)
www.panstanford.com

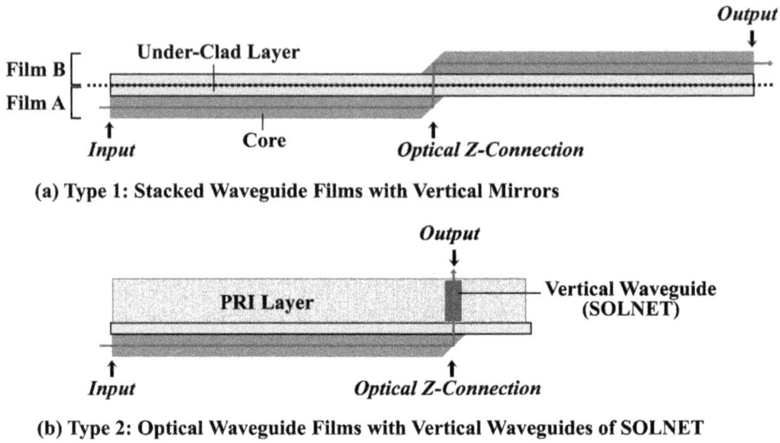

Figure 7.1. Schematic illustration of 3-D optical circuits. Reprinted with permission from T. Yoshimura, M. Miyazaki, Y. Miyamoto, N. Shimoda, A. Hori, and K. Asama, "Three-Dimensional Optical Circuits Consisting of Waveguide Films and Optical Z-Connections," *J. Lightwave Technol.* **24**, 4345–4352. © 2006 IEEE.

Type 2 is "Optical waveguide films with vertical waveguides of SOLNET." A vertical waveguide is constructed above a vertical mirror aperture in an optical waveguide film using SOLNET. SOLNET enables us to construct optical waveguides for optical vias in a region filled with a PRI material. When a write beam is introduced into Input, it is reflected into the PRI layer to grow a vertical waveguide at Optical Z-Connection. The vertical waveguide confines a reflected probe beam to guide it efficiently to Output at the PRI layer surface.

7.2 Type 1: Stacked Waveguide Films with Vertical Mirrors

7.2.1 *Demonstration of 3-D Optical Wiring*

In Fig. 7.2, a fabrication/measurement setup for 3-D optical circuits consisting of stacked waveguide films is shown. Film B put on an alignment stage was stacked on Film A with adjusting the positions of vertical mirrors in Film A and B to construct Optical Z-Connection.

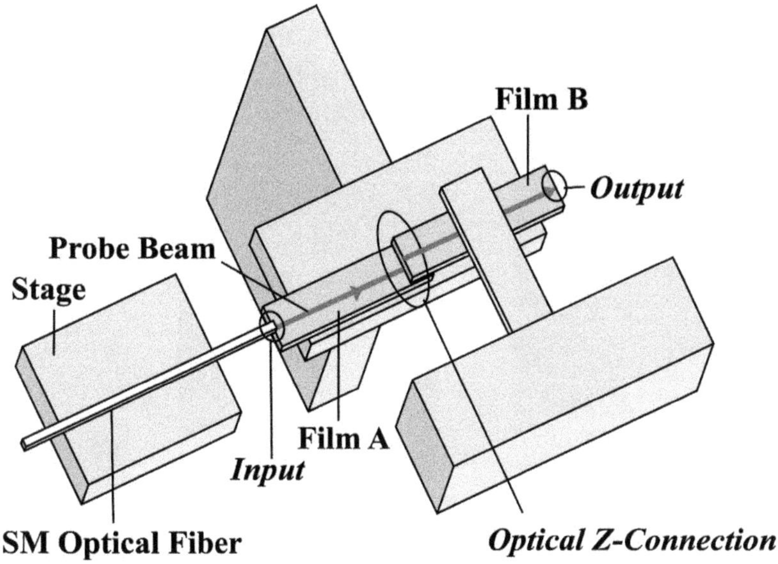

Figure 7.2. Experimental setup for demonstration of 3-D optical circuits of stacked waveguide films. Reprinted with permission from T. Yoshimura, M. Miyazaki, Y. Miyamoto, N. Shimoda, A. Hori, and K. Asama, "Three-Dimensional Optical Circuits Consisting of Waveguide Films and Optical Z-Connections," *J. Lightwave Technol.* **24**, 4345–4352. © 2006 IEEE.

A SM optical fiber for 1.3 μm with core diameter of 9.5 μm was connected to Input of a core in Film A by butt joint. A probe beam was introduced into Input and propagated to Output in Film B through Optical Z-Connection [1].

Figure 7.3 shows propagation of a 650 nm-wavelength probe beam before and after alignment of Film A and Film B. Before alignment, as shown in Fig. 7.3(a), since vertical mirror in Film B is not located above the vertical mirror in Film A, the probe beam reflected by the vertical mirror in Film A propagates through under-clad layers to reach a camera set above the vertical mirror. After alignment, as shown in Fig. 7.4, the two vertical mirrors are correctly aligned to form Optical Z-Connection. The reflected probe beam can be transmitted into Film B through Optical Z-Connection to reach Output (Fig. 7.3(b)). A $\sim$30 $\times$ 30 μm^2 NFP is observed at Output. Thus, 3-D optical wiring operation in the 3-D optical circuit is demonstrated successfully.

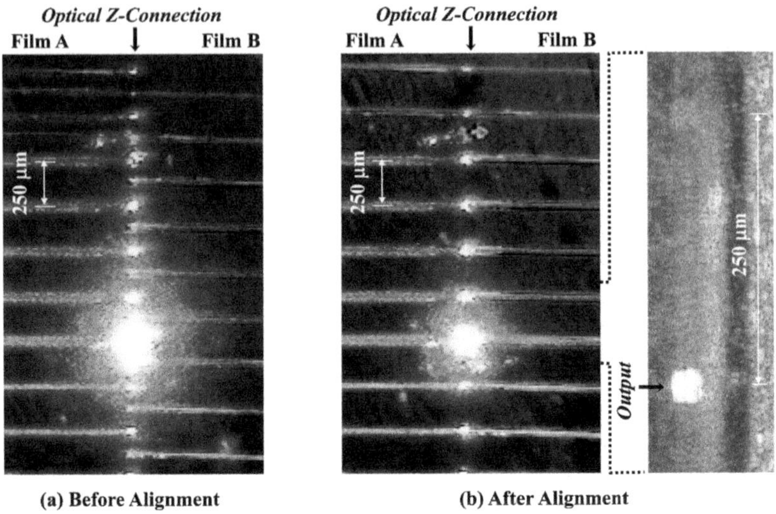

Figure 7.3. Probe beam propagation in the 3-D optical circuit with stacked waveguide films. Reprinted with permission from T. Yoshimura, M. Miyazaki, Y. Miyamoto, N. Shimoda, A. Hori, and K. Asama, "Three-Dimensional Optical Circuits Consisting of Waveguide Films and Optical Z-Connections," *J. Lightwave Technol.* **24**, 4345–4352. © 2006 IEEE.

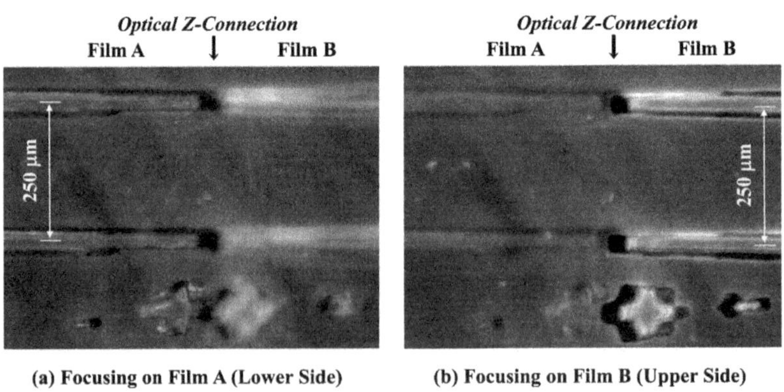

Figure 7.4. Photographs of Optical Z-Connection part. Reprinted with permission from T. Yoshimura, M. Miyazaki, Y. Miyamoto, N. Shimoda, A. Hori, and K. Asama, "Three-Dimensional Optical Circuits Consisting of Waveguide Films and Optical Z-Connections," *J. Lightwave Technol.* **24**, 4345–4352. © 2006 IEEE.

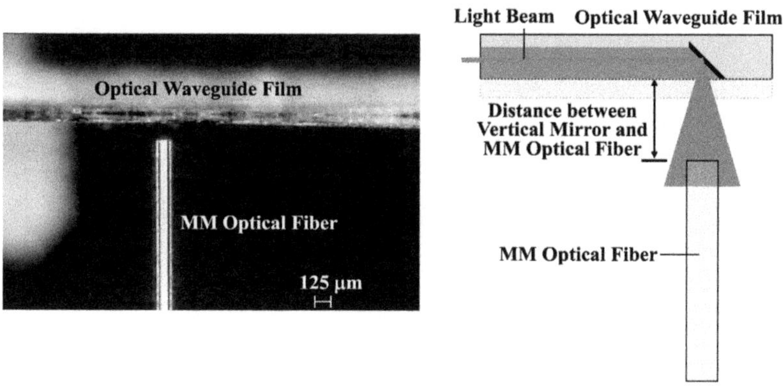

Figure 7.5. Setup for measurement of coupling efficiency from the optical waveguide film to the MM optical fiber.

The cutback method revealed that the propagation loss of the optical waveguide film is 0.8 dB/cm for 1.3 μm in wavelength. Reflection loss of the vertical mirror was estimated to be 2 dB by collecting the reflected probe beams by a MM optical fiber with core diameter of 50 μm as shown in Fig. 7.5. The measured loss is relatively large, which might be attributed to numerical aperture (NA) mismatch between the optical waveguide and the MM optical fiber.

7.2.2 *Loss Measurements at Optical Z-Connections*

7.2.2.1 Problems in measurements [2]

Figure 7.6 shows dependence of coupled power from the optical waveguide film to the MM optical fiber on distance between the vertical mirror and the MM optical fiber. The measurements were carried out by the setup depicted in Fig. 7.5. For air-clad type waveguide, where only one side of the core contacts a clad, the coupled power decreases with the distance rapidly. For over-clad type waveguide, where three sides of the core are covered by a clad, the rate of the coupled power decrease is low compared with the case of the air-clad type waveguide.

The results can be explained by the model schematically illustrated in Fig. 7.7. The air-clad type waveguide can guide

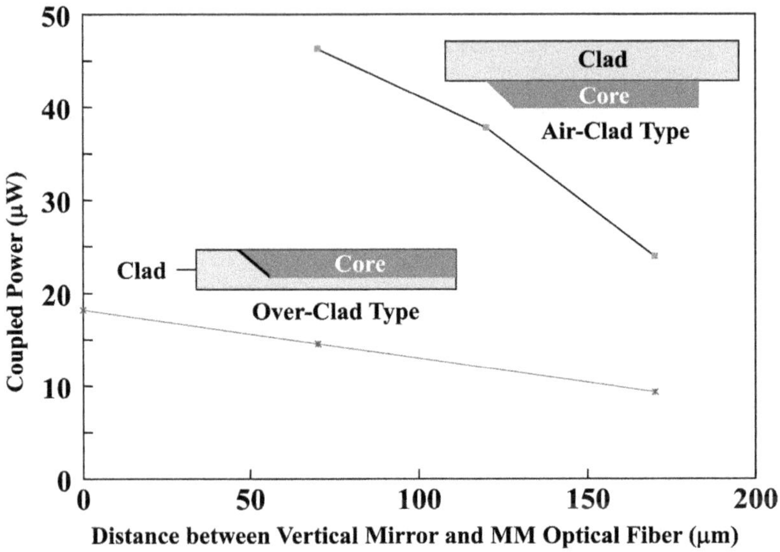

Figure 7.6. Dependence of coupled power from the optical waveguide film to the MM optical fiber on distance between the vertical mirror and the MM optical fiber.

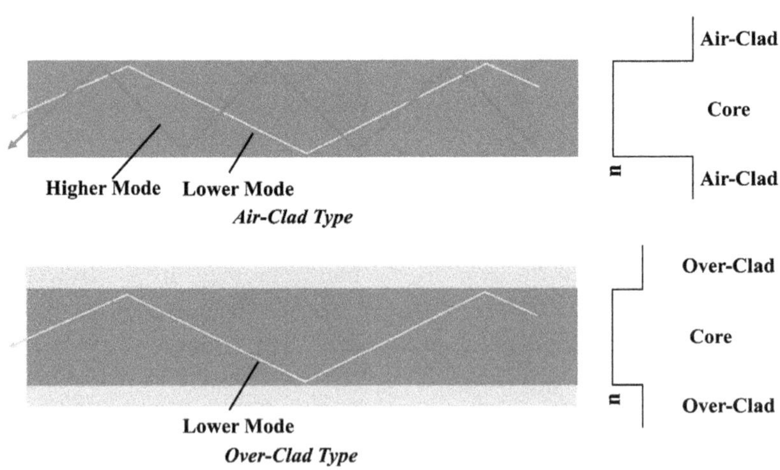

Figure 7.7. Schematic illustration of light beam confinement into an optical waveguide.

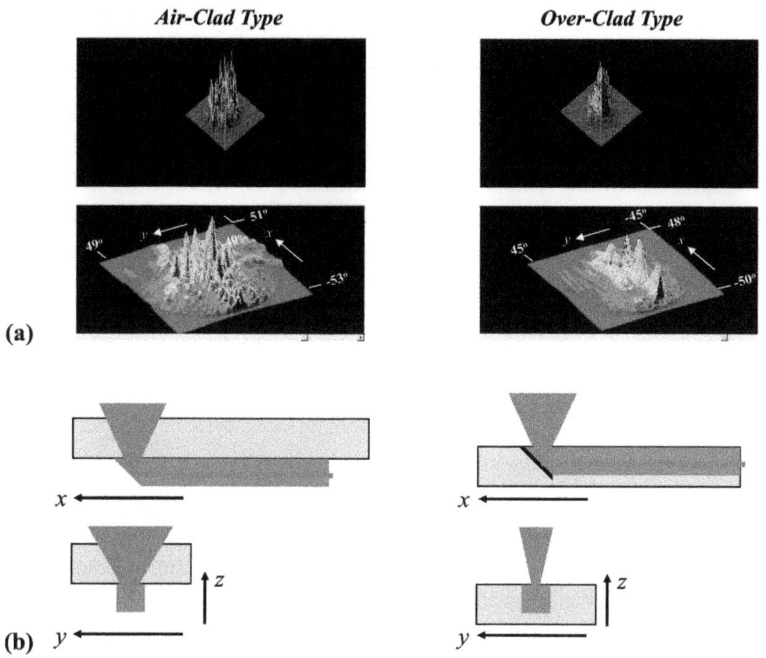

Figure 7.8. (a) NFPs and FFPs of reflected beams from vertical mirrors and (b) model for degree of spread of reflected beams.

light beams in higher modes comparing with the over-clad type waveguide because refractive index difference between the core and the clad is larger in the former than in the latter. This implies that NA of the air-clad type waveguide is larger than that of the over-clad type waveguide, causing larger emitting angles of light beams from the edge in the air-clad type waveguide.

NFPs and FFPs shown in Fig. 7.8(a) confirm the above-mentioned explanation. In the air-clad type waveguide, higher modes are excited and large beam spread along the y-direction, the direction in which the light beams are strongly confined by the air-clad regions of both sides, occurs. The larger the emitting angle becomes, the higher the rate of the coupled power decrease becomes. This results in the difference in the rate of the coupled power decrease shown in Fig. 7.6.

For beam spread along x-direction, no difference is observed between the air-clad type and over-clad type since in this direction the light beams are confined similarly by an air-clad region and a clad region in both types of optical waveguides.

The measured reflection loss of the vertical mirrors might be minimized as follows. In the present case, since the optical waveguide is an air-clad type, NA of the optical waveguide is larger than that of the MM optical fiber. The NA mismatch decreases the coupled efficiency, resulting in overestimation of the loss. The use of an optical fiber with larger core-clad refractive index differences or a coupling lens may reduce the overestimated value.

7.2.2.2 Loss at optical Z-Connections

In the present experiment, total propagation length from Input to Output is 5 cm, giving rise to a loss of 8 dB for the propagation loss. The insertion loss of the 3-D optical circuit, which involves the propagation loss and the reflection loss of two vertical mirrors at the Optical Z-Connection, is ~22 dB. These results suggest that the loss of the Optical Z-Connection is ~14 dB. The main origin of the loss might be the leakage of probe beams reflected from the vertical mirror with large diverging angles from the surface-normal direction.

In the present experiment, for easy handling of film samples, we used 70 μm-thick under-clad layers, resulting in a long free-space path length of ~170 μm for the reflected probe beams at Optical Z-Connection. So, only a limited portion of the reflected probe beams can be received by the vertical mirror in Film B, causing considerable probe beams passing through Film B even after alignment as can be seen in Fig. 7.3(b). Furthermore, in air-clad waveguides, light beams of higher modes can propagate in the core with a critical angle of ~40° to the core sidewall. As a result, the emitting angle range of the reflected probe beams from the vertical mirror in Film A is widely spread as shown in Fig. 7.8, increasing the loss at Optical Z-Connection.

A slight curvature implementation to vertical mirror surfaces may reduce the loss at the Optical Z-connection. Decreases in the under-clad layer thickness may also reduce the loss. Another promising challenge to reduce the loss at the Optical Z-Connection is optical via formation that is discussed in the following section.

7.3 Type 2: Optical Waveguide Films with Vertical Waveguides of SOLNET

As one of the ways for loss reduction at Optical Z-Connection, the implementation of vertical waveguides on an optical waveguide film is promising. The fabrication/measurement setup is shown in Fig. 7.9. On the back of an under-clad layer in an optical waveguide film, a PRI material layer of ~500 μm thickness was coated. For the PRI material, monomer/monomer type photo-polymer described in Section 5.1 was used.

A write beam of 405 nm blue LD was introduced from an optical fiber of 9.5 μm core diameter into Input of the optical waveguide film with power of ~100 μW or lesser. The propagated write beam was reflected to the surface-normal direction by a vertical mirror. The write beam passed through the under-clad layer and was injected into the PRI layer. Since refractive index of the PRI material increases with the write beam exposure, the write beam tends to be focused

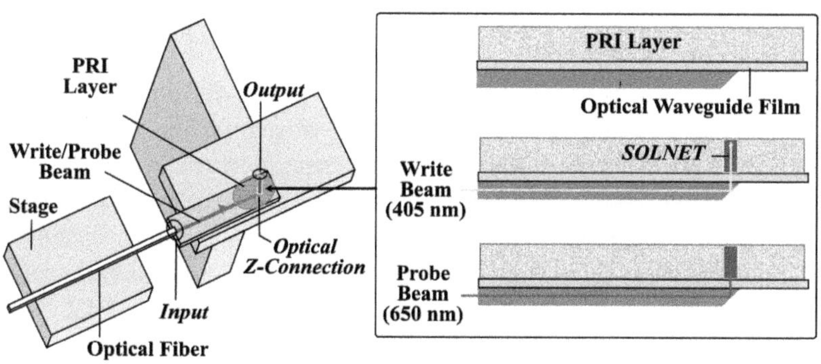

Figure 7.9. Experimental setup for demonstration of 3-D optical circuits of optical waveguide films with vertical waveguides of SOLNET. Reprinted with permission from T. Yoshimura, M. Miyazaki, Y. Miyamoto, N. Shimoda, A. Hori, and K. Asama, "Three-Dimensional Optical Circuits Consisting of Waveguide Films and Optical Z-Connections," *J. Lightwave Technol.* **24**, 4345–4352. © 2006 IEEE.

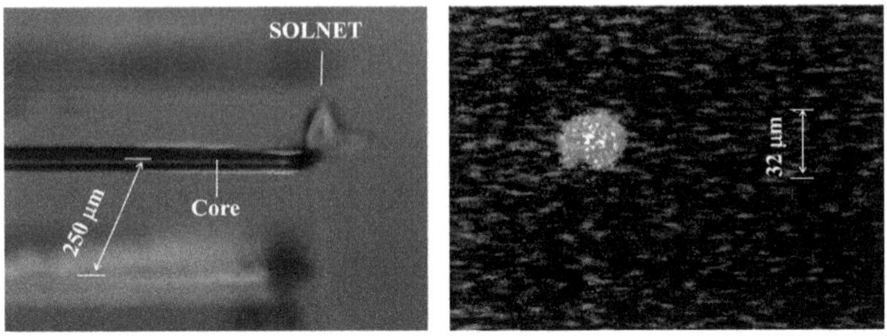

(a) Vertical Waveguide of SOLNET
(View from above with an angle of ~30°)

(b) Near Field Pattern

Figure 7.10. (a) The vertical waveguide of SOLNET grown above the 45° mirror of the optical waveguide film. (b) An NFP of the probe beam guided in the vertical waveguide at the PRI layer surface. Reprinted with permission from T. Yoshimura, M. Miyazaki, Y. Miyamoto, N. Shimoda, A. Hori, and K. Asama, "Three-Dimensional Optical Circuits Consisting of Waveguide Films and Optical Z-Connections," *J. Lightwave Technol.* **24**, 4345–4352. © 2006 IEEE.

along the center of the propagation axis. The self-focusing effect grew a vertical waveguide of SOLNET in the PRI layer [1]. Then, a probe beam of 650 nm in wavelength was introduced into Input of the optical waveguide film. The beam was guided in the vertical waveguide at Optical Z-Connection to reach Output at the PRI layer surface of the SOLNET.

Figure 7.10(a) shows SOLNET grown vertically on the under-clad layer. It is found that the vertical waveguide is constructed in the PRI layer above the vertical mirror aperture. An NFP of a guided probe beam at Output is shown in Fig. 7.10(b). An NFP of $\sim 30 \times 30\ \mu m^2$, which is approximately the same as the core size of the optical waveguide film, is observed. These results demonstrate optical via capability of the vertical waveguide of SOLNET for 3-D optical wiring.

It should be noted that SOLNET was formed by write beams from the under-clad layer, which is a free space with no beam confining structures. In other words, SOLNET can be formed even when some spacers are inserted between vertical mirrors and PRI layers.

A future challenge is the implementation of SOLNET inserted between two vertical mirrors by combining 3-D optical circuits of

Type 1 and 2 like the model shown in Fig. 10.5. By connecting two optical waveguide films with vertical waveguides of SOLNET, light beams may be guided from one vertical mirror to the other with strong confinement, reducing the leakage of the light beams at Optical Z-Connection.

References

1. T. Yoshimura, M. Miyazaki, Y. Miyamoto, N. Shimoda, A. Hori, and K. Asama, "Three-Dimensional Optical Circuits Consisting of Waveguide Films and Optical Z-Connections," *J. Lightwave Technol.* **24**, 4345–4352 (2006).
2. T. Yoshimura, *Molecular Nano Systems — Applications to Optoelectronic Computers and Solar Energy Conversion*, (Corona Publishing Co., Ltd., Tokyo, 2007) (in Japanese).

Chapter 8

Heterogeneous Thin-Film Device Integration

As mentioned in Chapter 2, in order to remove the optics excess, a new concept of OE interconnect hardware "S-FOLM" consisting of films with embedded thin-film devices was proposed [1, 2].

For OE module fabrication, so far, bulk chip packaging based on flip-chip bonding has been the major technology [3, 4]. However, the method is not suitable for S-FOLM, which has a 3-D stacked structure, due to its thick-bulk-characteristics. Furthermore, the method requires precise alignment of OE chips using chip-bonders and is material and space consuming, preventing the OE modules from a drastic cost reduction.

Meanwhile, unique device processing technology called "Epitaxial Lift-Off (ELO)" was developed by Yablonovitch et al. [5, 6]. ELO enables mounting thin-film OE devices on substrates, which realizes smart pixels [7, 8] and high-performance OE transceivers [9]. In conventional ELO, however, thin-film devices are placed on substrates one-by-one using the pick-and-place method, giving time-consuming processes. Therefore, the most critical issue in the ELO-assisted heterogeneous integration process is how to place thin-film devices on substrates at one time with desired distribution.

Optical Electronics: Self-Organized Integration and Applications
Tetsuzo Yoshimura

ISBN 978-981-4310-90-1 (Hardcover), 978-981-4364-08-9 (eBook)
www.panstanford.com

In this chapter, a heterogeneous integration process, "Photolithographic Packaging with Selectively Occupied Repeated Transfer**(PL-Pack with SORT)**," is reviewed [10–14]. PL-Pack with SORT is a resource-saving heterogeneous integration process with multi-step device transfers and is an important core technology in Optical Electronics. The process is based on photolithography, and is completed only by fabrication processes for the integrated circuits without chip assembly, enabling both cost and resource saving.

A process design for S-FOLM is presented as an example of PL-Pack with SORT. Preliminary estimation of III-V epitaxial material consumption and module cost is carried out. Process step counts for SORT and conventional pick-and-place method are compared. Advantages of PL-Pack with SORT against the flip-chip-bonding-based packaging are discussed from a viewpoint of cost/resource saving. Some examples of thin-film device transfers by SORT are presented as experimental demonstrations, including transfers of waveguide lenses, waveguide cores, and transfers of two kinds of model devices.

8.1 PL-Pack with SORT versus Flip-Chip-Bonding-Based Packaging

In Fig. 8.1, PL-Pack with SORT is compared with conventional flip-chip-bonding-based packaging for an example of assembling VCSEL and PD on optical waveguides. In the flip-chip-bonding-based packaging, VCSEL chip, PD chip, and driver/amplifier chip are mounted on a substrate with optical waveguides and vias/pads/electrical lines, providing conventional OE module.

In PL-Pack with SORT, first, VCSEL flake array and PD flake array are grown on a VCSEL wafer and a PD wafer, respectively. Driver/amplifier flake array is also grown on a driver/amplifier wafer. Next, individual VCSEL flakes and PD flakes in the arrays on the wafers are selectively transferred to appropriate sites on a substrate and embedded with optical waveguides and vias/pads/electrical lines. Driver/amplifier flakes are embedded in an interface film. Finally, the interface film is stacked on the substrate to complete S-FOLM. As mentioned in Section 2.2.2,

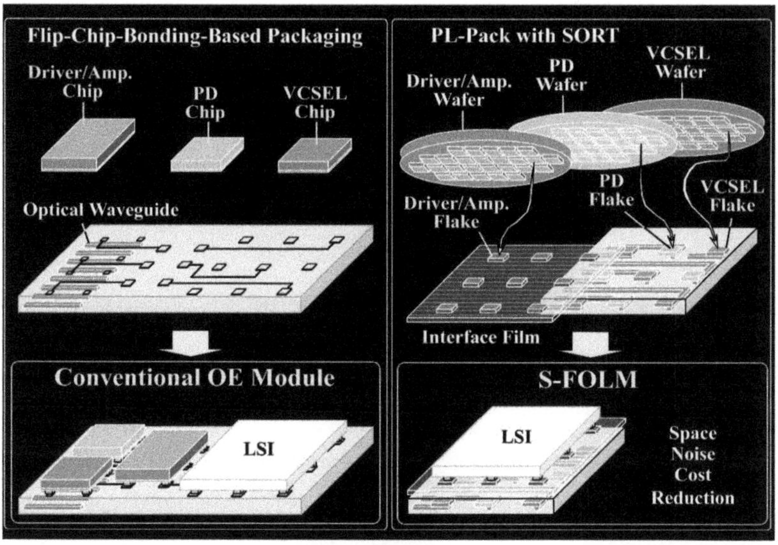

Figure 8.1. (a) Conventional OE module by flip-chip-bonding-based packaging and (b) S-FOLM by PL-Pack with SORT. See also Color Insert.

S-FOLM fabricated by PL-Pack with SORT has following advantages against conventional modules:

- Space saving (by embedding and stacking)
- Noise reduction (by eliminating long metal lines)
- Cost/Resource saving (to be described in Section 8.3.1)
- Thermal stress reduction (to be described in Section 8.3.2)

8.2 PL-Pack with SORT

8.2.1 *Process Flow of PL-Pack*

Total process flow of PL-Pack is shown in Fig. 8.2. A substrate (LSI, MCM, PCB, removable substrate, etc.) with pads/lines is prepared. Device flakes, which are prepared by epitaxial liftoff [5, 6], polishing, or something like that, are put on the pads of the substrate by SORT described in Section 8.2.2. After coating polymer films, the surface is planarized to embed the device flakes. Then, metallization for vias, pads, and electrical lines is carried out. By substrate removal,

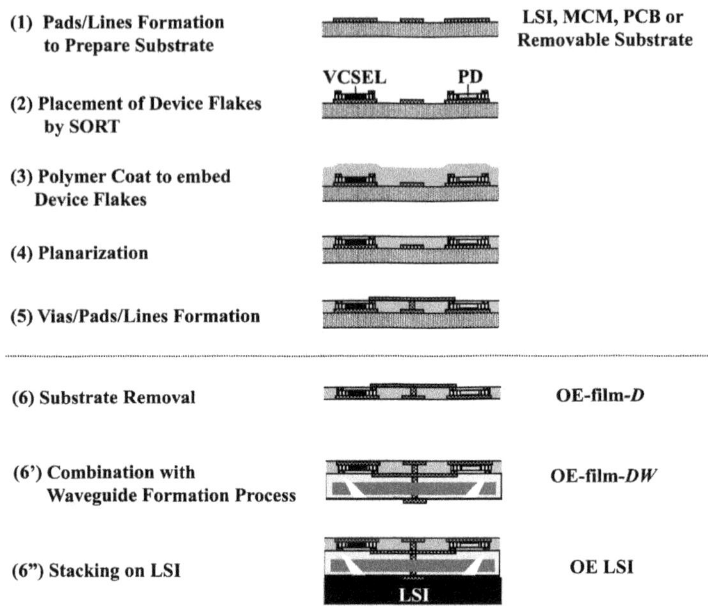

Figure 8.2. Total process flow of PL-Pack. Reprinted with permission from T. Yoshimura, J. Roman, Y. Takahashi, M. Lee, B. Chou, S. I. Beilin, W. V. Wang, and M. Inao, "Optoelectronic Scalable Substrates Based on Film/Z-Connection and Its Application to Film Optical Link Module (FOLM)," *Proc. SPIE* **3952**, 202–213 (2000).

a device-flakes-embedded OE film, which corresponds to OE-film-*D* described in Section 10.3.1, is obtained. By combining the process steps 1–5 with waveguide formation processes, a device-flakes-embedded optical waveguide film, which corresponds to OE-film-*DW* in Section 10.3.1, is obtained. By stacking the OE-film-*DW* on LSI, OE LSI is constructed.

Here, for steps 1, 3, 4, and 5, mature process technologies already exist. The critical part is step 2, device flake placement on the substrate by SORT, which is described below in detail.

8.2.2 *Process Flow of SORT*

Figure 8.3 shows the outline of SORT. For simplicity, two kinds of devices are considered, although, in general, many kinds of different devices are integrated. Device I flake array and Device II flake array are respectively grown with a pitch of d on growth wafer I and growth wafer II. The device flakes are obtained by dividing the grown thin film with boundary grooves. All the Device I flakes are picked up on a pickup substrate at one time by, for example, ELO, followed by pre-transfer to a supporting substrate Sub I. Similarly, the Device II flakes are pre-transferred to Sub II. The distribution patterns of the flakes on Sub I and Sub II are respectively designed to match the pad distribution patterns for Device I and Device II on the final substrate. The flakes left on the pick-up substrates are provided to another Sub I and Sub II. For final transfer, Sub I is attached on one of the final substrates, where catch-up sites with a pitch of p_P are formed, to place Devices I flakes at sites for Device I on the final substrate selectively. Then, Sub II is attached on the final substrate, to place Device II flakes at sites for Device II. The Sub I and Sub II are further attached to other final substrates or other places of the final substrate to provide Device I flakes and Device II flakes selectively. By repeating the procedure, all the flakes on Sub I and Sub II are transferred to final substrates. For the case shown in Fig. 8.3, there are $4 \times 4 = 16$ flakes corresponding to one catch-up site. This implies that one supporting substrate can provide flakes to 16 final substrates. Two supporting substrates are available for Device I and II respectively, indicating that 32 final substrates, on which Device I flakes and II flakes are distributed with a desired configuration, can be obtained from a growth wafer for Device I and a growth wafer for Device II.

For fabrication of OE LSIs, optical interconnect boards, or 3D stacked OE LSIs, Devices I and II could be VCSELs, light modulators, wavelength filters and PDs. For the fabrication of 3D-MOSS, Device I and II could be, for example, optical switches of PLZT films, electro-optic (EO) organic films or III-V epitaxial films, micro-reflectors of mirrors/wavelength filters or photonic crystals, waveguide lenses of Ti_xSi_yO thin films, and so on.

In addition to material consumption reduction, PL-Pack with SORT has another advantage, namely, it can remove the conventional flip-chip-bonding-based batch packaging, which is called "Jisso" in Japanese, to achieve heterogeneous device integration only by the semiconductor device fabrication process based on the

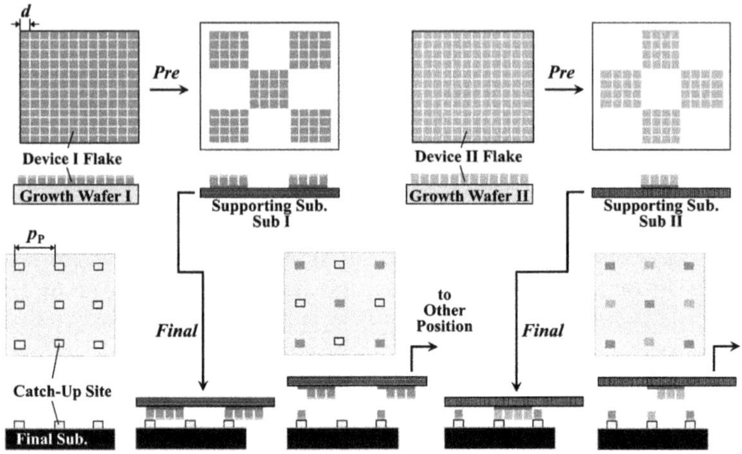

Figure 8.3. Outline of SORT for integration of two kinds of active elements with length/width of d. Reprinted with permission from T. Yoshimura, M. Ojima, Y. Arai, and K. Asama, "Three-Dimensional Self-Organized Micro Optoelectronic Systems for Board-Level Reconfigurable Optical Interconnects — Performance Modeling and Simulation," *IEEE J. Select. Topics Quantum Electron.* **9**, 492–511. © 2003 IEEE. See also Color Insert.

photolithography. These features are effective on cost reduction of OE systems.

An example of precise SORT flow is illustrated in Figs. 8.4 (a–e). For simplicity, two kinds of devices are considered.

(a) Device I flake array is grown on a growth wafer with a pitch of *d*. The value of *d* is assumed to be equal to device flake size by neglecting the boundary groove area. By ELO, the flake array is transferred to a pick-up substrate, Sub-I.1. Similarly, Device II flake array grown on a growth wafer is transferred to Sub-II.1.

(b) For pre-transfer, Sub-I.1 is attached on a supporting substrate, Sub-I.2, where adhesive layers are formed. The pattern is designed so that the adhesive regions correspond to Device I sites on a final substrate. Device I flakes are selectively placed on Sub-I.2 (1). The Device I flakes left on Sub-I.1 are placed on Sub-I.2 (2) to use all the Devices I flakes.

(c) With similar procedure, Device II flakes are selectively placed on Sub-II.2 (1) and (2).

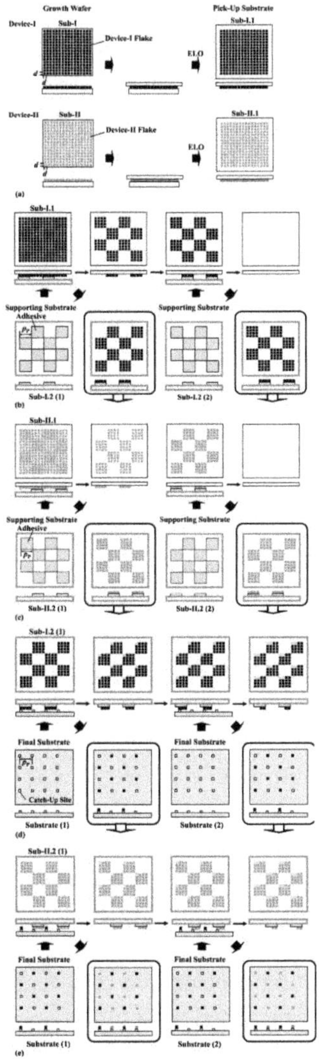

Figure 8.4. A SORT Process Example. (a) Transfer from wafer to pick-up substrate, (b) pre-transfer for Device-I flakes, (c) pre-transfer for Device-II flakes, (d) final-transfer for Device-I flakes, and (e) final-transfer for Device-II flakes. Reprinted with permission from T. Yoshimura, K. Kumai, T. Mikawa, and O. Ibaragi, "Photolithographic Packaging with Selectively Occupied Repeated Transfer (PL-Pack with SORT) for Scalable Optical Link Multi-Chip-Module (S-FOLM)," *IEEE Trans. Electron. Packag. Manufact.* **25**, 19–25. © 2002 IEEE.

(d) Final substrates, Substrate (1), (2), …, (32) are prepared. The substrates have catch-up sites with a pitch of p_P. For final-transfer, Sub-I.2 (1) is attached on Substrate (1). Device I flakes are selectively placed at sites for Device I on Substrate (1). Then, Sub-I.2 (1) is attached on Substrate (2), placing Device I flakes selectively on Substrate (2). Similar procedure is repeated to provide the device flakes to other final substrates. At last, Sub-I.2 (1) can make $4 \times 4 = 16$ final substrates of (1), (2), …, (16) with Device I flakes. Similarly, Sub-I.2 (2) makes another 16 final substrates of (17), (18), …, (32) with Device I flakes. Thus, 32 final substrates with Device I flakes are made.

(e) On above-described Substrate (1) with Device I flakes, Sub-II.2 (1) is attached. Device II flakes are selectively transferred at sites for Device II on Substrate (1). By repeating the procedure, finally, 32 final substrates, on which Device I flakes and Device II flakes are distributed with a desired configuration, are obtained.

A marker alignment procedure can be applied to the device flake transfer process by using an aligner. Therefore, SORT can be regarded as an all-photolithographic process, eliminating batch process in the conventional flip-chip-bonding-based packaging or "Jisso." In the present case, the supporting substrate and the final substrate have the same size. However, the model can work in the case that final substrate is larger than the supporting substrate just by replacing "Substrate (1)" and "Substrate (2)" with "Region (1)" and "Region (2)". Namely, device flakes are transferred from a supporting substrate to a plurality of regions in the same final substrate.

SORT can stably be performed by using adhesive, whose adhesion strength is controlled with "Adhesion Strength Hierarchy" or "Dynamic Adhesion Strength Control." In the former, the adhesion strength is increased with step progress, namely, in the order of Sub-I.1/Sub-II.1, Sub-I.2/Sub-II.2, and final substrate. In the latter, adhesion strength of the adhesive is controlled during the process by, for example, UV light exposure using UV curable materials, which corresponds to Light-Assisted SORT described in Section 8.6.1.

The SORT process described above is one of the examples. A lot of variations are possible. The essential point is that a number of device flakes are distributed on final substrates in desired configurations with minimum steps by successive device transfers using supporting substrate, where patterned adhesion regions are selectively formed.

8.3 Impact of PL-Pack with SORT

Schematic illustrations of structures obtained by PL-Pack with SORT and conventional flip-chip bonding are shown in Fig. 8.5 for the case of devices made of III-V materials, typically, VCSELs and PDs. In flip-chip bonding, a bulk chip of III-V material is put on a substrate. In PL-Pack with SORT, an OE film, in which III-V epitaxial material flakes with length and width of d are embedded with a pitch of p_P, is put on a substrate. The difference brings remarkable advantages to the structure built by PL-Pack with SORT over flip-chip bonding as described below.

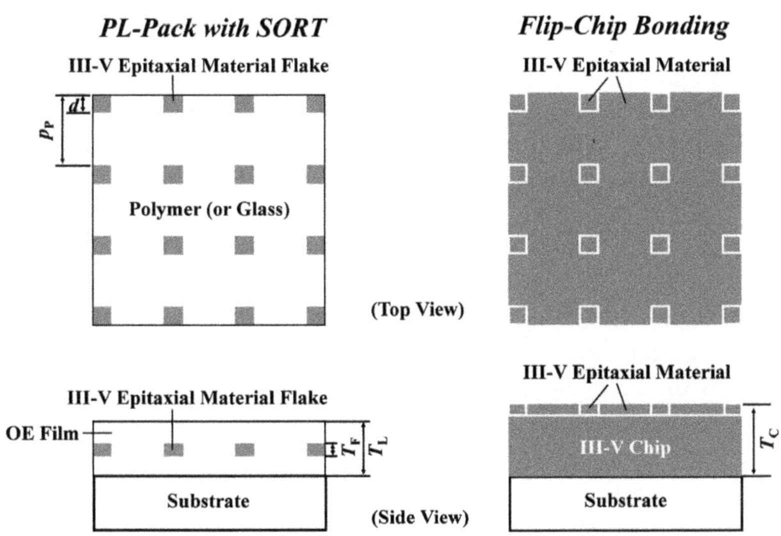

Figure 8.5. Schematic illustrations of structures obtained by PL-Pack with SORT and conventional flip-chip bonding. Reprinted with permission from T. Yoshimura, M. Ojima, Y. Arai, and K. Asama, "Three-Dimensional Self-Organized Micro Optoelectronic Systems for Board-Level Reconfigurable Optical Interconnects — Performance Modeling and Simulation," *IEEE J. Select. Topics Quantum Electron.* **9**, 492–511. © 2003 IEEE.

8.3.1 *Resource Consumption and Cost*

One of the advantages of PL-Pack with SORT is resource-saving characteristics. In flip-chip bonding, as shown in Fig. 8.5, whole area is occupied by expensive III-V epitaxial materials. In PL-Pack with SORT, on the other hand, the epitaxial materials exist only at necessary sites. The III-V epitaxial material saving effect can be measured by the material saving factor (MSF) of PL-Pack with SORT defined by

$$\mathrm{MSF} = \frac{[1/R_{\mathrm{SORT}}]}{[1/R_{\mathrm{FC\,Bonding}}]} = \left(\frac{p_{\mathrm{P}}}{d}\right)^2. \tag{8.1}$$

Here, R_{SORT} and $R_{\mathrm{FC\,Bonding}}$ represent III-V epitaxial material consumption in PL-Pack with SORT and in flip-chip bonding, respectively. Material flake length/width d and pitch p_{P} in the OE film in Fig. 8.5 respectively correspond to the device flake pitch on a growth wafer and the catch-up site pitch in Fig. 8.4. Figure 8.6(a) shows MSF as a function of d for various p_{P}. When p_{P} is 100 μm, MSF equals 1 for d of 100 μm. This implies that whole area of the OE film prepared by PL-Pack with SORT is occupied by III-V epitaxial materials since p_{P} is equal to d. MSF increases with an increase in p_{P} and with a decrease in d, reaching more than 100∼1000 typically.

The resource saving characteristics of PL-Pack with SORT results in cost reduction of OE systems. Based on "cost versus device pitch" characteristics shown in Fig. 8.7 for an OE module with VCSEL array [15], the impact of PL-Pack with SORT on cost can be estimated. As the 0th approximation, costs are assumed the same for PL-Pack with SORT and flip-chip bonding except for III-V epitaxial material cost. Due to the epitaxial-material saving effect in PL-Pack with SORT, module cost is reduced with a decrease in d. For example, in the case that p_{P} and d are 200 μm and 50 μm, respectively, cost reduction of ∼1/6 is expected. By further decrease in d, cost reduction of $<1/10$ might be expected.

Another cost reduction factor is process simplicity in heterogeneous integration of thin-film devices. In flip-chip-bonding-based packaging, generally, different kinds of OE devices should be fabricated in the same chip, which requires a sophisticated film growth method, raising fabrication cost. In PL-Pack with SORT, as

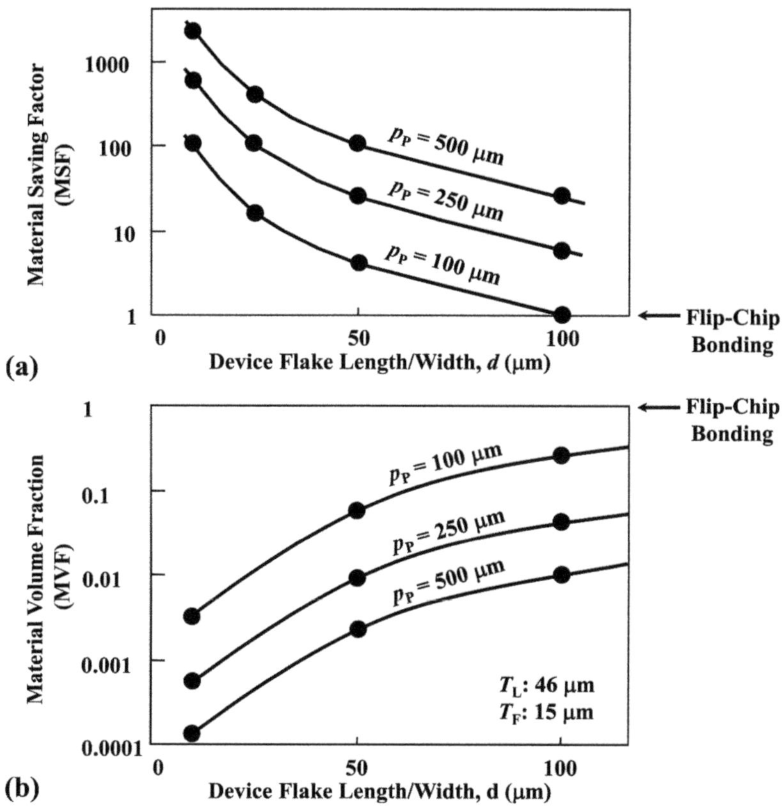

Figure 8.6. (a) Dependence of MSF on device flake length/width *d*. (b) Dependence of MVF on device flake length/width *d* for the case that TL is 46 μm and TF is 15 μm. Reprinted with permission from T. Yoshimura, M. Ojima, Y. Arai, and K. Asama, "Three-Dimensional Self-Organized Micro Optoelectronic Systems for Board-Level Reconfigurable Optical Interconnects — Performance Modeling and Simulation," *IEEE J. Select. Topics Quantum Electron.* **9**, 492–511. © 2003 IEEE.

shown in Fig. 8.1, heterogeneous integration can be performed by delivering OE device flakes from a set of wafers. On individual wafers, arrays consisting of a single kind of devices are grown by simple growth method, contributing to cost reduction.

For example, in WDM module fabrications, the above-mentioned process simplicity is very effective because many kinds of devices

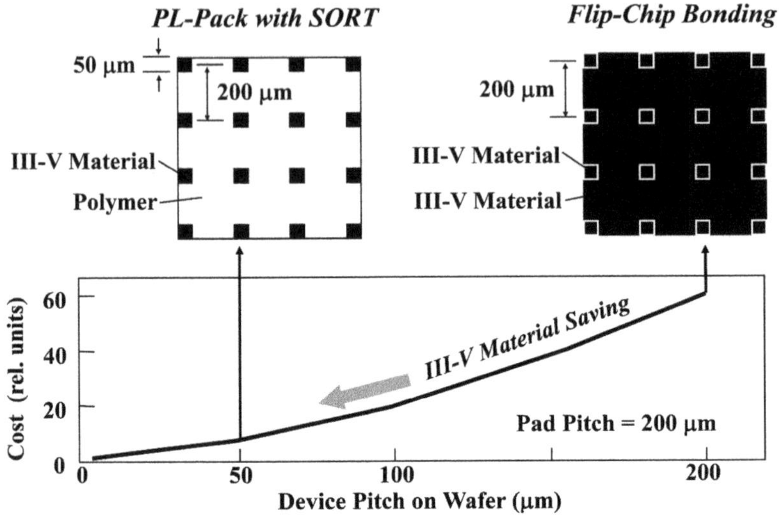

Figure 8.7. Cost versus device pitch characteristics for OE modules with VCSEL arrays. Reprinted with permission from T. Yoshimura, K. Kumai, T. Mikawa, and O. Ibaragi, "Photolithographic Packaging with Selectively Occupied Repeated Transfer (PL-Pack with SORT) for Scalable Optical Link Multi-Chip-Module (S-FOLM)," *IEEE Trans. Electron. Packag. Manufact.* **25**, 19–25. © 2002 IEEE.

should be embedded in the module. When a 10-λ system is built, 10 kinds of laser diodes (LDs) or VCSELs are implemented in the OE system. In flip-chip-bonding-based packaging, 10-different-wavelength OE devices should be formed in a single chip. In PL-Pack with SORT, on the other hand, WDM integration can be achieved by delivering the 10-different-wavelength OE devices from 10 wafers for individual wavelengths to a substrate. Each wafer has same-wavelength devices made by a simple low-cost growth method. The sophisticated epitaxial growth method for integration of the OE devices of different wavelengths into one chip is not necessary.

8.3.2 *Mechanical Properties*

PL-Pack with SORT has a potentiality to reduce thermal stress. As an example, consider a case that a two-dimensional VCSEL array of

L^2 μm^2, in which VCSELs are located with a pitch of p_P μm, is put on a plastic substrate (Fig. 8.5). For flip-chip bonding, an L^2 μm^2 bulk III-V chip of T_C μm thick is mounted on a plastic substrate. For PL-Pack with SORT, an L^2 μm^2 OE film of T_L μm thick, where T_F μm-thick III-V epitaxial material flakes with d^2 μm^2 area are embedded with a pitch of p_P μm, is mounted on a plastic substrate. In the flip-chip bonding case, due to the large difference in CTE between III-V material and plastics as well as due to the large thickness of III-V chip with large Young's modulus, a large thermally induced stress between the chip and the substrate is induced as well known. In PL-Pack with SORT, compared with the flip-chip bonding, thickness of the III-V material flake is reduced by a factor of T_F/T_C and the area by a factor of d^2/L^2, indicating that the III-V material volume is reduced by a factor of $(T_F/T_C)(d^2/L^2)$. For a case that $T_F = 15$ μm, $T_C =$ 500 μm, $d = 20$ μm, and $L = 1$ cm, roughly speaking, the power inducing the stress is reduced by seven orders of magnitude in PL-Pack with SORT.

Figure 8.6(b) shows material volume fraction (MVF) of III-V material in an OE film, which is defined by

$$\mathrm{MVF} = \frac{(d^2 T_F)}{P_P^2 T_L} \tag{8.2}$$

For a case that $d = 20$ μm, $T_F = 15\,\mu m$, $p_P = 200\,\mu m$, and $T_L = 46\,\mu m$, MVF is calculated to be 0.003, suggesting that CTE of the OE films is close to that of the host matrix polymer with a low Young's modulus. Thus, stress between the OE film and the plastic substrate is drastically reduced comparing with that between the III-V chip and the plastic substrate. Furthermore, basically, no large thermal expansion difference arises between OE films because the CTE of the film is almost determined by the host matrix polymer. Then, it is expected that thermally induced displacement at optical Z-connections may be neglected in S-FOLM where films are pinned by electrical pad connections.

The arguments mentioned in the present subsection are applicable to other active elements such as light modulators, optical switches, micro reflectors, wavelength filters, micro lenses/prisms, photonic crystal devices, ICs, dielectric films like lithium niobate and PLZT, organic compounds, glasses, and so on.

8.3.3 *Step Count*

Main concern in SORT is step count. It can be calculated using a model shown in Fig. 8.8(a). Two kinds of devices, Device I and Device II, are considered, namely, number of device types, M, is 2. The Device I flakes and Device II flakes are assumed to be placed alternately on a final substrate.

Device I flakes are on a pick-up substrate with a pitch of d. The effective area of the substrate is $L \times L$. The Device I flakes are transferred to M supporting substrates by pre-transfers. Similarly, Device II flakes on another pick-up substrate are transferred to M more supporting substrates. Thus, M^2 steps are involved in

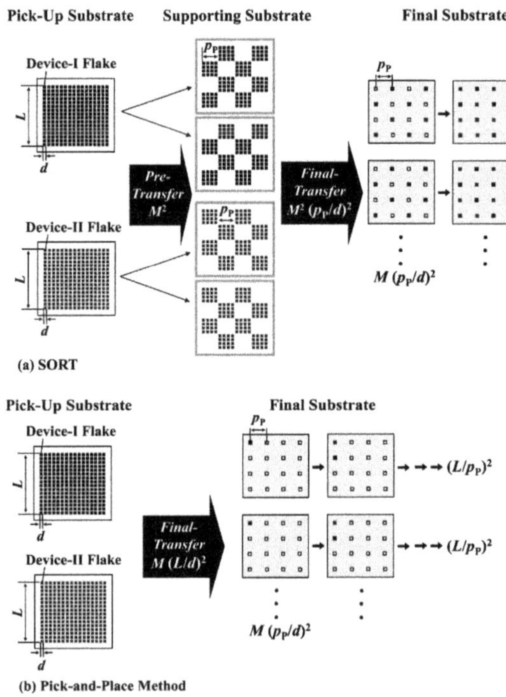

Figure 8.8. Model for device flake placement steps. (a) SORT and (b) pick-and-place method. Reprinted with permission from T. Yoshimura, K. Kumai, T. Mikawa, and O. Ibaragi, "Photolithographic Packaging with Selectively Occupied Repeated Transfer (PL-Pack with SORT) for Scalable Optical Link Multi-Chip-Module (S-FOLM)," *IEEE Trans. Electron. Packag. Manufact.* **25**, 19–25. © 2002 IEEE.

the pre-transfer for M kinds of devices. In the present example, $M = 2$, so, M^2 is 4. By final transfers, individual Device I flakes are placed on $M\ (p_P/d)^2$ final substrates, where pads are formed with a pitch of p_P. For example, when $p_P = 200\ \mu m$, $d = 50\ \mu m$ and $M = 2$, $M(p_P/d)^2 = 32$. Next, Device II flakes are placed on the final substrates. Consequently, $M(p_P/d)^2$ final substrates, on which Device I flakes and Device II flakes are placed alternately, are obtained. Step count involved in the final transfer for M kinds of devices is $M^2(p_P/d)^2$. Therefore, total step count necessary for SORT is found to be

$$\text{Step Count (SORT)} = M^2(1 + (p_P/d)^2) \qquad (8.3)$$

In the present example, 68 steps are required to make 32 final substrates.

For comparison, step count in conventional pick-and-place method was calculated using a model shown in Fig. 8.8(b). Device I flakes and Device II flakes are transferred from pick-up substrates to $M(p_P/d)^2$ final substrates one by one. In order to complete one final substrate, $(L/p_P)^2$ steps are necessary. Therefore, total step count in pick-and-place method is expressed as

$$\text{Step Count (Pick and Place)} = M(p_P/d)^2(L/p_P)^2 = M(L/d)^2 \qquad (8.4)$$

When $p_P = 200\ \mu m$, $d = 50\ \mu m$, and $L = 1$ cm, $M\ (L/d)^2 = 80000$ to make 32 final substrates. It is found that the step count in SORT is smaller by three orders of magnitude than that in pick-and-place method.

Figure 8.9 shows step counts in SORT and pick-and-place method for $p_P = 250\ \mu m$ and $d = 25\ \mu m$. Horizontal axis represents number of device types, M. For example, $M = 10$ implies 10-wavelength WDM communication using VCSEL-$\lambda_1 \sim \lambda_{10}$. In pick-and-place method, the step count increases with an increase in effective area of ELO substrates because the number of device flakes contained in the pick-up substrate is proportionally increased with the effective area. In SORT, on the other hand, step count is constant regardless of the effective area due to its "at-once transfer" characteristics. In a range of $L = 1 \sim 5$ cm, and $M = 2$–100, step count in SORT is smaller by factor of (1/10)–(1/10000) than that in pick-and-place method.

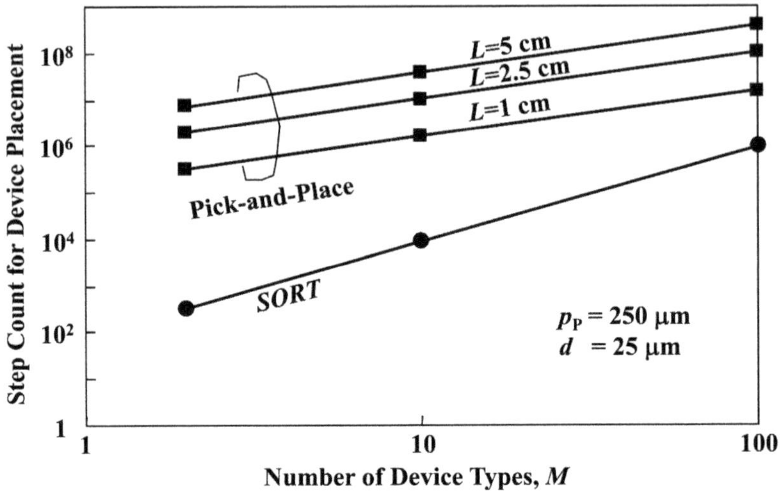

Figure 8.9. Step count for device placement. Reprinted with permission from T. Yoshimura, K. Kumai, T. Mikawa, and O. Ibaragi, "Photolithographic Packaging with Selectively Occupied Repeated Transfer (PL-Pack with SORT) for Scalable Optical Link Multi-Chip-Module (S-FOLM)," *IEEE Trans. Electron. Packag. Manufact.* **25**, 19–25. © 2002 IEEE.

8.3.4 *Device Flake Placement Density*

In SORT, as can be seen in Fig. 8.3, there are no restrictions on device flake placement density in principle. In pick-and-place method, on the other hand, slope limit occurs, which can be estimated using a model shown in Fig. 8.10. Device flakes to be placed are on a polyimide film with a pitch of d. By deforming the polyimide film, the device flakes are transferred one by one to pads with a pitch of p_P on a final substrate. In some conditions, device flakes that were already put on the neighboring pads prevent the device flakes from being placed on unoccupied pads. The slope limit condition is expressed as

$$\mathrm{Slope} = t/(p_\mathrm{P} - d) < \tan\theta_\mathrm{d} \qquad (8.5)$$

where, t is device thickness and tan θ_d a deformation limit of the polyimide film.

In Fig. 8.11, slopes necessary for device flake placement are plotted as a function of pad pitches on the final substrate for

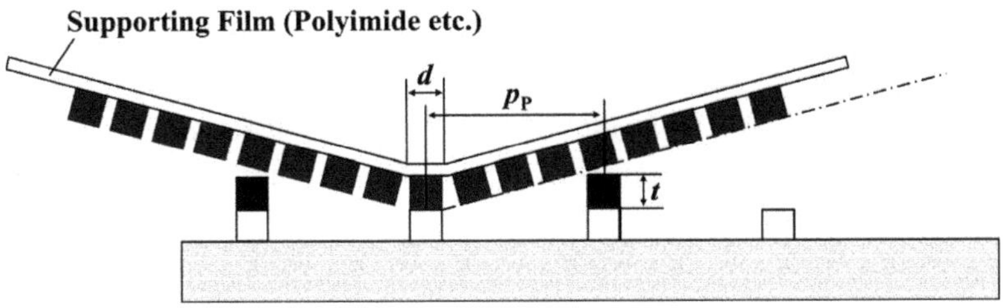

Figure 8.10. Model for slope limit. Reprinted with permission from T. Yoshimura, K. Kumai, T. Mikawa, and O. Ibaragi, "Photolithographic Packaging with Selectively Occupied Repeated Transfer (PL-Pack with SORT) for Scalable Optical Link Multi-Chip-Module (S-FOLM)," *IEEE Trans. Electron. Packag. Manufact.* **25**, 19–25. © 2002 IEEE.

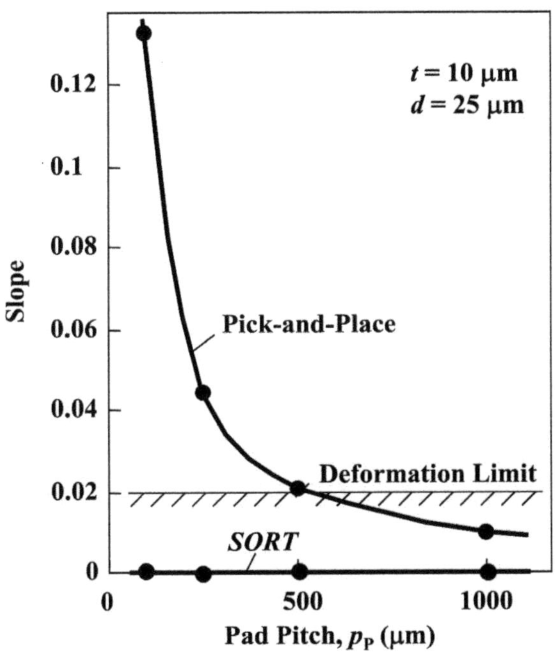

Figure 8.11. Slope versus pad pitch characteristics. Reprinted with permission from T. Yoshimura, K. Kumai, T. Mikawa, and O. Ibaragi, "Photolithographic Packaging with Selectively Occupied Repeated Transfer (PL-Pack with SORT) for Scalable Optical Link Multi-Chip-Module (S-FOLM)," *IEEE Trans. Electron. Packag. Manufact.* **25**, 19–25. © 2002 IEEE.

$t = 10\ \mu m$ and $d = 25\ \mu m$. In pick-and-place method, required slope rapidly increases with decreasing pad pitch. If, for example, $\tan\theta_d = 0.02$ is assumed, a slope limit arises around pad pitch of 500 μm, ruling out high-density placement. In SORT, on the other hand, there is no limit, implying a possibility of high-density device flake placement.

8.4 SORT of Polymer Waveguide Lenses

SORT process for polymer waveguide lenses is shown in Fig. 8.12 [16]. On a glass substrate, Al patterns of 50 nm thickness with apertures of waveguide lens shapes were formed to make a built-in mask, followed by spin-coating a PVA film of 10 μm thickness. After drying the PVA in ambient atmosphere at room temperature for 1 h, an acrylic photo-definable material film was spin-coated on it with a thickness of 40 μm. The film was exposed to UV light (from a

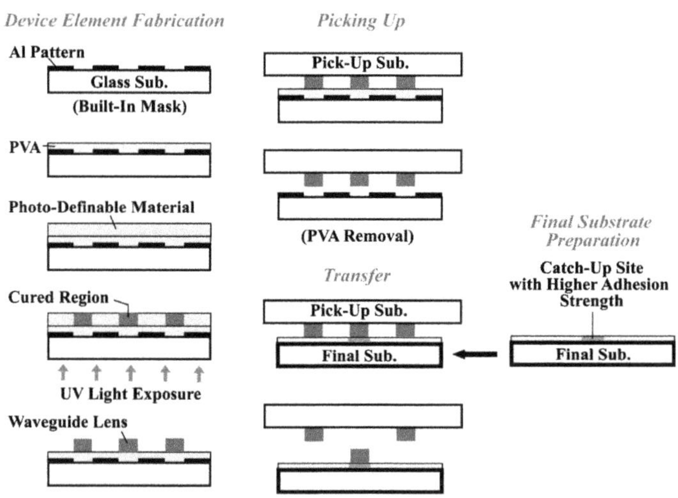

Figure 8.12. SORT process for polymer waveguide lenses using "built-in mask" method. Reprinted with permission from T. Yoshimura, D. Tamaki, S. Kawamura, Y. Arai, and K. Asama, "A Device Transfer Process "Selectively Occupied Repeated Transfer (SORT)" for Resource-Saving Integration of Polymer Micro Optics Fabricated by "Built-in Mask" Method," *IEEE Trans. Comp. Packag. Technol.* **27**, 468–471. © 2004 IEEE.

high-pressure mercury lump) of 400 mJ/cm^2 through the built-in mask to form a polymer waveguide lens array by developing. The waveguide lens diameter is 300 μm, and the pitch in the array is 450 μm. For picking up, a pick-up substrate, which is a glass substrate covered with an adhesive film (GelPak film-×8 provided by Gel-Pak), was attached on the top of the polymer waveguide lens array. By solving the PVA film in water, the polymer waveguide lens array was picked up to the pick-up substrate at one time. The photograph in Fig. 8.13 (top) represents the polymer waveguide lens array on the adhesive film of the pick-up substrate.

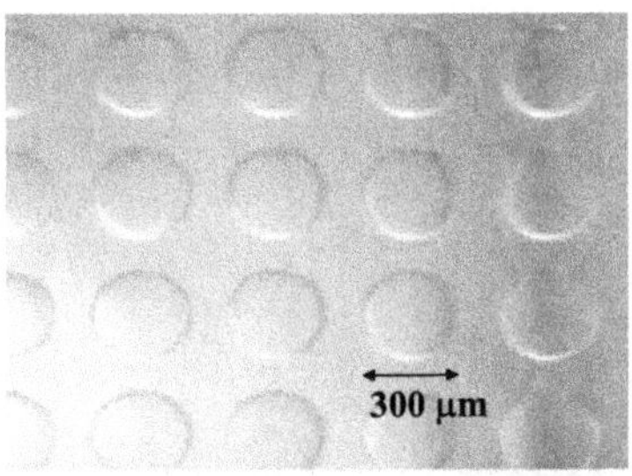

Transfer onto Pick-Up Substrate

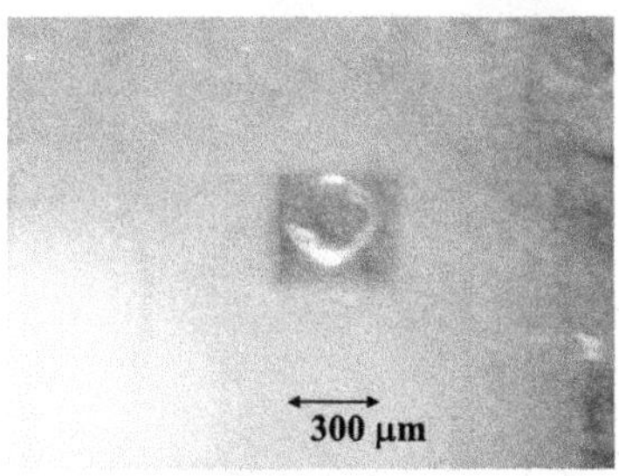

Transfer onto Final Substrate

Figure 8.13. Microscopic photographs for polymer waveguide lens array picked-up on a pick-up substrate (top) and a polymer waveguide lens transferred onto a final substrate (bottom). The dark square observed under the polymer waveguide lens in the bottom photograph is photomask pattern for catch-up site formation. Reprinted with permission from T. Yoshimura, D. Tamaki, S. Kawamura, Y. Arai, and K. Asama, "A device transfer process "Selectively Occupied Repeated Transfer (SORT)" for Resource-Saving Integration of Polymer Micro Optics Fabricated by "Built-in Mask" Method," *IEEE Trans. Comp. Packag. Technol.* **27**, 468–471. © 2004 IEEE.

A final substrate was prepared by exposing a glass substrate covered with a UV-curable-material-coated film (UV tape UE-111AJ provided by Nitto Denko) to UV light (from a high-pressure mercury lump) of 180 mJ/cm^2 through a photo-mask. The adhesion strength of the exposed region decreased by about two orders of magnitude compared with the unexposed region, providing a final substrate with catch-up sites. For transfer, the polymer waveguide lens array on the pick-up substrate was attached onto the final substrate, and then, the two substrates were detached from each other. As shown in the photograph in Fig. 8.13 (bottom), the polymer waveguide lens was transferred selectively onto the catch-up site of the final substrate. This demonstrates the feasibility of SORT process.

For device element transfers, the SORT equipment shown in Fig. 8.14 was used. The equipment was proposed by the author and

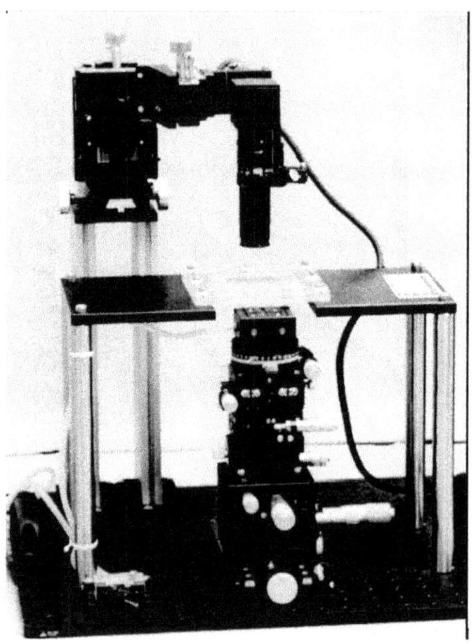

Figure 8.14. SORT equipment. Reprinted with permission from T. Yoshimura, K. Ogushi, Y. Kitabayashi, K Naito, Y. Miyamoto, and M. Miyazaki, "Optical Waveguide Films with Two-Layer Skirt-Type Core End Facets for Beam Leakage Reduction at 45° Mirrors," *Proc. SPIE* **6899**, 689913-1-11 (2008).

was assembled in Suruga Seiki Co., Ltd. using components for fiber optics with positional accuracy in the order of 0.1~1 μm. Alignment between two substrates was achieved by optical stages, on which a lower substrate was set, below the upper substrate holder. One problem of the equipment is that it has one-eye alignment system, which reduces angular alignment accuracy. The problem can be solved by implementing two-eye alignment system into the equipment. Improved SORT equipment with two-eye alignment system is commercially available from Suruga Seiki as "Wafer Bonding Unit."

8.5 SORT of Waveguide Cores

In order to reduce the material consumption of optical waveguides as well as to realize all-air-clad waveguides for strong guided beam confinement, we developed a material-saving waveguide fabrication process and demonstrated transfers of cores made of photo-definable polymers onto a substrate to fabricate optical waveguides [17].

8.5.1 *Material-Saving Process for Waveguide Fabrication*

Figure 8.15(a) shows a concept of material-saving optical waveguide fabrication using SORT. On a pick-up substrate, cores for optical waveguides are placed in an array with high density. On final substrates, catch-up sites are formed. By attaching the pick-up substrate to a final substrate and detaching them, the cores located at sites corresponding to the catch-up sites are selectively transferred onto the final substrate to construct optical waveguides. In the case shown in Fig. 8.15, since core pitch on the final substrate is six times larger than that on the picking-up substrate, one pick-up substrate can provide cores to six final substrates. This implies that core materials are saved by six times.

In usual optical waveguides for chip-scale optical interconnects, devices such as vertical mirrors and wavelength filters with metal/dielectric film coating on core end facets coexist. To make the devices, complicated deposition/photolithography processes are

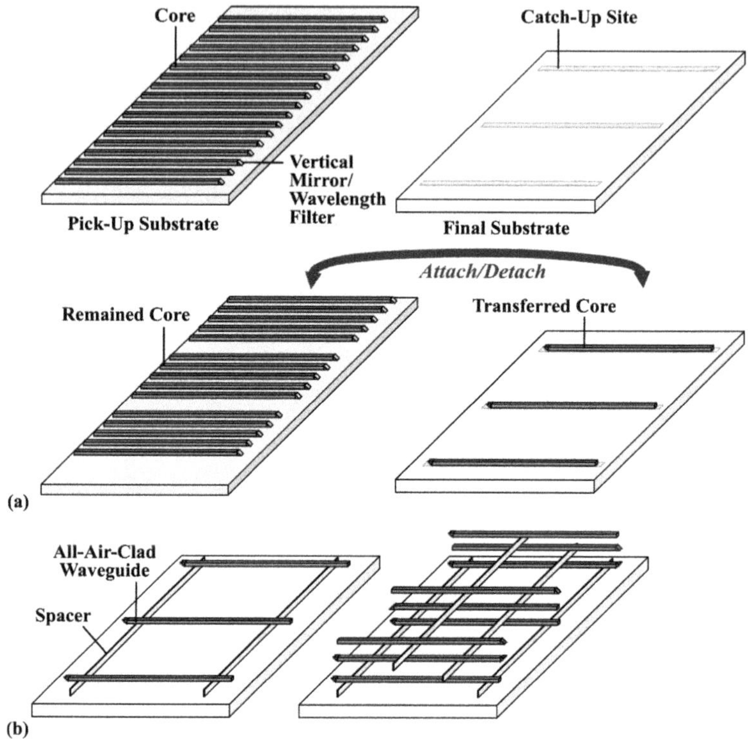

Figure 8.15. Concept of material-saving waveguide fabrication using SORT. (a) Cores are selectively transferred from a pick-up substrate with a high-density core array onto a final substrate with catch-up sites to construct optical waveguides. (b) All-air-clad waveguides for fine-pattern optical interconnects. Reprinted with permission from T. Yoshimura, K. Ogushi, Y. Kitabayashi, K Naito, Y. Miyamoto, and M. Miyazaki, "Optical Waveguide Films with Two-Layer Skirt-Type Core End Facets for Beam Leakage Reduction at 45° Mirrors," *Proc. SPIE* **6899**, 689913-1-11 (2008).

involved. The material-saving waveguide fabrication process shown in Fig. 8.15 can also reduce the process efforts for fabricating the devices by six times.

8.5.2 *All-Air-Clad Waveguides*

SORT is applicable to the fabrication of all-air-clad waveguides shown in Fig. 8.15(b). For intra-chip optical interconnects, fine-pattern optical circuits with light-beam-confined sizes of less than ~1 μm are necessary. To realize this, refractive index contrast between cores and clad regions should be high. For the purpose, all-air-clad waveguides, where all sides of cores are surrounded by air, are desirable. By transferring cores on spacers of final substrates, all-air-clad waveguides are obtained. Furthermore, by building spacers/cores up in multi-layer configurations, 3-D optical wiring consisting of the all-air-clad waveguides will be fabricated.

The all-air-clad structure is a kind of a goal to be achieved in future. In this case, the losses caused by the spacer edge scattering will be a serious concern. The use of low-index materials for spacers and high-index materials for cores as well as narrowing the spacers may contribute to the scattering loss reduction.

Core transfers onto a glass substrate with catch-up sites were attempted. A linear array of 3 cm-long cores with 32 × 32 μm^2 cross sections was formed with 250 μm pitch on a built-in mask using a photo-definable material with refractive index of 1.58. The linear array was transferred to a pick-up substrate from the built-in mask. Meanwhile, catch-up sites were formed with 500 μm pitch on a final substrate using UV-curable-material-coated films. In the case that one kind of device elements are arranged in a linear array, direct transfers from the pick-up substrate to the final substrate are possible without the supporting substrate. By attaching/detaching of the two substrates, cores were transferred to the final substrate selectively.

Figures 8.16(a) and (b), respectively, show the cores on the pick-up substrate and the final substrate after selective transfer. It can be seen that a core is transferred onto a catch-up site of the final substrate while a missing core site is found on the pick-up substrate. As shown in Fig. 8.16(c), a guided beam of 650 nm in wavelength is clearly observed at the core edge. These results confirm that SORT can be applied to the transfer/integration of micro-scale optical circuits, providing the proof-of-concept of the material-saving waveguide fabrication process.

Figure 8.17 shows an all-air-clad waveguide made of fluorinated photo-definable material developed by DIC. Core formed by the built-in mask method was transferred on a GelPak film. The film has a rough surface, which can be regarded as a surface with natural

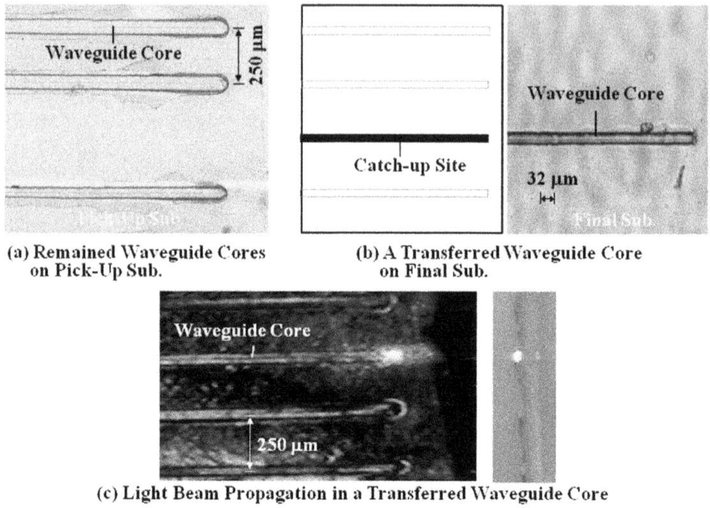

Figure 8.16. (a) Remaining cores on the pick-up substrate and (b) transferred cores on the final substrate. (c) Guided beams in an optical waveguide with a transferred core. Reprinted with permission from T. Yoshimura, K. Ogushi, Y. Kitabayashi, K Naito, Y. Miyamoto, and M. Miyazaki, "Optical Waveguide Films with Two-Layer Skirt-Type Core End Facets for Beam Leakage Reduction at 45° Mirrors," *Proc. SPIE* **6899**, 689913-1-11 (2008).

spacers to provide all-air-clad waveguides. A 650 nm guided beam and an emitted beam from the core edge are observed. The cutback method revealed that the propagation loss is ~0.5 dB/cm.

Figure 8.17. An all-air-clad waveguide on a GelPaK film.

8.6 Transfers of Two Kinds of Model Devices

8.6.1 *Light-Assisted SORT (LA-SORT)*

Figure 8.18 shows process flow of light-assisted SORT (LA-SORT) [14, 17]. Catch-up sites of a supporting substrate are prepared by exposing an UV curable material layer on the substrate to a patterned UV light. Unexposed regions have stronger adhesion strengths comparing with the exposed regions, forming catch-up sites. For catch-up sites of final substrates, any materials that induce adhesion strength contrasts can be used. One candidate is an UV curable material similar to the catch-up site material for the supporting substrate. In this case, however, there is a concern about reliability, that is, adhesion strength between device flakes and catch-up sites decreases gradually with time. To avoid this problem, photo-definable materials are promising for the catch-up sites. When the device flakes are cores of optical waveguides made of a photo-definable material, the device flakes preserve strong adhesion strength to the catch-up sites made of another

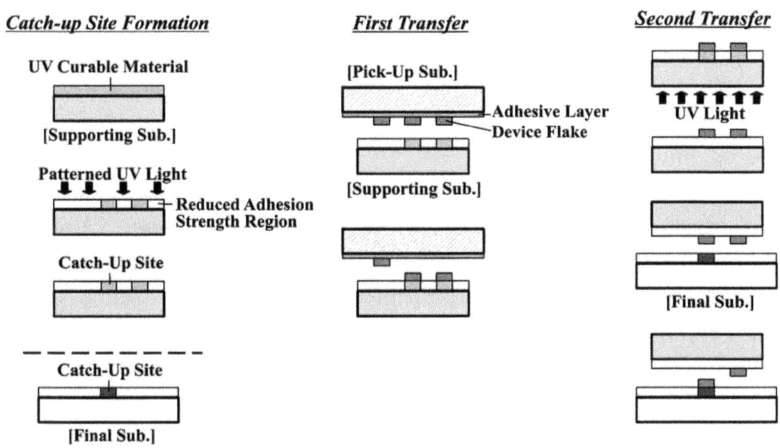

Figure 8.18. Process flow of LA-SORT. Reprinted with permission from T. Yoshimura, K. Ogushi, Y. Kitabayashi, K Naito, Y. Miyamoto, and M. Miyazaki, "Optical Waveguide Films with Two-Layer Skirt-Type Core End Facets for Beam Leakage Reduction at 45° Mirrors," *Proc. SPIE* **6899**, 689913-1-11 (2008).

photo-definable material after curing. Pads of solder, diffusible metal, and polymer can also be applied to the catch-up sites with the assistance of thermal, super-sonic, or optical excitation. When Au and InSn are used for the diffusible metal, bonding of Au-InSn catch-up sites to Au pads of device flakes can occur by alloying of InSn with Au [7]. When polyimide is used for the polymer pad, polyimide bonding can be performed.

For the first transfer, device flakes on a pick-up substrate are attached to the supporting substrate to be transferred onto the catch-up sites. For the second transfer, the catch-up sites are exposed to blanket UV light from the back of the substrate, which reduces the adhesion strength of all the catch-up sites. Then, the device flakes on the supporting substrate are attached to the final substrate and detached from it to distribute the flakes on the final substrate.

8.6.2 *Experimental Demonstration of LA-SORT for Two Kinds of Devices*

A fabrication process of polymer device flakes using the built-in mask method is shown in Fig. 8.19. The process is similar to the process shown in Fig. 8.12. An array of $100 \times 130\ \mu m^2$ rectangular polymer device flakes (Device I flakes) with 1000 μm pitch was transferred to the pick-up substrate GelPak film-×8. Device II flakes were obtained by depositing Au thin films on the Device I flakes by the vacuum evaporation.

Using the Device I flakes and Device II flakes as model devices, we carried out the process shown in Fig. 8.18 for LA-SORT demonstrations. To form catch-up sites of supporting and final substrates, the UV-curable-material-coated films were used. The UV light exposure energy for catch-up site formation was 270 mJ/cm^2. For device flake transfers, the SORT equipment shown in Fig. 8.14 was used.

In the first transfer, after the device flakes on the pick-up substrate were aligned to the corresponding catch-up sites on the supporting substrate, the pick-up substrate was attached to the supporting substrate, and then, detached from it, resulting in selective transfer of the device flakes onto the supporting substrate as shown in Fig. 8.20(a). In the second transfer, after the

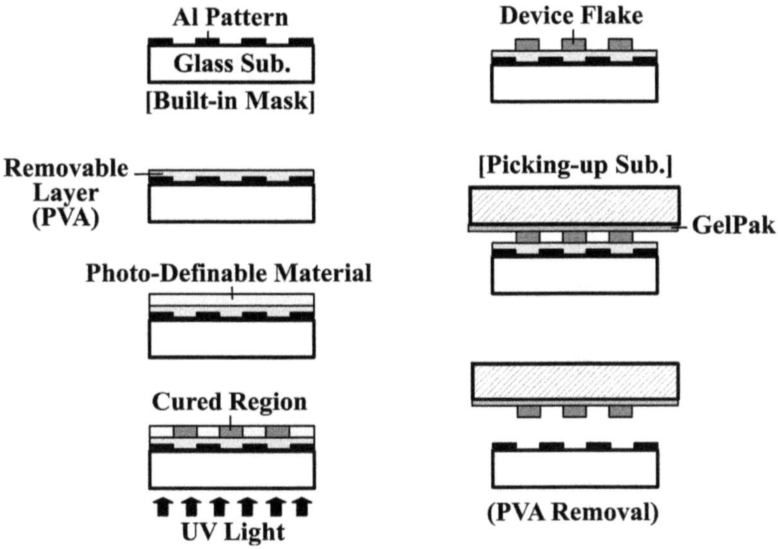

Figure 8.19. Fabrication process for polymer device flakes using the built-in mask method. Reprinted with permission from T. Yoshimura, K. Ogushi, Y. Kitabayashi, K Naito, Y. Miyamoto, and M. Miyazaki, "Optical Waveguide Films with Two-Layer Skirt-Type Core End Facets for Beam Leakage Reduction at 45° Mirrors," *Proc. SPIE* **6899**, 689913-1-11 (2008).

UV-curable-material-coated film, on which transferred device flakes are, was exposed to UV light from the back, the supporting substrate was attached to the final substrate and detached from it, resulting in selective transfer of the device flake onto the final substrate. Figure 8.20(b) shows an example of heterogeneous integration of Device I flake and Device II flake on the final substrate. It should be noted that cross-shaped markers consisting of two lines with 10 μm width and 100 μm length were transferred on substrates, suggesting that the alignment accuracy of the SORT equipment is at least in the order of a few μm.

These results indicate the feasibility of LA-SORT process for heterogeneous integration in μm-scale resolutions. The procedure described above for polymer devices might be applied to other kinds of devices made of semiconductor, glass, and dielectric materials with process condition tuning.

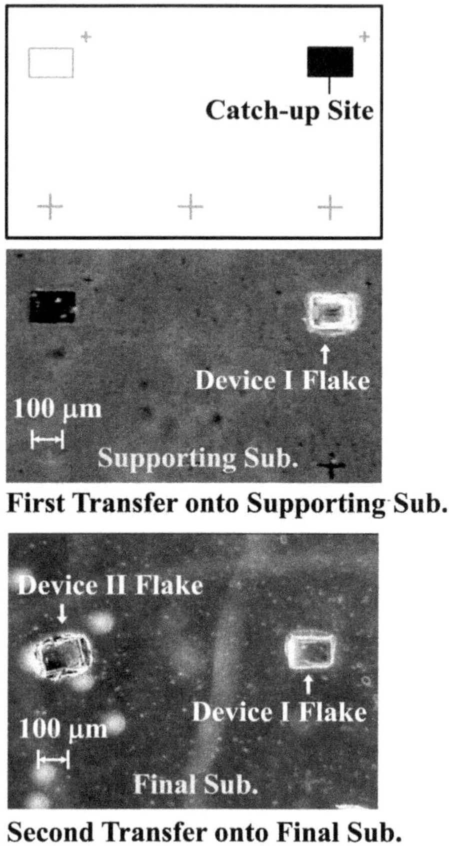

Figure 8.20. (a) Selective transfer of 100 × 130 μm^2 rectangular polymer device flakes onto a supporting substrate by the first transfer and (b) heterogeneous transfer of Devices I and II flakes onto a final substrate by the first transfer and the second transfer. Reprinted with permission from T. Yoshimura, K. Ogushi, Y. Kitabayashi, K Naito, Y. Miyamoto, and M. Miyazaki, "Optical Waveguide Films with Two-Layer Skirt-Type Core End Facets for Beam Leakage Reduction at 45° Mirrors," *Proc. SPIE* **6899**, 689913- 1-11 (2008).

8.7 Concept of SORT for Nano-Scale Heterogeneous Integration

Figure 8.21 shows the concept of nanoscale integration for HIC waveguides [18]. The HIC waveguides are picked up onto a

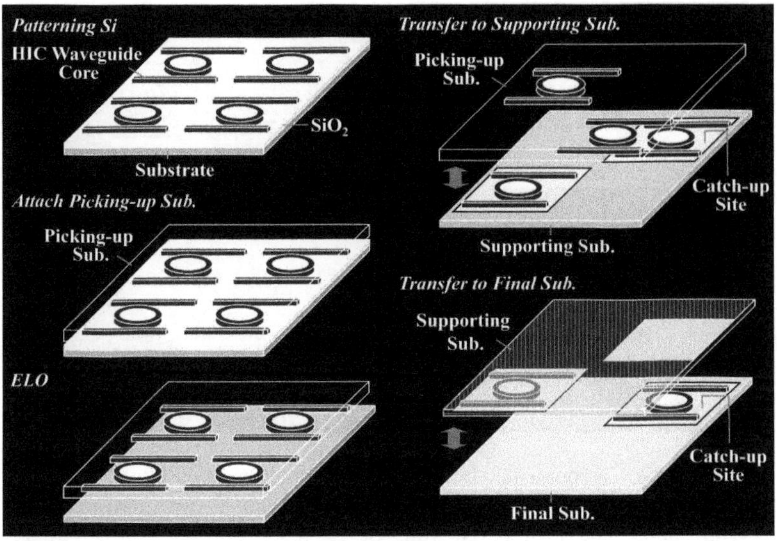

Figure 8.21. Nano-scale integration: SORT of HIC waveguides. Reprinted from T. Yoshimura, *Trends in Thin Solid Films Research* (ed. A. R. Jost), Ch. 5 "Self-Organized Growth of Polymer Wire Networks with Designed Molecular Sequences for Wavefunction-Controlled Nano Systems," Nova Science Publishers, New York (2007). With permission from Nova Science Publishers, Inc.

pick-up substrate by ELO and are transferred to a final substrate via a supporting substrate using SORT. In Fig. 8.22, photonic crystals (PhCs) and HIC waveguides, which respectively correspond to Device I flakes and Device II flakes in Fig. 8.3, are integrated in an optical waveguide film. To realize such nano-scale transfers, improvement of alignment accuracy of SORT equipment to sub-micron ranges is necessary.

Nano-scale optical circuits are expensive due to the ultra-fine fabrication processes using submicron photolithography. Meanwhile, in optical interconnects, the nano-scale optical circuits are placed with pitches of $\sim$100 μm corresponding to I/O pads of chips. SORT enables us to deliver the nano-scale optical circuits from one growth substrate to a lot of final substrates, say, 100 substrates, saving material and process steps for the nano-scale optical circuit fabrication. Therefore, SORT may greatly contribute to reduction of

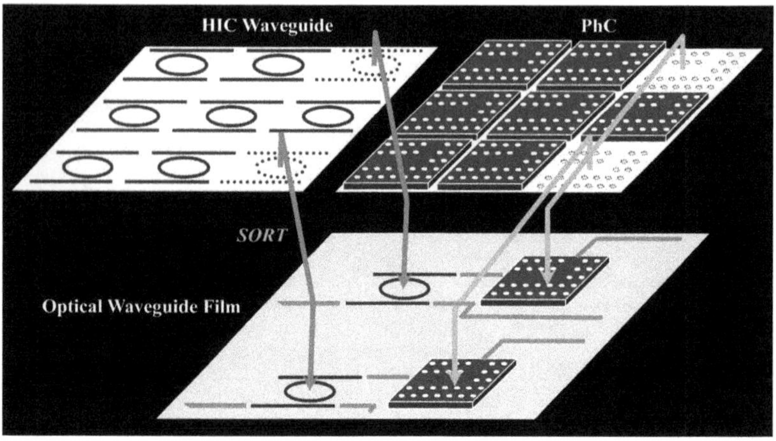

Figure 8.22. Nano-scale heterogeneous integration: Photonic Crystals/HIC waveguides using SORT. Reprinted from T. Yoshimura, *Trends in Thin Solid Films Research* (ed. A. R. Jost) Ch. 5, "Self-Organized Growth of Polymer Wire Networks with Designed Molecular Sequences for Wavefunction-Controlled Nano Systems," Nova Science Publishers, New York (2007). With permission from Nova Science Publishers, Inc.

material consumption and process efforts in the optical interconnect systems.

References

1. T. Yoshimura, J. Roman, Y. Takahashi, M. Lee, B. Chou, S. I. Beilin, W. V. Wang, and M. Inao, "Optoelectronic Scalable Substrates Based on Film/Z-Connection and Its Application to Film Optical Link Module (FOLM)," *Proc. SPIE* **3952**, from Photonics West 2000, San Jose, California, 202–213 (2000).
2. T. Yoshimura, J. Roman, Y. Takahashi, W. Wang, M. Inao, T. Ishitsuka, K. Tsukamoto, S. Aoki, K. Motoyoshi, and W. Sotoyama, "Self-Organizing Lightwave Network (SOLNET) and Its Application to Film Optical Circuit Substrates," *IEEE Trans. Comp. Packag. Technol.* **24**, 500–510 (2001).
3. M. E. Robertsson, O.-J. Hagel, G. Gustafsson, A. Dabec, M. Popall, L. Cergel, P. Wennekers, P. Kiely, M. Lebby, and T. Lindahl, "O/e MCM Packaging with New, Patternable Dielectric and Optical Materials," *Proc. 48th Electron. Components Technol. Conf. (ECTC)*, 1423–1421 (1998).

4. S. Koike, F. Shimokawa, T. Matsuura, and H. Takahara, "New Optical and Electrical Packaging Techniques Using Optical Waveguides for Optoelectronic Multichip Modules," *IEEE Trans. Comp. Hybrids Manuf. Technol.* **19**, 124–130 (1996).
5. E. Yablonovitch, T. Gmitter, J. P. Harbison, and R. Bhat, "Extreme Selectivity in the Lift-Off of Epitaxial GaAs Films," *Appl. Phys. Lett.* **51**, 2222–2224 (1987).
6. W. Chang, G. Pike, C. Kao, E. Yablonobitch, MRS 1997 Fall Meeting, Abstracts, H10.1, 210 (1997).
7. R. Pu, C. Duan, and C. Wilmsen, "Hybrid Integration of VCSEL's to CMOS Integrated Circuits," *IEEE J. Select. Topics in Quantum Electron.* **5**, 201–208 (1999).
8. J. J. Chang, S. Jung, M. Vrazel, K. Jung, M. Lee, M. A. Brooke, N. M. Jokerst, and S. Wills, "A Hybrid Optically Interconnected Microprocessor: An InP I-MSM Integrated onto a Mixed Signal CMOS Analog Optical Receiver with a Digital CMOS Microprocessor," *Proc. SPIE* **4089**, 708–714 (2000).
9. C. B. Morrison, R. L. Strijek, and J. H. Bechtel, "New Processing Approach for Grafting Optoelectronic Devices and Applications to Multichip Modules," *Proc. SPIE* **3005**, 120–127 (1997).
10. T. Yoshimura, K. Kumai, T. Mikawa, and O. Ibaragi, "Photolithographic Packaging with Selectively Occupied Repeated Transfer (PL-Pack with SORT) for Scalable Optical Link Multi-Chip-Module (S-FOLM)," *IEEE Trans. Electron. Packag. Manuf.* **25**, 19–25 (2002).
11. T. Yoshimura, M. Ojima, Y. Arai, and K. Asama, "Three-Dimensional Self-Organized Micro Optoelectronic Systems for Board-Level Reconfigurable Optical Interconnects — Performance Modeling and Simulation-," *IEEE J. Select. Topics Quantum Electron.* **9**, 492–511 (2003).
12. T. Yoshimura, J. Roman, Y. Takahashi, M. Lee, B. Chou, S. I. Beilin, W. V. Wang, and M. Inao, "Proposal of Optoelectronic Substrate with Film/Z-Connection Based on OE-Film," *Proc. 3rd IEMT/IMC Symp.*, 140–145 (1999).
13. T. Yoshimura, J. Roman, W-C. Wang, M. Inao, M. T. McCormack, "Device Transfer Method," U.S. Patent 6,669,801 B2 (2003).
14. T. Yoshimura, Y. Arai, "3D Optoelectronic Micro System," U.S. Patent 7,387,913 B2 (2008).
15. C. Stirk, Q. Liu, and M. Ball, "Manufacturing Cost Analysis of Integrated Photonic Packages," *Proc. SPIE* **3631**, 224–233 (1999).

16. T. Yoshimura, D. Tamaki, S. Kawamura, Y. Arai, and K. Asama, "A Device Transfer Process "Selectively Occupied Repeated Transfer (SORT)" for Resource-Saving Integration of Polymer Micro Optics Fabricated by 'Built-In Mask' Method," *IEEE Trans. Comp. Packag. Technol.* **27**, 468–471 (2004).
17. T. Yoshimura, T. Kofudo, T. Kashiwazako, K. Naito, K. Ogushi, and Y. Kitabayashi, "A Material-Saving Optical Waveguide Fabrication Process with Selective Transfers of Cores," *Opt. Eng.* **47**, 014601-1-7 (2008).
18. T. Yoshimura, Y. Suzuki, N. Shimoda, T. Kofudo, K. Okada, Y. Arai, and K. Asama, "Three-Dimensional Chip-Scale Optical Interconnects and Switches with Self-Organized Wiring Based on Device-Embedded Waveguide Films and Molecular Nanotechnologies," *Proc. SPIE* **6126**, from Photonics West 2006, San Jose, California, 612609-1-15 (2006).

Chapter 9

Optical Switches

High-speed/small-size light modulators and optical switches are the challenging core technology for Optical Electronics. In order to improve their performance, both of device structures and electro-optic (EO) materials should be improved.

In Fig. 9.1, promising candidates of light modulators/optical switches are shown, such as the Variable Well Optical IC (VWOIC) [1–4], the waveguide prism deflector (WPD) optical switch [2–10], the total internal reflection (TIR) optical switch, the digital optical switch, and optical switches using ring resonators and photonic crystals. VWOIC, WPD optical switch, TIR optical switch, and digital optical switch are phase-insensitive flux-controlled devices that are suitable for large-scale integration.

Examples of EO material candidates for the light modulators/optical switches are listed in Table 9.1. Materials having the Pockels effect are $LiNbO_3$ (LN), III-V semiconductor quantum dot (QD) [11, 12], styrylpyridinium cyanine dye (SPCD) [13], which is a molecular crystal, and polymer multiple quantum dot (MQD) [14–18], which is an artificial material consisting of polymer wires with MQDs. EO property of the polymer MQDs was theoretically predicted [16]. EO material having the Kerr effect is lead lanthanum zirconate titanate (PLZT) [19].

Optical Electronics: Self-Organized Integration and Applications
Tetsuzo Yoshimura

ISBN 978-981-4310-90-1 (Hardcover), 978-981-4364-08-9 (eBook)
www.panstanford.com

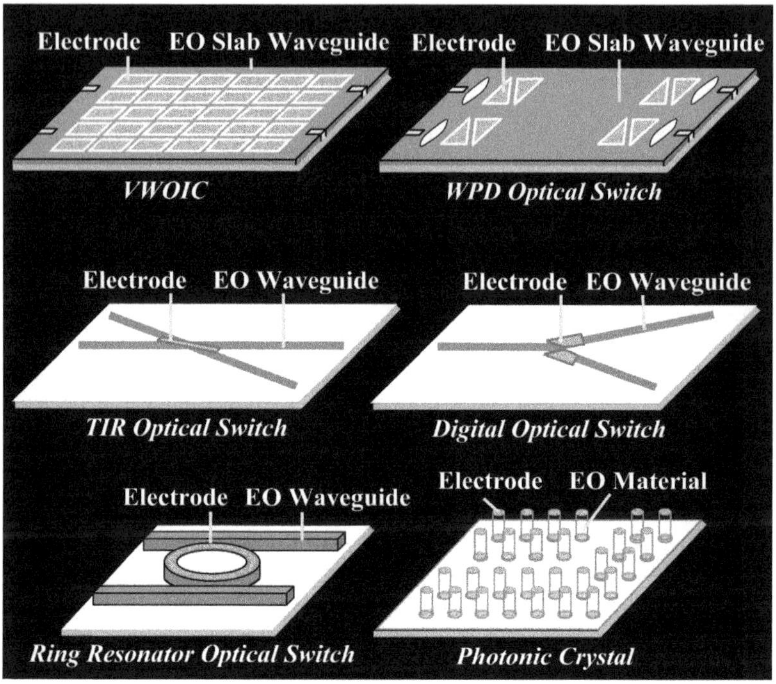

Figure 9.1. Structures of light modulators/optical switches. Reprinted from T. Yoshimura, *Trends in Thin Solid Films Research* (ed. A. R. Jost) Ch. 5 "Self-Organized Growth of Polymer Wire Networks with Designed Molecular Sequences for Wavefunction-Controlled Nano Systems." Nova Science Publishers, New York (2007). With permission from Nova Science Publishers, Inc.

Table 9.1. EO material candidates for light modulators/optical switches

Pockels effect					Kerr effect	
	LN	III-V QD	SPCD	polymer MQD		PLZT
n	2.2	3.5	1.55	2	n	2.48
r (pm/V)	31	26[11]	430[13]	1000*[16]	R (m^2/V^2)	5.7×10^{-10}[19]

*Molecular Orbital Calculation

Electric-field-induced refractive index change, Δn, is calculated by following expressions:

$$\text{Pockels Effect:} \Delta n = \frac{1}{2} n^3 r E \quad (9.1)$$

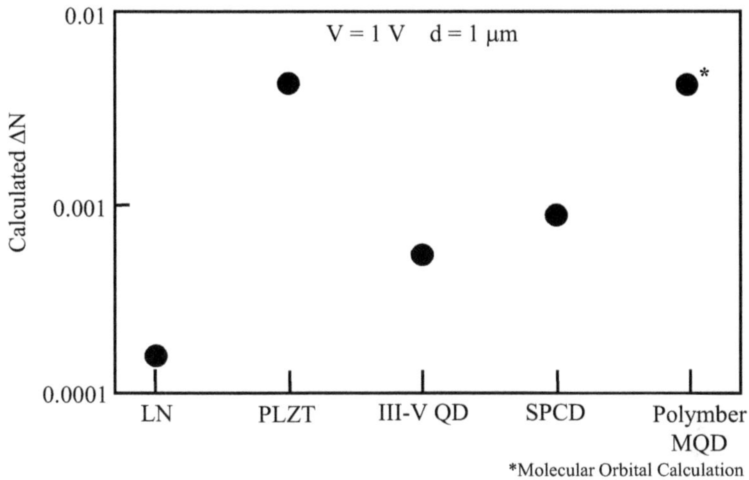

Figure 9.2. Electric-field-induced refractive index change Δn calculated for typical EO materials.

$$\text{Kerr Effect:}\Delta n = \frac{1}{2}n^3 R E^2 \tag{9.2}$$

Here, n and E represent refractive index of the EO materials and applied electric field in the materials. r and R are EO coefficients for the Pockels effect and the Kerr effect, respectively. Figure 9.2 shows Δn calculated by assuming that 1 V is applied between electrodes on top and bottom surfaces of 1 μm-thick EO materials. It is found that III-V semiconductor QD and SPCD exhibit large Δn compared with LN. PLZT and polymer MQD exhibit extremely large Δn. With applied voltages of a few volts, Δn more than 0.01 is expected.

In this chapter, first, design and predicted performance of VWOICs and WPD optical switches that the author previously proposed are reviewed. The VWOIC and the WPD optical switch are especially superior as EO switches for integrated optical circuits with a large number of elements. Ring resonator optical switches that are expected as ultra-small optical switches are briefly described. In addition, band width limit in photonic crystal waveguides is discussed.

Since light modulators and optical switches have same structures in many cases, light modulators are included into optical switches later in this chapter.

9.1 Variable Well Optical ICs (VWOICs) and Waveguide Prism Deflectors (WPDs)

Structures of VWOICs [1–4] and WPD optical switches [2–10] are schematically shown in Fig. 9.3. In the former, rectangular electrodes are formed on an EO slab waveguide. In the latter, prism-shaped cascading electrodes are formed on an EO slab waveguide. The slab waveguides consist of three layers of clad/core/clad placed between counter electrodes and rectangular or prism-shaped electrodes. The core is made of EO materials, and the refractive index changes by electric fields. Interface films, in which thin-film interface ICs are embedded, are stacked on the surface for performing selective voltage application to the electrodes to drive the individual optical switches.

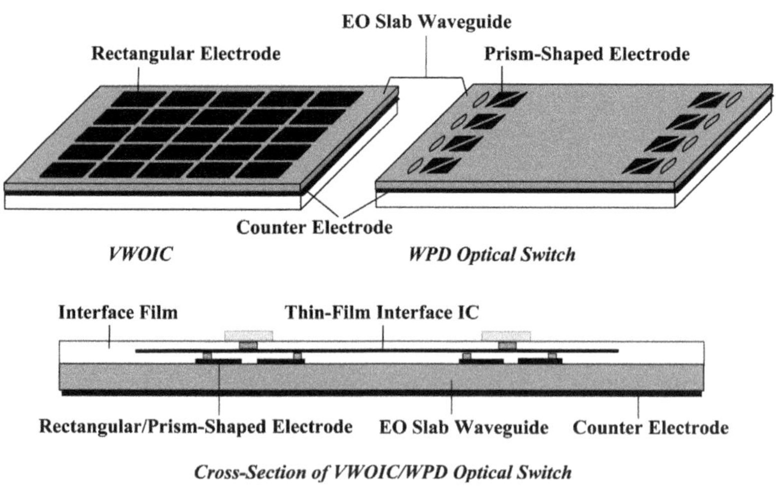

Figure 9.3. Structures of the VWOIC and the WPD optical switch. Reprinted with permission from T. Yoshimura, S. Tsukada, S. Kawakami, Y. Arai, H. Kurokawa, and K. Asama, "3D Micro Optical Switching System (3D-MOSS) Architecture," *Proc. SPIE* **4653**, 62–70 (2002).

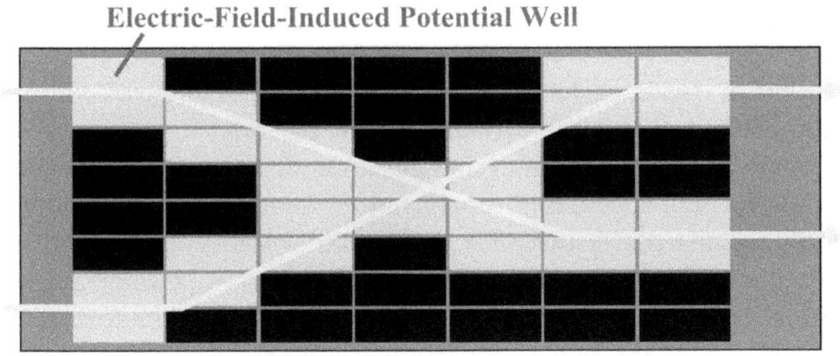

Figure 9.4. Operation principle of the VWOIC. Reprinted with permission from T. Yoshimura, S. Tsukada, S. Kawakami, Y. Arai, H. Kurokawa, and K. Asama, "3D Micro Optical Switching System (3D-MOSS) Architecture," *Proc. SPIE* **4653**, 62–70 (2002).

As shown in Fig. 9.4, in VWOIC, by selectively applying voltage to the EO waveguide through the patterned electrodes, refractive index change is induced to form potential wells for light waves. Then, electric-field-induced dynamic optical waveguides are constructed all over the slab waveguide to switch guided light beams. In WPD optical switches, as shown in Fig. 9.5, electric-field-induced dynamic waveguide prisms, which deflect guided light beams introduced from input waveguides, are constructed, enabling us to switch the light beams through two-dimensional (2-D) free space in the slab waveguide like "search lights." Waveguide lenses for light beam

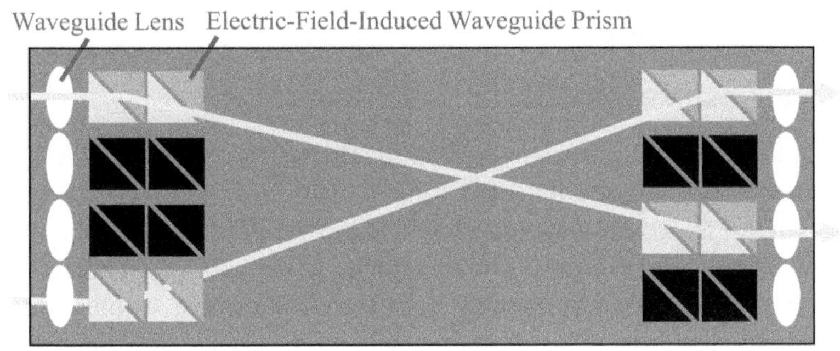

Figure 9.5. Operation principle of the WPD optical switch. Reprinted with permission from T. Yoshimura, S. Tsukada, S. Kawakami, Y. Arai, H. Kurokawa, and K. Asama, "3D Micro Optical Switching System (3D-MOSS) Architecture," *Proc. SPIE* **4653**, 62–70 (2002).

collimation are formed to suppress spreading due to diffraction of the propagated light beams in the free space region.

Optical switches for large-scale optical interconnects within computers and optical switching systems, where considerable thermal and device shape/configuration disturbance is caused, should be high-speed, thin, small, and at the same time, should be stable against the disturbance. In ordinary Mach–Zehnder optical switches and directional coupler optical switches, a small phase difference is used for the switching operation. Therefore, the dependence of the device operation on the temperature and the processing accuracy is substantial. In fact, in a $LiNbO_3$ matrix optical switch, individual precise adjustment is required for each element even in the 8×8 scale.

On the contrary, VWOICs and WPD optical switches are phase-insensitive flux-controlled devices. Namely, electric-field-induced differences in refractive index in the slab waveguide are used to switch the optical path directly. The refractive index difference is insensitive to the temperature changes compared with the refractive index itself. Furthermore, for processing accuracy, as far as the edge angles are maintained, the size of each part does not largely affect the switching characteristics. Therefore, VWOICs and WPD optical switches are expected to exhibit stable operation compared to ordinary optical switches, and are suited to large-scale optical systems such as integrated optical interconnects within boxes and the three-dimensional micro optical switching systems (3D-MOSSs) described in Chapter 10.

9.1.1 *Design of VWOIC*

Switching characteristics were simulated by the BPM for VWOIC with 12×7 electrodes of 2 μm × 100 μm. The results are shown in Fig. 9.6 for wavelength of 1.3 μm. Refractive index change induced by voltage application is assumed to be 0.005. Two adjacent electrodes are simultaneously selected for voltage application, which corresponds to 4 μm-wide waveguide construction. Selected regions are shifted by 2 μm along a direction perpendicular to light propagation direction in order to achieve quasi-continuous index variation. It can be seen that light beams are switched in a length of 700 μm.

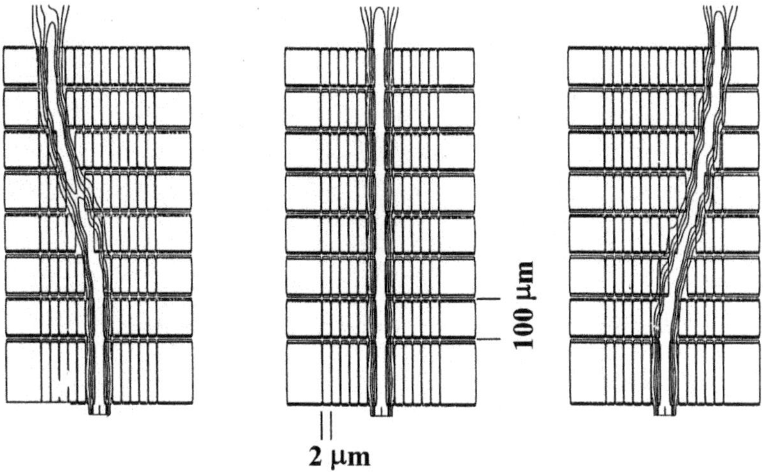

Figure 9.6. Switching characteristics of the VWOIC simulated by BPM. [T. Yoshimura, S. Tsukada, S. Kawakami, Y. Arai, H. Kurokawa, and K. Asama, "3D Micro Optical Switching System (3D-MOSS) Architecture," *Proc. SPIE* **4653**, 62–70 (2002).

9.1.2 *Design of WPD Optical Switch Utilizing the Pockels Effect*

9.1.2.1 Simulation procedure

Figure 9.7 shows a preliminary model of WPD optical switch for structural design [8]. The pitch of the input and output waveguides with aperture of D is set to P μm, and the number of steps of the prism electrode pair to N. The prism electrode has a right-angle triangle shape with a length and an aperture of D_P.

When designing the WPD optical switches, the following two points are kept under consideration:

(1) Reduction of the prism size and the inter-channel pitch
(2) Suppression of the beam spreading accompanying the free space propagation in the slab waveguide

These two points have a reciprocal relationship. If the prism size is reduced, the prism aperture decreases and the diffraction angle increases. Consequently, the beam width increases. So it is important

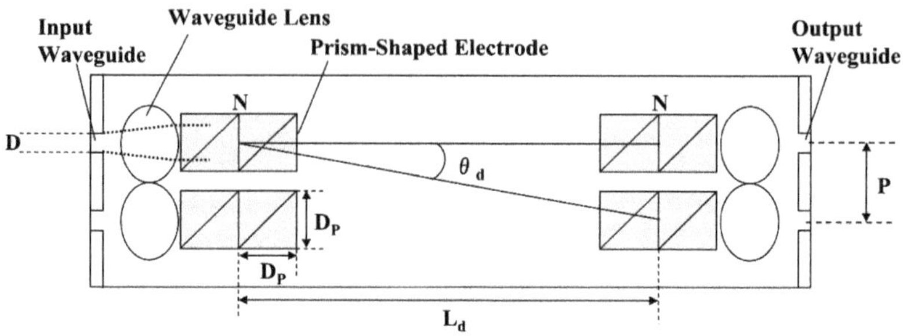

Figure 9.7. Preliminary model of WPD optical switch.

to find a practical solution for the WPD optical switch structure within this trade-off.

As shown in Fig. 9.8, the prism electrode is formed by a pair. When a negative voltage is applied to one electrode and positive voltage to the other, prism-shaped refractive index changes of $-\Delta n$ and $+\Delta n$ are induced in the slab waveguide under the electrodes. As a result, a waveguide prism is produced, and the light beam is deflected. The deflection angle θ_d is given by the following equation using the angle of incidence θ_1 and the angle of refraction θ_2.

$$\theta_d = \theta_1 - \theta_2 \tag{9.3}$$

By lining up multiple prism electrode pairs in a cascade, a larger deflection angle can be achieved. The voltage pattern of the prism-shaped electrode at the output has its polarity reversed. Thus, the deflected light beam returns in the horizontal direction and can

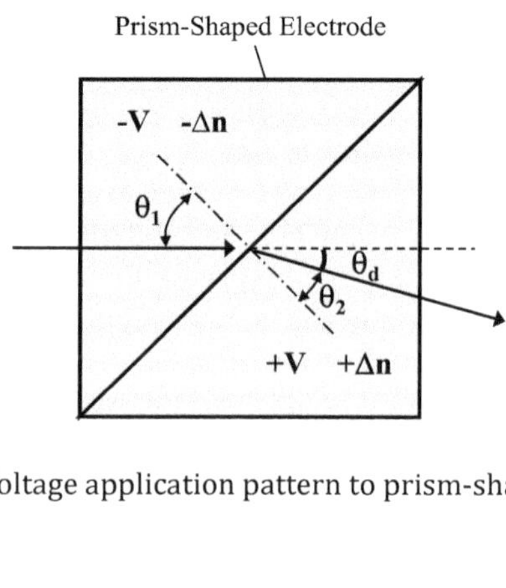

Figure 9.8. Voltage application pattern to prism-shaped electrodes.

smoothly couple to the output waveguide. The light beam introduced from the input waveguide spreads as it propagates in the slab waveguide. As illustrated in Fig. 9.5, waveguide lenses are inserted to suppress this spread.

The switching characteristics are simulated by the two-dimensional analysis of the wide-angle approximate BPM using the finite difference method. The propagated light beam has wavelength λ of 1.3 μm and is in TE mode. Δz step in the light beam propagation direction is 0.5 μm and Δx step in the perpendicular direction is $\lambda/30$. The core refractive index of the input and output waveguides is equal to the core refractive index of the slab waveguide, that is, the refractive index of the EO material with no applied voltage. The difference in the refractive index between the core and the clad is 0.03. The electric field distribution of the input light beam is that of the input waveguide fundamental mode. The integrated intensity of the input light beam is 1.0.

For the present simulation, EO material is assumed to be a 400 nm-thick SPCD film with $r = 430$ pm/V and $n = 1.55$. When 6.6 V is applied to the 400 nm-thick EO waveguide, $\Delta n \approx 0.013$ is expected.

9.1.2.2 Preliminary settings of the general structural model

In Fig. 9.7, the distance L_d between the prisms required for switching is estimated. For convenience, L_d is defined as the distance in the center of the multistage prisms at the input and the output. The refractive index n_1 in the negative voltage application unit and the refractive index n_2 of the positive voltage application unit are expressed as follows:

$$n_1 = n_0 - \Delta n \tag{9.4}$$

$$n_2 = n_0 + \Delta n \tag{9.5}$$

where n_0 is the refractive index when no voltage is applied. From Snell's law at the prism electrode boundary, the relationship between the angle of incident θ_1 and the angle of refraction θ_2 is given by

$$\theta_2 = \sin^{-1}\left(\frac{n_1}{n_2}\sin\theta_1\right) \tag{9.6}$$

The deflection angle θ_d is determined from Eqs. (9.6) and (9.3). For an N-stage cascade, N iterations of the calculation in Eq. (9.6) determine the final θ_d. The value of L_d in the 0th approximation is expressed in terms of θ_d and P by

$$L_d = \frac{P}{\tan\theta_d} \tag{9.7}$$

In WPD optical switch, since light beams propagate in 2-D free space, the beam diameter spreads by diffraction. The diffraction angle increases with decreasing D. The beam diameter must be less than or equal to the waveguide pitch P. Therefore, restrictions develop in L_d depending on D.

9.1.2.3 Performance evaluation by simulation

A. 2 x 2WPD Optical Switch From preliminary tests, $N = 4$ and $V = 11$ V was extracted as the appropriate condition for WPD optical switches with SPCD. A 2 × 2 WPD optical switch model is created with this condition as shown in Fig. 9.9. Here, $P = 20$ μm, $D = 4$ μm, and $D_P = 16$ μm. The diameter and the refractive index of the lens are 40 μm and 2.0, respectively. The optical switch length is 390 μm.

Figure 9.10 shows the BPM simulation results of the Input 1 → Output 1 switching and the Input 2 → Output 1 switching. Light beam deflection by the waveguide prism and smooth switching operation are seen. For Input 1 → Output 1, the cross talk is –12 dB and the insertion loss, 0.44 dB. For Input 2 → Output 1, the cross talk is –14 dB and the insertion loss, 0.96 dB. Thus, the overall WPD

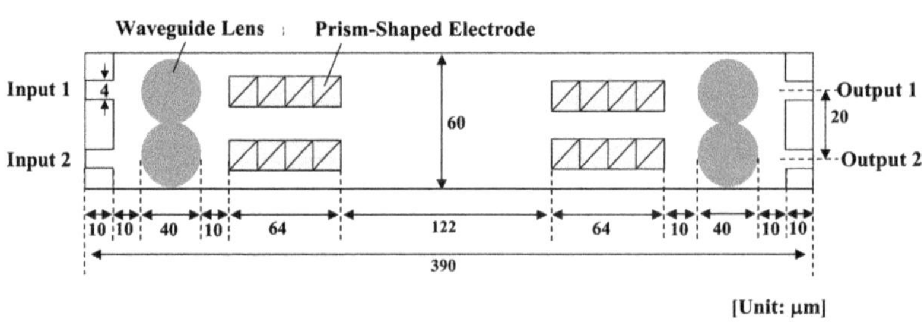

Figure 9.9. 2 × 2 WPD optical switch model with an EO slab waveguide of SPCD.

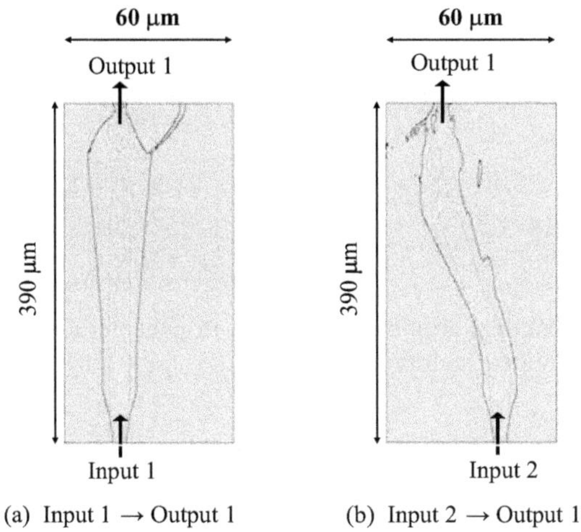

Figure 9.10. Switching characteristics of the 2 × 2 WPD optical switch with SPCD simulated by BPM.

optical switch characteristics are the cross talk ≤ −12 dB and the insertion loss ≤ 0.96 dB.

B. 3 x 3 WPD Optical Switch Figure 9.11 shows a 3 × 3 WPD optical switch model. Figure 9.12 shows the voltage application pattern, and Fig. 9.13 shows the BPM simulation results of the Input 1 → Output 3 switching and the Input 1→ Output 2 switching. The cross talk is −10 ∼ −12 dB, insertion loss is <1.2 dB, and the length is 400 μm.

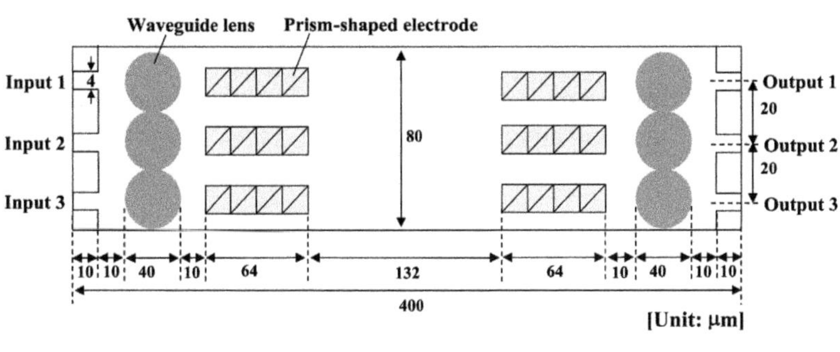

Figure 9.11. A 3 × 3 WPD optical switch model with an EO slab waveguide of SPCD.

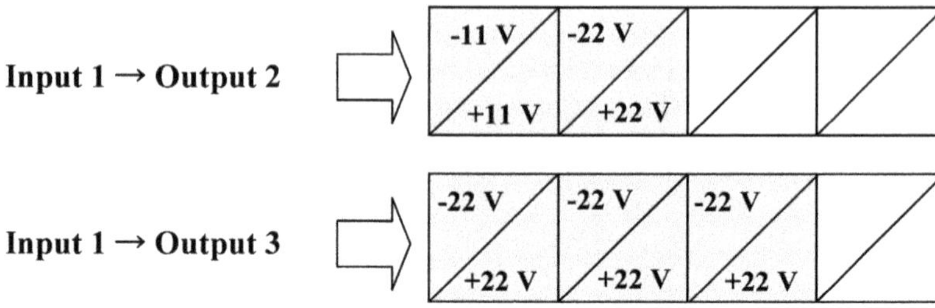

Figure 9.12. Voltage-application pattern to prism-shaped electrodes in the 3 × 3 WPD optical switch model.

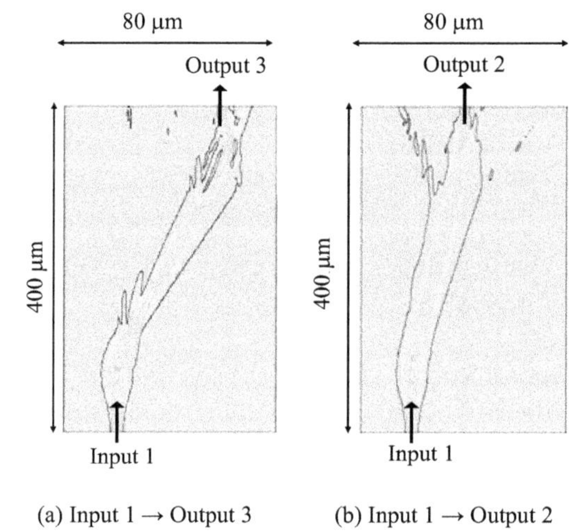

Figure 9.13. Switching characteristics of the 3 × 3 WPD optical switch with SPCD simulated by BPM.

C. Switching Time The switching time τ of WPD optical switch is given by the following equation:

$$\tau = R_{\mathrm{ON}} C \ [s] \tag{9.8}$$

where, the ON resistance of the driver circuit is R_{ON} [Ω] and the capacitance between the prism electrode and the opposite electrode is C [F]. When the area of the prism-shaped electrode is S [m^2], the distance between electrodes is d [m], and dielectric constant of the

EO material is ε, C is given by

$$C = \frac{\varepsilon S}{d} [F] \tag{9.9}$$

By substituting $S = (1/2) \times (16 \times 10^{-6}) \times (16 \times 10^{-6}) = 128 \times 10^{-12}$ m^2, $d = 4 \times 10^{-7}$ m and the assumed dielectric constant of SPCD $\varepsilon \approx 4\varepsilon_0$ into (9.9), $C = 11$ fF is obtained. Then, from Eq. (9.8), τ is calculated to be 1.1 ps for $R_{ON} = 100\ \Omega$. We see that the speed of the switch is sufficiently fast.

Power consumption for a charge/discharge cycle is given by

$$\text{Power} = FCV^2 \tag{9.10}$$

where F is switching rate. When $F = 5 \times 10^9$ 1/s and $V = 11$ V, Power = 6.9 mW is obtained.

9.1.3 *Design of WPD Optical Switch Utilizing the Kerr Effect*

As shown in Fig. 9.2, PLZT is expected to induce a large electric-field-induced refractive index change due to its large Kerr coefficient. In the present subsection, design, simulated switching characteristics, and a preliminary experimental result of WPD optical switches made with EO slab waveguides of PLZT thin films are reviewed.

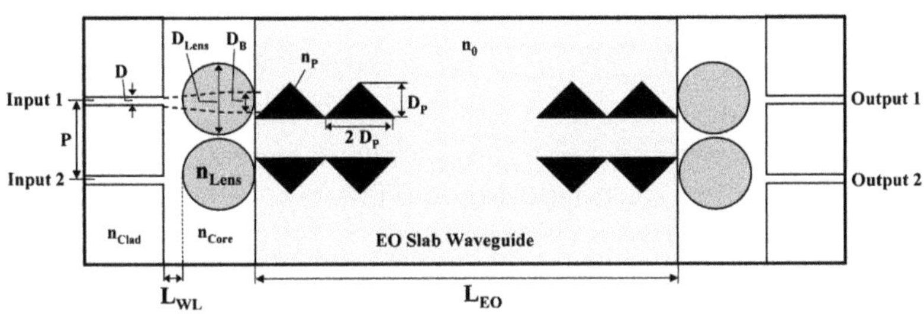

Figure 9.14. Preliminary model of 2 × 2 WPD optical switch with an EO slab waveguide of PLZT. Reprinted with permission from T. Yoshimura, M. Ojima, Y. Arai, N. Fujimoto, and K. Asama, "Simulation on Cross-Talk Reduction in Multi-Mode-Waveguide-Based Micro Optical Switching Systems Using Single Mode Filters," *Appl. Opt.* **43**, 1390–1395 (2004).

9.1.3.1 Simulation procedure

A preliminary model of 2 × 2 WPD optical switch with an EO slab waveguide of PLZT is shown in Fig. 9.14. Electric-field-induced dynamic waveguide prisms deflect optical signals from input waveguides to switch them through the two-dimensional slab waveguide.

Light wavelength λ is assumed 1.3 μm. Input and output waveguides consist of channel waveguide with core width D of 4 μm, core refractive index n_{Core} of 1.52, and clad refractive index n_{Clad} of 1.50, which permits to propagate the fundamental mode of $m = 0$ and the first excited mode of $m = 1$. Channel distance P is 20 μm. Waveguide lens diameter D_{Lens} is 40 μm. The lens expands light beam size to D_{B}. Right-angle prism-shaped electrodes with an aperture D_{P} of 16 μm and length $2D_{\mathrm{P}}$ are formed on the slab waveguide with a separation d of 400 nm from a bottom electrode beneath the PLZT film. Two-step prism cascading configuration is considered. Reflections from the integrated components such as waveguide lenses are not taken into account. AR coating on the component edge surfaces using the atomic layer deposition (ALD) or the molecular layer deposition (MLD), which can grow ultra-thin conformal films on surfaces with 3-D structures such as trenches, might make the model viable.

9.1.3.2 Preliminary settings of the general structural model

In order to determine a rough sketch of WPD optical switch structure, estimation for EO slab waveguide length L_{EO} is carried out. A detail of the electric-field-induced waveguide prism is shown in Fig. 9.15.

Refractive index change Δn is induced in the EO slab waveguide by a driving voltage V according to Eq. (9.2). Waveguide prism refractive index n_{P} is given by

$$n_P = n_0 + \Delta n \tag{9.11}$$

Here, n_0 is refractive index of the EO slab waveguide with no applied voltage. Deflection angle θ_{d} can be calculated by the following

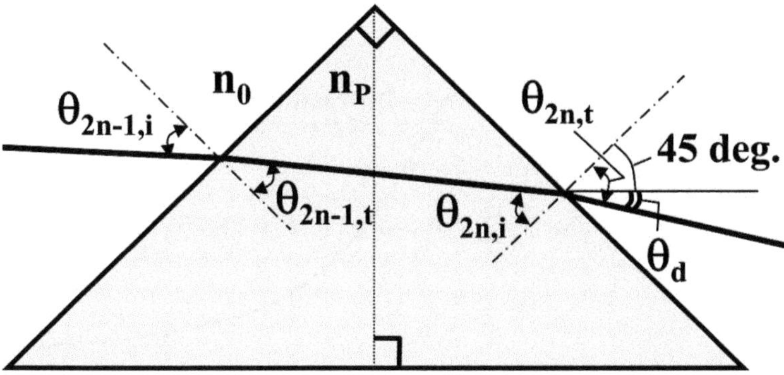

Figure 9.15. Detail of the electric-field-induced waveguide prism in the WPD optical switch. Reprinted with permission from T. Yoshimura, M. Ojima, Y. Arai, N. Fujimoto, and K. Asama, "Simulation on Cross-Talk Reduction in Multi-Mode-Waveguide-Based Micro Optical Switching Systems Using Single Mode Filters," *Appl. Opt.* **43**, 1390–1395 (2004).

equations:

$$\begin{aligned} n_0 \sin\theta_{2n-1,i} &= n_P \sin\theta_{2n-1,t} \\ n_P \sin\theta_{2n,i} &= n_0 \sin\theta_{2n,t} \\ \theta_{2n,i} &= \frac{\pi}{2} - \theta_{2n-1,t} \\ \theta_{2n+1,i} &= \frac{\pi}{2} - \theta_{2n,t} \\ (n &= 1, 2, \ldots, N) \\ \theta_d &= \theta_{2N,t} - \frac{\pi}{4} \end{aligned} \tag{9.12}$$

As the 0th approximation, neglecting the size of prism-shaped electrodes, L_{EO} is given by

$$L_{\text{EO}} = P/\tan\theta_{\text{d}} \tag{9.13}$$

Diffraction angle of light beams propagating in the slab waveguide is expressed as $\theta = \lambda/(\pi n_0 D_B)$. Light beam width should be less than P at the end of the EO slab waveguide after propagation. Then, L_{EO} should satisfy

$$L_{\text{EO}} < \frac{P}{2}/\tan\theta \tag{9.14}$$

In Fig. 9.16, V dependence of L_{EO} derived from (9.2) and (9.11–9.13) using parameters for PLZT shown in Table 9.1 is shown for

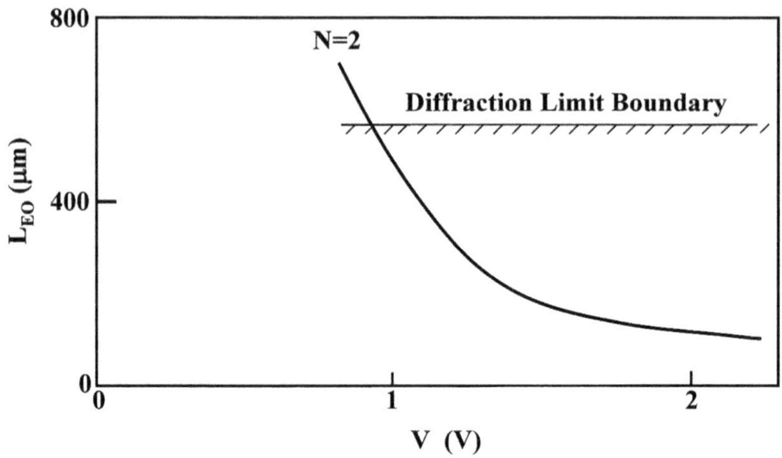

Figure 9.16. V dependence of L_{EO} and diffraction limit for D_B of 5 μm. Reprinted with permission from T. Yoshimura, M. Ojima, Y. Arai, N. Fujimoto, and K. Asama, "Simulation on Cross-Talk Reduction in Multi-Mode-Waveguide-Based Micro Optical Switching Systems Using Single Mode Filters," *Appl. Opt.* **43**, 1390–1395 (2004).

two-step prism cascading configuration ($N = 2$). Diffraction limit boundary obtained by (9.14) is also shown for D_B of 5 μm. It can be seen that lowest possible driving voltage is around 1 V, predicting L_{EO} of around 500 μm. In the present simulation, $V = 1.2$ V is assumed, which generates waveguide prism refractive index $n_{P(On)}$ of 2.52 derived from Eqs. (9.2) and (9.11). Waveguide prism refractive index $n_{P(Off)}$ for 0 V application is equal to n_0 of 2.48.

Based on the parameters mentioned above, waveguide lens refractive index n_{Lens}, distance between waveguide lens and input/output waveguide L_{WL}, and L_{EO} are optimized. Figure 9.17 shows the BPM simulation results for n_{Lens} of 1.8, 2.0, and 2.5. For n_{Lens} = 1.8, collimation of the light beam in the EO slab waveguide is insufficient, while for n_{Lens} of 2.5, focusing effect is too strong so that large beam size expansion occurs. n_{Lens} of 2.0 is proper condition to produce collimated light beams, resulting in smooth coupling to output waveguide in a distance of 310 μm from input waveguide.

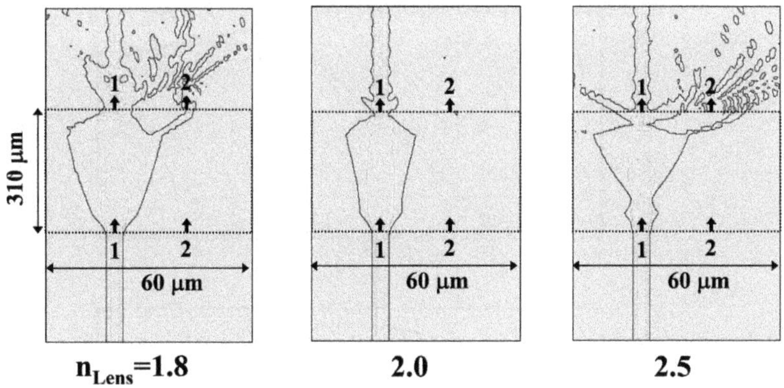

Figure 9.17. BPM simulation of light beam collimation for various n_{Lens}. Reprinted with permission from T. Yoshimura, M. Ojima, Y. Arai, N. Fujimoto, and K. Asama, "Simulation on Cross-Talk Reduction in Multi-Mode-Waveguide-Based Micro Optical Switching Systems Using Single Mode Filters," *Appl. Opt.* **43**, 1390–1395 (2004).

Figure 9.18(a) shows L_{WL} dependence of transmission efficiency from the input waveguide into the output waveguide. The optimum L_{WL} is found to be 20 μm. For L_{EO}, as shown in Fig. 9.18(b), the optimum is 190–200 μm. From a viewpoint of miniaturization of optical switches, 190 μm should be the optimum L_{EO}. Thus, the total length of the WPD optical switch is calculated to be 310 μm by adding $2(D_{\text{Lens}} + L_{\text{WL}})$ to L_{EO}.

Design parameters determined for the WPD optical switch with PLZT are summarized in Table 9.2 and the optimized structure is shown in Fig. 9.19. The slab waveguide is divided into two regions. One is a region with waveguide lenses. The waveguide lens refractive index of 2.0 is available by using $\text{Ti}_x\text{Si}_y\text{O}$, for example. The other is a region with prisms where whole slab waveguide is made of 400 nm-thick PLZT. Right-angle prism-shaped electrodes with an area of 16×16 μm^2 are formed on the slab waveguide. Size of the main part of the optical switch is 310×20 μm. When channel separation is widened to 60 μm by bending waveguides of 2.5 mm radius at the input and output ports, total switch length L_{SW} and width W_{SW} become 1190 μm and 100 μm, respectively.

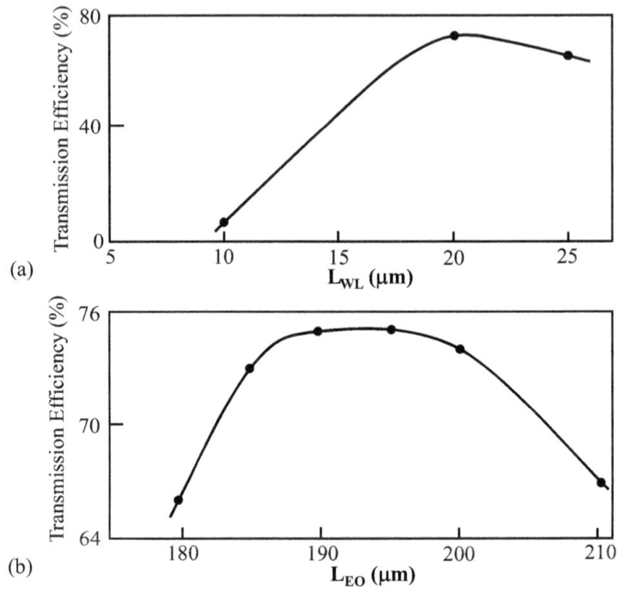

Figure 9.18. (a) L_{WL} and (b) L_{EO} dependence of transmission efficiency into the output waveguide. Reprinted with permission from T. Yoshimura, M. Ojima, Y. Arai, N. Fujimoto, and K. Asama, "Simulation on Cross-Talk Reduction in Multi-Mode-Waveguide-Based Micro Optical Switching Systems Using Single Mode Filters," *Appl. Opt.* **43**, 1390–1395 (2004).

9.1.3.3 Performance evaluation by simulation

Figure 9.20 shows results of the BPM simulation of switching characteristics for the 2 × 2 WPD optical switch with PLZT. When no voltages are applied to the prism-shaped electrodes, a light beam propagates from Input 1 to Output 1, achieving "Bar" condition. By 1.2 V application to the electrodes, a light beam propagates from Input 1 to Output 2, achieving "Cross" condition. For both "Bar" and "Cross", cross talk is less than −20 dB. Transmission efficiency is 80% and 69% for "Bar" and "Cross", respectively. Maximum insertion loss within the WPD optical switch is 1.7 dB including losses in bending waveguides.

Switching speed and power consumption can be estimated based on a charge-discharge model for a prism-shaped capacitor as the

Table 9.2. Design parameters for WPD optical switch with PLZT

Optical signal	
Wavelength: λ	1.3 μm
Channel Waveguide	
Width: D	4 μm
Channel separation: P	20 μm
Core refractive index: n_{Core}	1.52
Clad refractive index: n_{Clad}	1.50
Waveguide Lens	
Lens-Input/Output waveguide distance: L_{WL}	20 μm
Diameter: D_{Lens}	40 μm
Refractive index: n_{Lens}	2.0
EO Slab Waveguide	
Length: L_{EO}	190 μm
Electrode separation: d	400 nm
Driving voltage: V	1.2 V
Cascaded prism count: N	2
Kerr coefficient: R_{Kerr}	5.7×10^{-16} m^2/V^2
Refractive index: n_0	2.48
Prism refractive index (Off): $n_{\mathrm{P(Off)}}$	2.48
Prism refractive index (On): $n_{\mathrm{P(On)}}$	2.52
Structure	
Total Length:	310 μm

Source: Reprinted with permission from T. Yoshimura, M. Ojima, Y. Arai, N. Fujimoto, and K. Asama, "Simulation on Cross-Talk Reduction in Multi-Mode-Waveguide-Based Micro Optical Switching Systems Using Single Mode Filters," *Appl. Opt.* **43**, 1390–1395 (2004).

0th approximation. A waveguide prism deflector is regarded as a capacitor with S [m^2] area, d [m] thickness, and ε dielectric constant. The capacitance C, switching speed τ, and power consumption Power are given by Eqs. (9.9), (9.8), and (9.10), respectively. In the model shown in Fig. 9.19, $S = 256 \times 10^{-12}$ m^2, $d = 4 \times 10^{-7}$ m, and the dielectric constant of PLZT $\varepsilon = 1000\varepsilon_0$. Then, C, τ, and Power are respectively calculated to be 5.7 pF and 570 ps for $R_{\mathrm{ON}} = 100\ \Omega$, and 41 mW for $F = 5 \times 10^9$ 1/s and $V = 1.2$ V.

9.1.4 *WPD Optical Switch with ADD Function*

For advanced optical switching systems, optical switches with ADD function are sometimes required. WPD can provide such optical

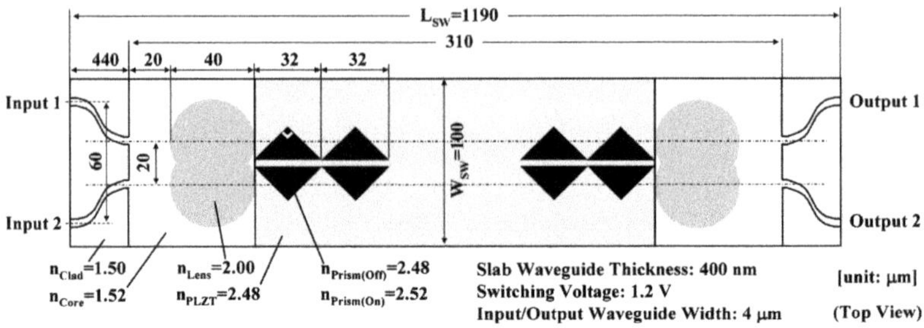

Figure 9.19. Optimized structure of 2 × 2 WPD optical switch using PLZT for EO material. Reprinted with permission from T. Yoshimura, M. Ojima, Y. Arai, and K. Asama, "Three-Dimensional Self-Organized Micro Optoelectronic Systems for Board-Level Reconfigurable Optical Interconnects — Performance Modeling and Simulation," *IEEE J. Select. Topics Quantum Electron.* **9**, 492–511. © 2003 IEEE.

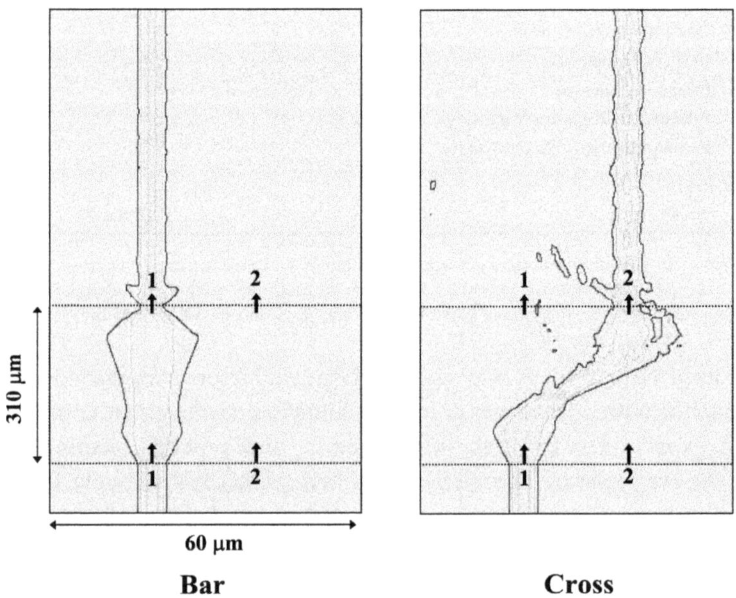

Figure 9.20. Switching characteristics of the 2 × 2 WPD optical switch with PLZT simulated by BPM. Reprinted with permission from T. Yoshimura, M. Ojima, Y. Arai, and K. Asama, "Three-Dimensional Self-Organized Micro Optoelectronic Systems for Board-Level Reconfigurable Optical Interconnects — Performance Modeling and Simulation," *IEEE J. Select. Topics Quantum Electron.* **9**, 492–511. © 2003 IEEE.

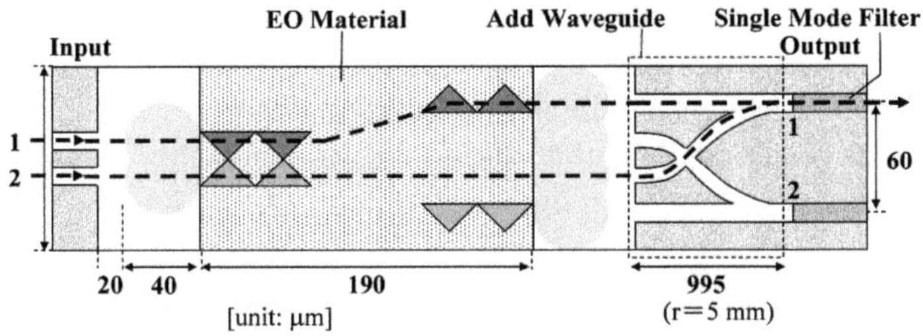

Figure 9.21. Model of WPD optical switch with ADD function. See also Color Insert.

switches as Fig. 9.21 shows [20]. The structure is based on the 2 × 2 WPD optical switch. To give the ADD function to the optical switch, Add waveguides with curvature of 5 mm radius are inserted. Single mode filters [9], which extract the fundamental mode from the optical signals, are placed after the Output.

By applying voltage to four prism-shaped electrodes in the upper-half region, optical signals in Input 1 are connected to Output 1, and at the same time, optical signals in Input 2 are also connected to Output 1, accomplishing the ADD function. Similarly, by applying four electrodes in the lower-half region, optical signals in Input 1 and Input 2 are connected to Output 2 simultaneously.

Figure 9.22 shows the switching characteristics of the WPD optical switch with ADD function simulated by BPM. Smooth adding

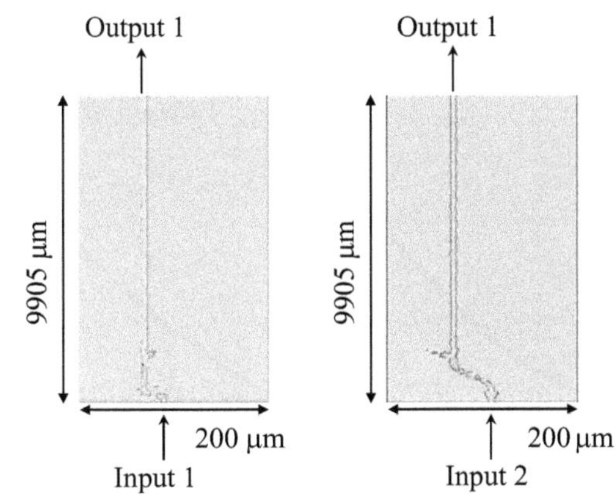

Figure 9.22. Switching characteristics of the WPD optical switch with ADD function simulated by BPM.

can be seen. The insertion loss and cross talk are respectively 1.4 dB and −22 dB. After passing through the single mode filter, they respectively become 5.4 dB and −21 dB.

These results suggest that the WPD optical switches will realize various kinds of functional optical circuits.

9.1.5 *Impact of Polymer Multiple Quantum Dots (MQDs) on Optical Switches*

As shown in Table 9.1 and Fig. 9.2, the polymer MQD [14–18] is expected to exhibit large EO effects and low dielectric constant, say, around $4\varepsilon_0$, which will improve performance of optical switches. Figure 9.23 shows dependences of switching operation of WPD optical switches on EO materials simulated by the BPM. When the polymer MQD described in Section 12.2 is used, large deflection angles induced by small driving voltages are expected.

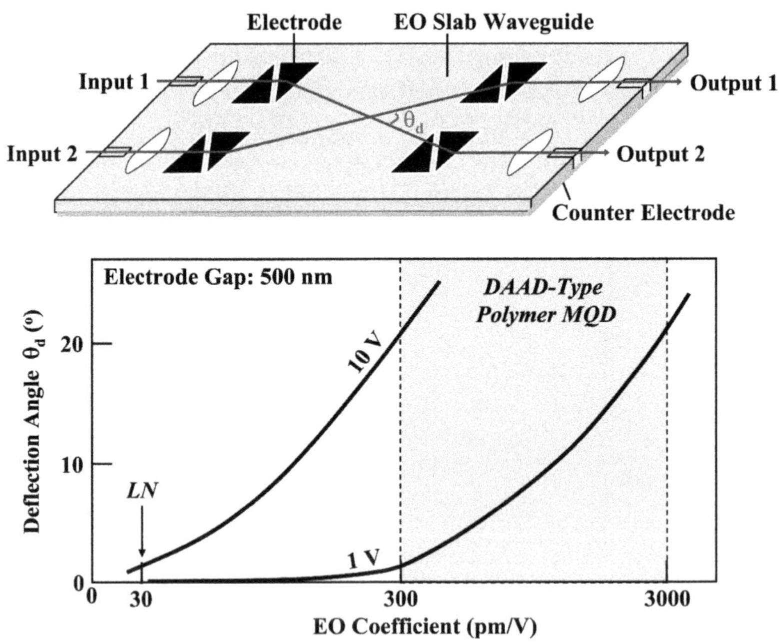

Figure 9.23. Deflection angles in WPD optical switches as a function of EO coefficient.

The dielectric constant for electrical signals is expected much smaller in the polymer MQD than in PLZT because the EO effect in the polymer wires arises from electronic polarization. This feature of the polymer MQD is desirable to achieve high-speed and low power operations. When ε for the polymer MQD is $4\varepsilon_0$ and for PLZT $1000\varepsilon_0$, τ and Power for the polymer MQD can be 1/250 of those for PLZT. Thus, the polymer MQD is expected to remove the drawbacks of slow switching time and large power consumption in PLZT. The polymer MQD is an ideal EO material for the large-scale integrated optical interconnects and optical switching systems such as the 3D-MOSS described in Chapter 10.

9.1.6 *Future Integration Issues*

The present simulation clearly shows the potential performance of the switching operations of 2 × 2 and 3 × 3 WPD optical switches having lengths of 100 μm scales. In addition to the optical switch itself, the method for connecting and integrating the driver circuits, waveguide systems, and optical switches is also a serious problem in the 3-D OE platform based on S-FOLM. The connecting and the integrating will be achieved by using PL-Pack with SORT described in Chapter 8, which transplants multiple different thin-film device flakes and material flakes together. EO thin films of the optical switches may easily be integrated with optical waveguides, waveguide lenses, waveguide filters, and waveguide mirrors by PL-PACK with SORT. For integration of EO thin films made of the polymer MQD, the molecular nano duplication shown in Chapter 12 will be applicable. SOLNET described in Chapters 4 and 5 is useful to connect the EO thin films to optical waveguides and other optical devices with self-aligned manners.

9.1.7 *Experimental Demonstration of WPD utilizing PLZT*

As a preliminary demonstration, we fabricated WPD using a PLZT thin film. A patterned indium tin oxide (ITO) transparent electrode was formed on a glass substrate. Then, a thin film of PLZT was coated all over the surface by the sol-gel method. The PLZT film thickness was 250 nm. On the film surface Al electrode was deposited by the

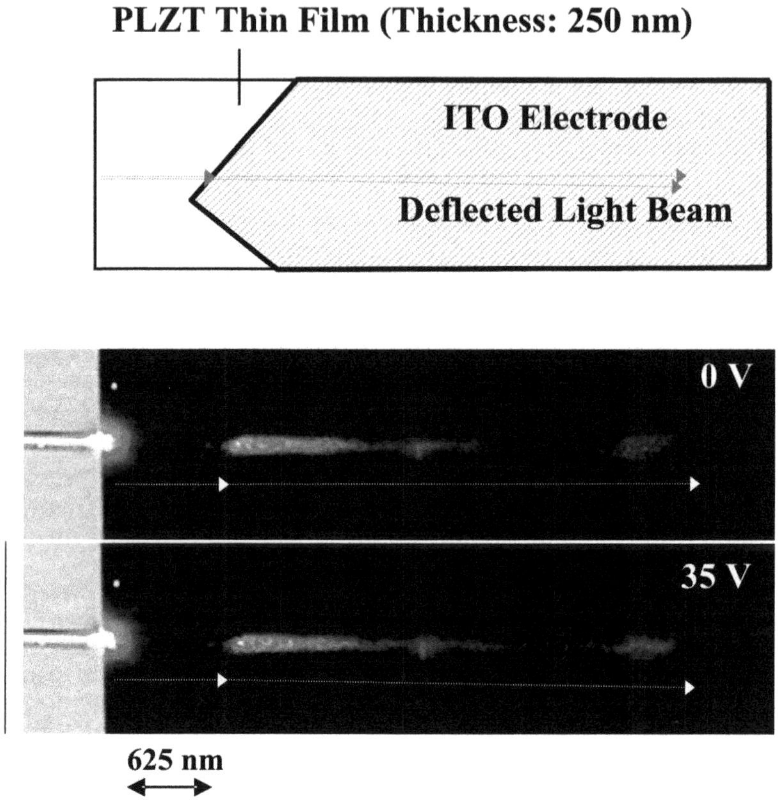

Figure 9.24. Experimentally observed light beam deflection in WPD with PLZT.

vacuum evaporation. Figure 9.24 shows guided beam deflection by applying 35 V between the ITO and Al electrodes. Deflection angle of 1.1° was observed.

The future challenges are optimization of the PLZT WPD optical switches and replacement of PLZT with organic EO materials such as SPCD or the polymer MQD.

9.2 Ring Resonator Optical Switches

Although the thermal and dimensional stability is usually insufficient, ring resonators consisting of HIC waveguides might still be

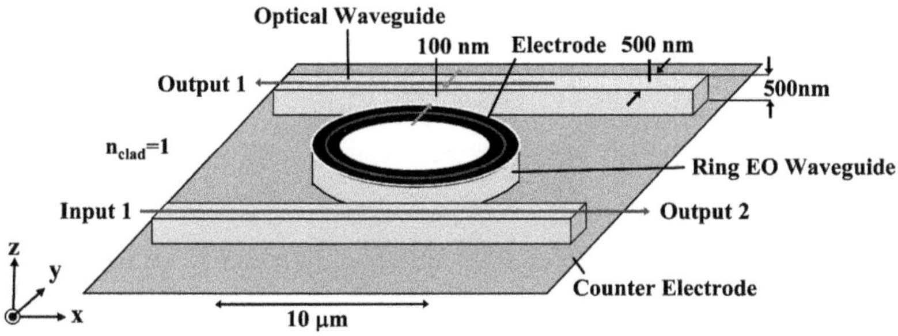

Figure 9.25. Model of a ring resonator optical switch.

promising for ultra small optical switches in precisely-controlled circumstance. In the present section, light beam propagation and switching operation in these devices are simulated by the FDTD method and investigate the possibility of optical switches.

Ring resonator optical switches have been developed by many researchers. The typical structural model is shown in Fig. 9.25. Between two straight waveguides a ring EO waveguide is inserted. Core width of these optical waveguides is usually submicrons. Figure 9.26 shows propagation of a light pulse in a ring resonator

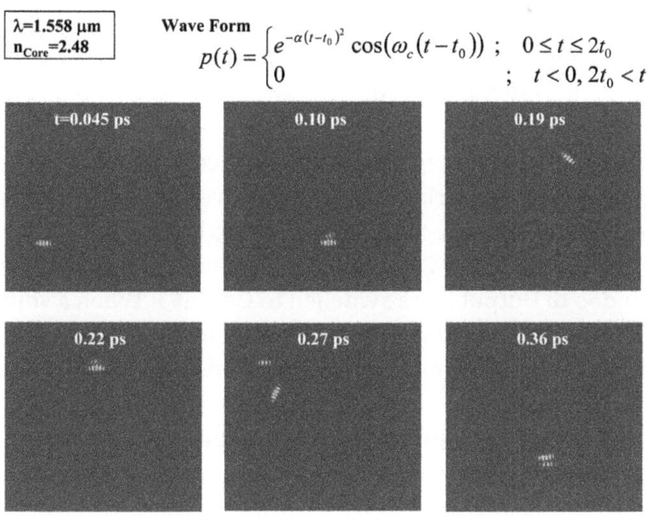

Figure 9.26. Propagation of a light pulse in a ring resonator optical switch simulated by FDTD method.

$$p(t) = \begin{cases} \frac{1}{2}(1-\cos\frac{\pi t}{T_\omega})\sin\omega_0 t \ ; & 0 \le t \le T_\omega \\ \sin\omega_0 t & ; \ t \ge T_\omega \end{cases}$$

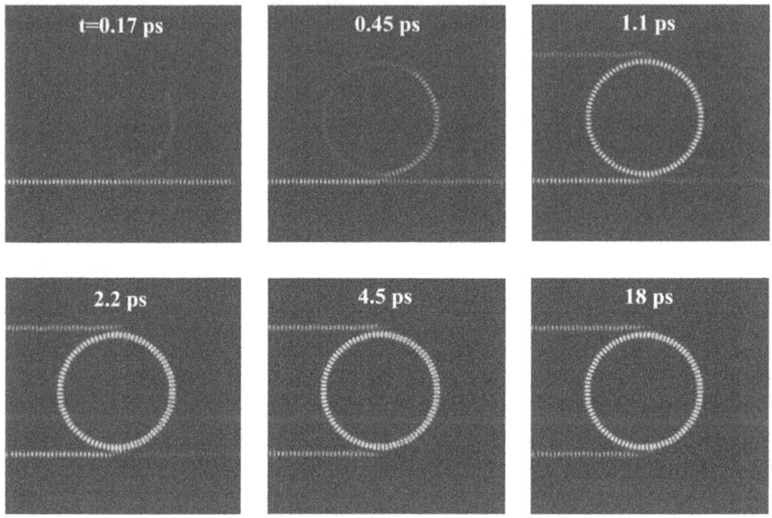

Figure 9.27. Resonance of light beams in a ring resonator optical switch simulated by FDTD method.

optical switch consisting of PLZT. A light pulse introduced from Input 1 is partially transferred between the straight waveguides and the ring EO waveguide.

When the device configuration and refractive index of the optical waveguides are appropriately tuned for a resonance condition, cw light beams gradually become localized into the ring EO waveguide as can be seen in Fig. 9.27. At the same time, the light beams that were guided to Output 2 are switched to Output 1. When a voltage is applied to the ring EO waveguide, the resonance condition is broken to switch the light beams from Output 1 to Output 2 as shown in Fig. 9.28.

The resonant condition varies with wavelength of the guided beams. Therefore, the ring resonator acts as a wavelength filter. By applying a voltage to the ring waveguide, the wavelength in a resonant condition is controlled, enabling tunable wavelength filter operation for WDM optical interconnects and switching systems.

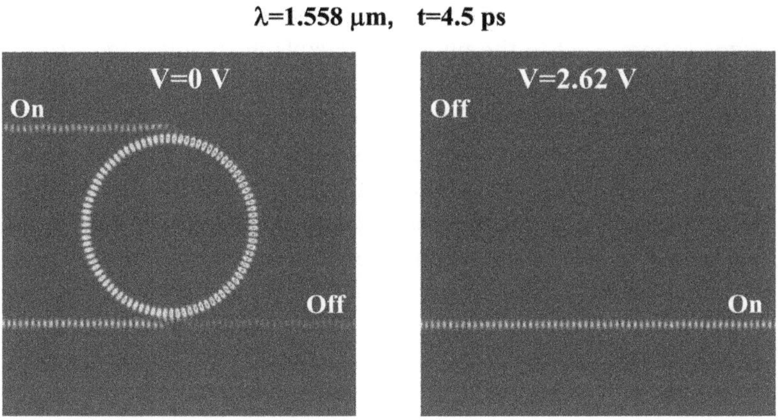

Figure 9.28. Switching operation of a ring resonator optical switch simulated by FDTD method.

9.3 Band Width Limit in Photonic Crystal Waveguides

The photonic crystal, whose concept was proposed by Yablonovitch [21], has periodic variations of refractive index like dielectric multilayer wavelength filters. As the energy bands for electrons in crystals that has periodic potential variations, energy bands for photons appear in photonic crystals as shown in Fig. 9.29. When the

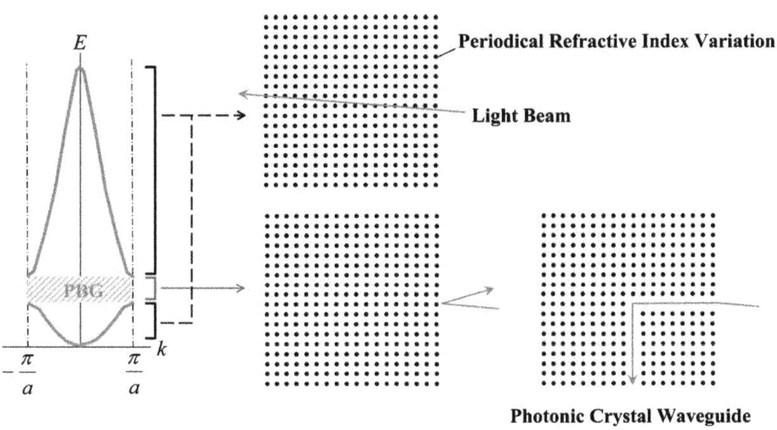

Figure 9.29. Band Diagrams for a photonic crystal and light beam propagation in it.

energy of photons is in the photonic band gap (PBG) region, light beams cannot exist in the photonic crystals in steady states. When the energy of photons is out of the PBG region, light beams can exist in the photonic crystals.

In the case that some imperfections are generated in a photonic crystal, light beams with photon energy within PBG can exist in the imperfection parts. By creating line-shaped imperfections as shown in Fig. 9.29, light beams are confined in the imperfection parts because they cannot exist in the surrounding perfect photonic crystal parts. When the line width of the imperfection is adjusted to be submicrons, nano-scale waveguides can be constructed in the photonic crystal. By the implementation of EO materials into the photonic crystal, ultra-small optical switches might be available.

Simulation for light beam propagation in a two-dimensional photonic crystal waveguide was carried out using FDTD method. The results are shown in Figs. 9.30 and 9.31. 104 nm-radius holes are

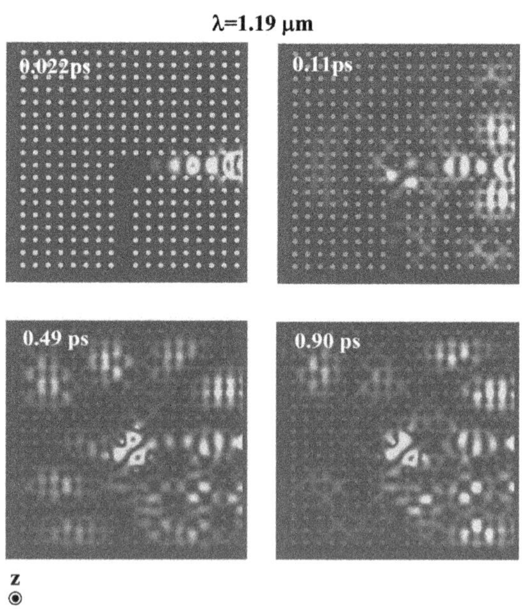

Figure 9.30. Light beam propagation in a photonic crystal waveguide for a wavelength out of PBG simulated by FDTD method.

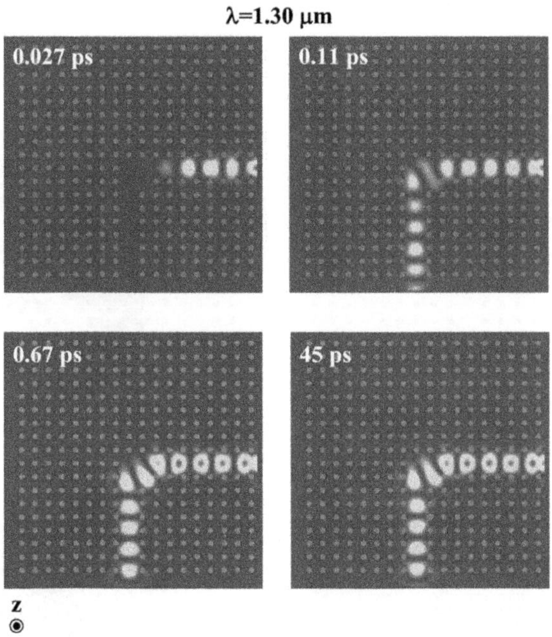

Figure 9.31. Light beam propagation in a photonic crystal waveguide for a wavelength within PBG simulated by FDTD method.

arranged with a 521 nm pitch in a medium with refractive index of 3.38. Light polarization is parallel to the *z*-axis. For wavelength $\lambda = 1.19$ μm, which corresponds to photon energy out of PBG, light beams are leaked from the optical waveguide (Fig. 9.30). For wavelength $\lambda = 1.30$ μm, which corresponds to photon energy near the center of PBG, light beams are strongly confined in the optical waveguide (Fig. 9.31).

Here, however, it is found that at the initial stage of light incidence, namely, at time duration (t) of 0.11 ps, confined light intensity is weak, and with time it gradually increases to reach a saturated state. Figure 9.32 show a result for the transient light beam propagation in the photonic crystal waveguide [22]. The drawing contrast is adjusted to enhance leaked light beams in the surrounding perfect crystal region. It is found that at the initial stage, considerable leakage of light beams into the surrounding perfect crystal region is observed. After 0.6 ps, the leakage is suppressed

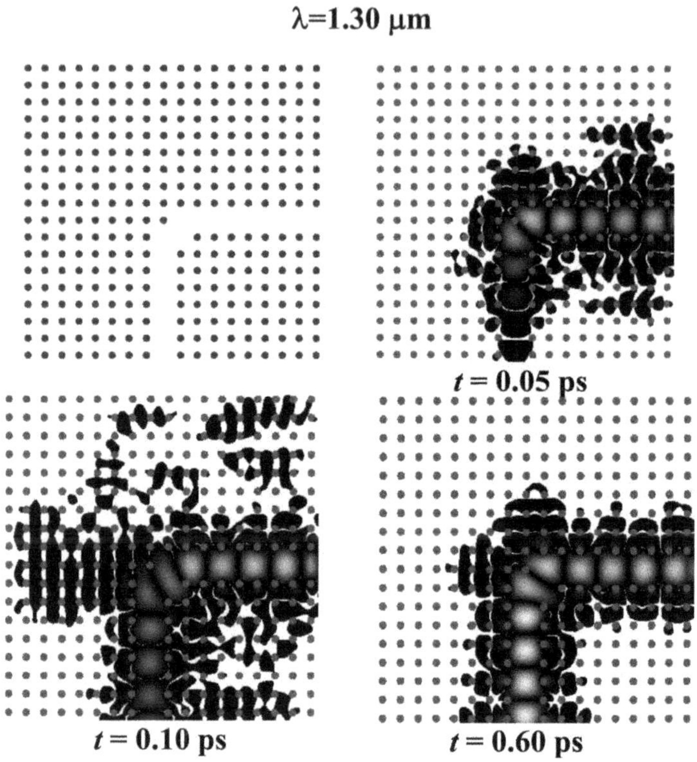

Figure 9.32. Transient light beam propagation in a photonic crystal waveguide simulated by FDTD method. Reprinted with permission from T. Yoshimura, Y. Suzuki, N. Shimoda, T. Kofudo, K. Okada, Y. Arai, and K. Asama, "Three-Dimensional Chip-Scale Optical Interconnects and Switches with Self-Organized Wiring Based on Device-Embedded Waveguide Films and Molecular Nanotechnologies," *Proc. SPIE* **6126**, from Photonics West 2006, San Jose, California, 612609-1-15 (2006).

and a strong confinement of light beams in the optical waveguide is achieved.

Such delay before settling waveguide functions arises from the fact that the photonic crystal device uses interference of light beams at a steady state. In the model shown in Fig. 9.32, at $t = 0.05$ and 0.1 ps, only a part of reflected light beams exist in the photonic crystal, namely, light beam interference in the photonic crystal is not completed yet. At $t = 0.6$ ps, since 0.6 ps corresponds to a

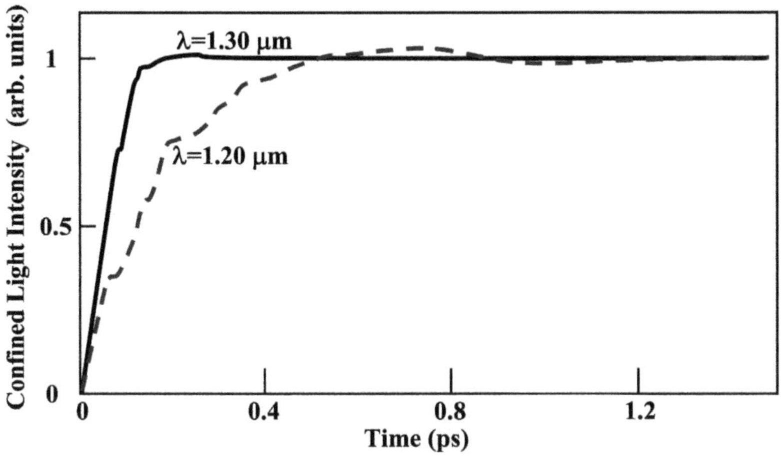

Figure 9.33. Transient responses of light beam confinement in a photonic crystal waveguide simulated by FDTD method.

light beam propagation distance of ~150 μm that is 17 times of the photonic crystal size in the present model, the interference is built-up completely, namely superposition of all the reflected light beams is completed in whole of the photonic crystal, to reach the steady state. Thus, the leakage is suppressed, resulting in perfect light beam confinement in the optical waveguide.

Transient responses of light beam confinement in a photonic crystal waveguide are shown in Fig. 9.33. For wavelength $\lambda = 1.20$ μm, which corresponds to photon energy at the edge of PBG, a long delay of ~0.5 ps is observed before the confined light intensity reaches its steady state. For wavelength $\lambda = 1.30$ μm located near the center of PBG, on the other hand, the confined light intensity rapidly increases to its saturated value with a short delay less than 0.2 ps. As can be seen in Fig. 9.34, light beam distribution after reaching steady state is wider for $\lambda = 1.20$ μm than for $\lambda = 1.30$ μm, indicating that in the case of $\lambda = 1.20$ μm the light beams penetrate into the deep part of the surrounding perfect crystal region, and, consequently, the interference occurs in a large region comparing with the case of $\lambda = 1.30$ μm. From these results, it is suggested that the delay before the waveguide function appears will cause problems when photonic crystals are used in high speed systems.

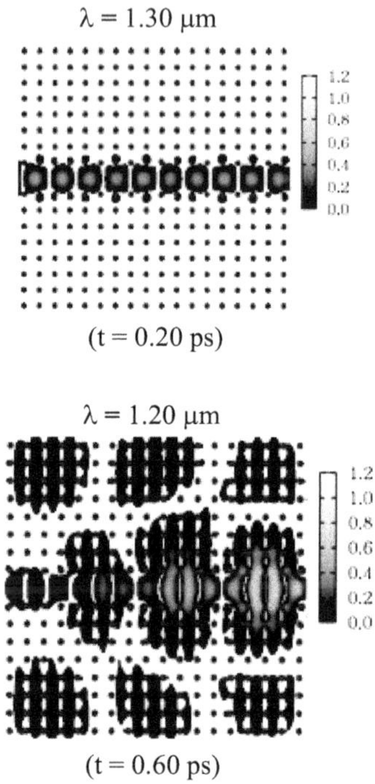

Figure 9.34. Light beam distribution in photonic crystal waveguides simulated by FDTD method. Reprinted with permission from T. Yoshimura, M. Ojima, Y. Arai, and K. Asama, "Three-Dimensional Self-Organized Micro Optoelectronic Systems for Board-Level Reconfigurable Optical Interconnects — Performance Modeling and Simulation," *IEEE J. Select. Topics Quantum Electron.* **9**, 492–511. © 2003 IEEE.

References

1. T. Yoshimura, S. Tsukada, S. Kawakami, Y. Arai, H. Kurokawa, and K. Asama, "3D Micro Optical Switching System (3D-MOSS) — Packaging Design," 2001 IEEE LEOS Annual Meeting Conference Proceedings, San Diego, California, 499 (2001).
2. T. Yoshimura, S. Tsukada, S. Kawakami, Y. Arai, H. Kurokawa, and K. Asama,, "3D Micro Optical Switching System (3D-MOSS) Architecture," *Proc. SPIE* **4653**, from Photonics West 2002, San Jose, California, 62–70 (2002).

3. T. Yoshimura, "Waveguide-Type Electro-Optic Devices," Japanese Patent Tokukai Hei 4-181231 (1992).
4. T. Yoshimura, "Optical Integrated Circuit Devices," Japanese Patent Tokukai Hei 4-204633 (1992).
5. T. Yoshimura, M. Ojima, Y. Arai, and K. Asama, "Three-Dimensional Self-Organized Micro Optoelectronic Systems for Board-Level Reconfigurable Optical Interconnects — Performance Modeling and Simulation," *IEEE J. Select. Top. Quantum Electron.* **9**, 492–511 (2003).
6. T. Yoshimura, S. Tsukada, S. Kawakami, M. Ninomiya, Y. Arai, H. Kurokawa, and K. Asama, "Three-Dimensional Micro-Optical Switching System Architecture Using Slab-Waveguide-Based Micro-Optical Switches," *Opt. Eng.* **42**, 439–446 (2003).
7. T. Yoshimura, "Design of Organic Nonlinear Optical Materials for Electro-Optic and All-Optical Devices by Computer Simulation," *FUJITSU Sc. Tech.* J. **27**, 115–131 (1991)
8. M. Ninomiya, Y. Arai, T. Yoshimura, H. Kurokawa, and K. Asama, "Characteristics Evaluation for Waveguide-Prism-Deflector Type Micro Optical Switches (WPD-MOS)," *Electron. Commun. Jpn.*, Part 2 **86**, 38–48 (2003) (Translated from *Denshi Joho Tsushin Gakkai Ronbunshi* **J85-C**, 1192–1201 (2002)).
9. T. Yoshimura, M. Ojima, Y. Arai, N. Fujimoto, and K. Asama, "Simulation on Cross-Talk Reduction in Multi-Mode-Waveguide-Based Micro Optical Switching Systems Using Single Mode Filters," *Appl. Opt.* **43**, 1390–1395 (2004).
10. T. Yoshimura, Y. Arai, "3D Optoelectronic Micro System," U.S. Patent 7,387,913 B2 (2008).
11. O. Qasaimeh, K. Kamath, P. Bhattacharya, and J. Phikkips, "Linear and Quadratic Electro-Optic Coefficients of Self-Organized $In_{0.4}Ga_{0.6}As$/GaAs Quantum Dots," *Appl. Phys. Lett.* **72**, 1275–1277 (1998).
12. T. Yoshimura and T. Futatsugi., "Non-Linear Optical Device Using Quantum Dots," U.S. Patent 6,294,794 (2001).
13. T. Yoshimura, "Characterization of the Electro-Optic Effect in Styrylpyridinium Cyanine Dye Thin-Film Crystals by an ac Modulation Method," *J. Appl. Phys.* **62**, 2028–2032 (1987).
14. T. Yoshimura, S. Tatsuura, W. Sotoyama, A. Matsuura, and T. Hayano, "Quantum Wire and Dot Formation by Chemical Vapor Deposition and Molecular Layer Deposition of One-Dimensional Conjugated Polymer," *Appl. Phys. Lett.* **60**, 268–270 (1992).

15. T. Yoshimura, A. Oshima D. Kim and Y. Morita "Quantum Dot Formation in Polymer Wires by Three-Molecule Molecular Layer Deposition (MLD) and Applications to Electro-Optic/Photovoltaic Devices," *ECS Trans.* **25** (4) "Atomic Layer Deposition Applications 5," from the Vienna, Austria Meeting, 15–25 (2009).
16. T. Yoshimura, "Enhancing Second-Order Nonlinear Optical Properties by Controlling the Wave Function in One-Dimensional Conjugated Molecules," *Phys. Rev. B* **40**, 6292–6298 (1989).
17. T. Yoshimura, R. Ebihara, and A. Oshima, "Polymer Wires with Quantum Dots Grown by Molecular Layer Deposition of Three Source Molecules for Sensitized Photovoltaics," *J. Vac. Sci. Technol. A* **29**, 051510-1-6 (2011).
18. T. Yoshimura, *Thin-Film Organic Photonics: Molecular Layer Deposition and Applications*, CRC/Taylor & Francis, Boca Raton, Florida, (2011).
19. T. Aizawa, Furuuchi Chemical Corporation Technical Articles for PLZT Shutter Arrays.
20. N. Fujimoto, H. Tomita, and T. Yoshimura, "A Path-Independent-Loss Type Optical Cross-Connect System Using High-Speed Waveguide Prism Micro-Optical Switches in WDM Networks," *Research Reports of the School of Engineering*, Kinki University, **40**, 81–85 (2006).
21. E. Yablonovitch, "Inhibited Spontaneous Emission in Solid State Physics and Electronics," *Phys. Rev. Lett.* **58**, 2059–2062 (1987).
22. T. Yoshimura, Y. Suzuki, N. Shimoda, T. Kofudo, K. Okada, Y. Arai, and K. Asama, "Three-Dimensional Chip-Scale Optical Interconnects and Switches with Self-Organized Wiring Based on Device-Embedded Waveguide Films and Molecular Nanotechnologies," *Proc. SPIE* **6126**, from Photonics West 2006, San Jose, California, 612609-1-15 (2006).

Chapter 10

OE Hardware Built by Optical Electronics

Optical Electronics will realize a variety of self-organized and 3-D OE hardware combining the core technologies such as SOLNET, 3-D optical circuits with stacked waveguide films, PL-Pack with SORT, and high-speed/small-size optical switches.

In this chapter, some of the examples of the OE hardware are presented, including "optical solder," "optical wiring in free spaces," "optical interconnects within boxes," and "optical switching systems."

10.1 Optical Solder

10.1.1 *One-Beam-Writing SOLNET*

The capability of self-aligned optical coupling in SOLNET realizes "optical solder" [1–3]. Two optical devices can be connected by putting the optical solder betweent them.

Figure 10.1 shows examples of the optical solder using one-beam-writing SOLNET. For optical waveguide-PD coupling in a conventional configuration without SOLNET, as shown in Fig. 10.1(a), light beams coming out from a waveguide core are reflected by a 45° mirror to be led to PD through a free space. This causes

Optical Electronics: Self-Organized Integration and Applications
Tetsuzo Yoshimura

ISBN 978-981-4310-90-1 (Hardcover), 978-981-4364-08-9 (eBook)
www.panstanford.com

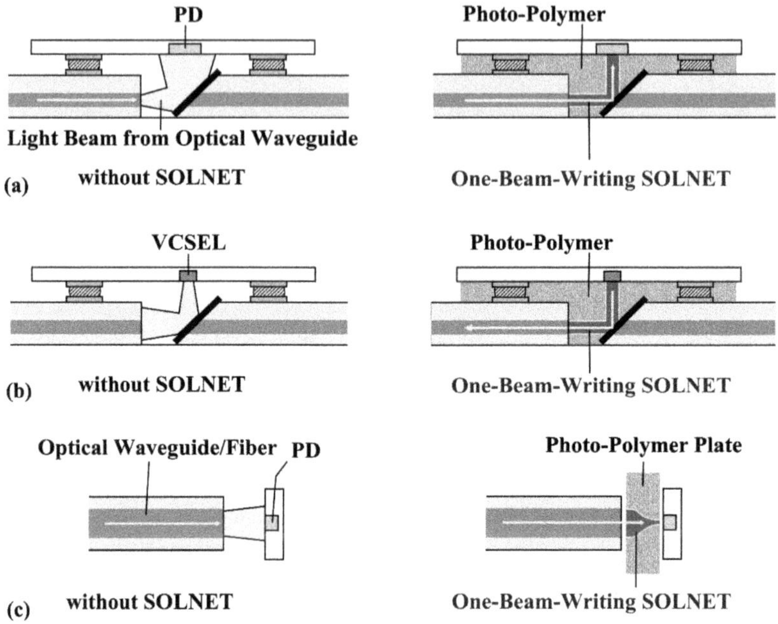

Figure 10.1. Optical solder using one-beam-writing SOLNET. Reprinted with permission from T. Yoshimura, J. Roman, Y. Takahashi, W. V. Wang, M. Inao, T. Ishitsuka, K. Tsukamoto, K. Motoyoshi, and W. Sotoyama, "Self-Organizing Waveguide Coupling Method "SOLNET" and Its Application to Film Optical Circuit Substrates," *Proc. 50th Electron. Components Technol. Conf. (ECTC)*, 962–969. © 2000 IEEE.

the expansion of light beam diameter, which sometimes makes it difficult to expose the PD active area to the light beams efficiently. By using one-beam-writing SOLNET, a write beam emitted from the waveguide core constructs a 3-D waveguide in the free space. The constructed light path will suppress the beam expansion to increase the coupling efficiency.

Similarly, as shown in Fig. 10.1(b), in optical waveguide-VCSEL coupling, light beams coming out from a VCSEL are reflected by a 45° mirror to be led to an optical waveguide through a free space. One-beam-writing SOLNET constructed by a write beam emitted from the VCSEL suppresses the beam expansion. In this case, the photo-polymer should be sensitized by dye molecules for the wavelength of the emitted light beams from the VCSEL.

Figure 10.1(c) shows optical waveguide (or fiber)-PD connector having a photo-polymer plate for optical solder. When an optical waveguide with a large core size is connected to a small size PD for high-speed operation, it may cause a decrease in coupling efficiency. By using one-beam-writing SOLNET, the beams can be shrunk down to a size less than PD, enabling a coupling efficiency increase and a cross talk reduction. The photo-polymer plate is detachable, so, it can be changed to new one when a new combination of optical devices is required.

10.1.2 *Two-Beam-Writing SOLNET*

Figure 10.2 shows optical solder using two-beam-writing SOLNET [1–3]. For optical waveguide-VCSEL coupling shown in Fig. 10.2(a), light beams emitted from a VCSEL are led to an optical waveguide by a 45° mirror. In the case that misalignment exists between the optical waveguide and the VCSEL, one-beam-writing SOLNET does not work. To solve the problem, two-beam-writing SOLNET is

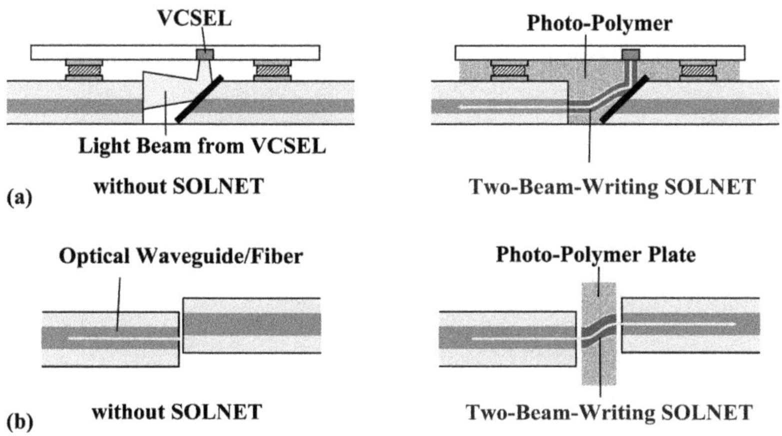

Figure 10.2. Optical solder using two-beam-writing SOLNET. Reprinted with permission from T. Yoshimura, J. Roman, Y. Takahashi, W. V. Wang, M. Inao, T. Ishitsuka, K. Tsukamoto, K. Motoyoshi, and W. Sotoyama, "Self-Organizing Waveguide Coupling Method "SOLNET" and Its Application to Film Optical Circuit Substrates," *Proc. 50th Electron. Components Technol. Conf. (ECTC)*, 962–969. © 2000 IEEE.

effective because of its automatic self-aligned light path construction capability between optical devices. The two-beam-writing SOLNET can be formed by emitting write beams from both of the optical waveguide and the VCSEL.

Figure 10.2(b) shows a schematic illustration of an optical connector with a detachable photo-polymer plate for connections between optical waveguides and optical fibers. Misalignment between the two optical devices causes loss and cross talk. By inserting a photo-polymer plate between them, a self-aligned light path of two-beam-writing SOLNET connecting the two optical devices can be made.

10.1.3 *R-SOLNET*

Figure 10.3 shows examples of optical solders using R-SOLNET for optical couplings involving devices that cannot emit write beams

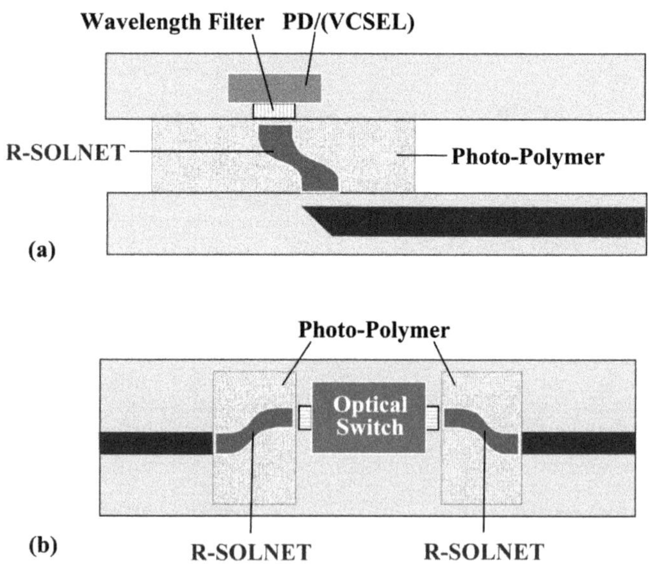

Figure 10.3. Optical solder using R-SOLNET. Reprinted with permission from T. Yoshimura, A. Hori, Y. Yoshida, Y. Arai, H. Kurokawa, T. Namiki, and K. Asama, "Coupling Efficiencies in Reflective Self-Organized Lightwave Network (R-SOLNET) Simulated by the Beam Propagation Method," *IEEE Photon. Technol. Lett.* **17**, 1653–1655. © 2005 IEEE.

such as PDs, VCSELs, and optical switches [4–6]. Wavelength filters are formed on PD surfaces, VCSEL surfaces, and waveguide edges of optical switches to pull the SOLNET to the wavelength filter locations.

10.2 Optical Wiring in Free Spaces

10.2.1 *Free-Space Optical Interconnects*

Figure 10.4(a) shows an example of waveguide construction in a 2-D free space by one-beam-writing SOLNET. Wavelength filters are put in a free space with photo-polymer between optical waveguides. Write beams are introduced from optical waveguides to construct

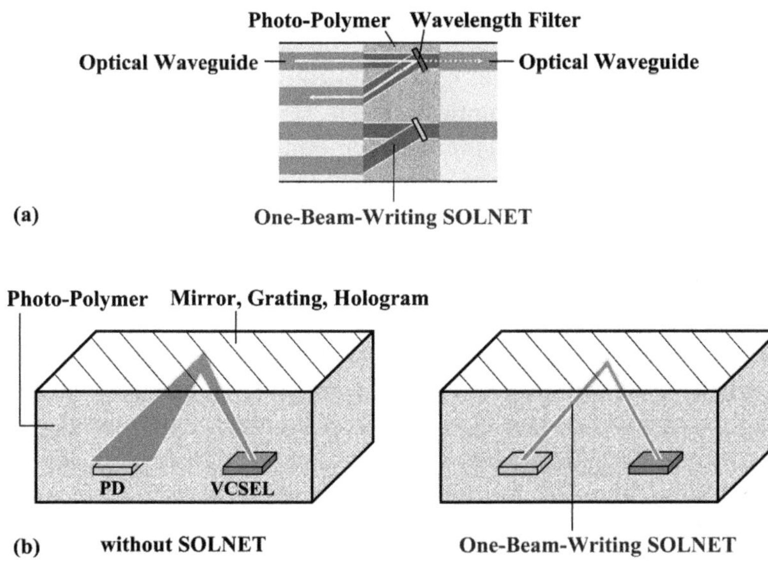

Figure 10.4. Free space optical wiring using one-beam-writing SOLNET in (a) an optical circuit with wavelength filters and (b) a free-space optical interconnects. Reprinted with permission from T. Yoshimura, J. Roman, Y. Takahashi, W. V. Wang, M. Inao, T. Ishitsuka, K. Tsukamoto, K. Motoyoshi, and W. Sotoyama, "Self-Organizing Waveguide Coupling Method "SOLNET" and Its Application to Film Optical Circuit Substrates," *Proc. 50th Electron. Components Technol. Conf. (ECTC)*, 962–969. © 2000 IEEE.

one-beam-writing SOLNET. By partial reflection of the write beams at the wavelength filters a prularity of SOLNET grow from the wavelength filters. This enables us to insert optical devices such as filters into optical circuits without cutting or etching processes.

In Fig. 10.4(b), an application of one-beam-writing SOLNET to free-space optical interconnects is shown [1, 2]. In conventional free-space systems, light beams propagate in a free space with a path control by mirrors, gratings, and holograms as well as relay lenses. In this case, sophisticated assembly of the optical devices is necessary for coupling efficiency increases and cross talk reduction. If 3-D waveguide networks are formed by SOLNET in a free space, relay lenses, which keep the diameters of propagating light beams small enough, can be removed, building simple 3-D optical interconnect systems.

10.2.2 *Optical Z-Connections in 3-D Optical Circuits*

Figure 10.5 shows free space optical wiring in an optical Z-connection of a 3-D optical circuit [7–11]. Photo-polymer is placed in a free space region between optical waveguide films. By introducing a write beam from one of the vertical mirrors as shown in Fig. 10.5(a), one-beam-writing SOLNET is formed, connecting the upper and the lower optical waveguides. When write beams are introduced from both vertical mirrors as shown in Fig. 10.5(b), two-beam-writing SOLNET is formed. In this case, even if some misalignment exists between the upper and the lower optical waveguides, a self-aligned vertical waveguide can be formed. By depositing a wavelength filter at a vertical mirror apperture of one of the optical waveguide films, R-SOLNET can be formed as shown in Fig. 10.5(c). The wavelength filter selectively reflects write beams (usually wavelength is in a range of 0.35–0.8 μm) and transmits signal beams (usually is in a range of 0.4–1.5 μm). This enables us to construct self-aligned vertical waveguide between the two stacked waveguides in one-beam writing configurations. The R-SOLNET is especially useful when one of the coupled waveguides cannot emit write beams, for example, optical waveguides in inner regions of optical circuits. The theoretical analysis of a model similar to that shown in Fig. 10.5(c) is described in Section 4.4.5.

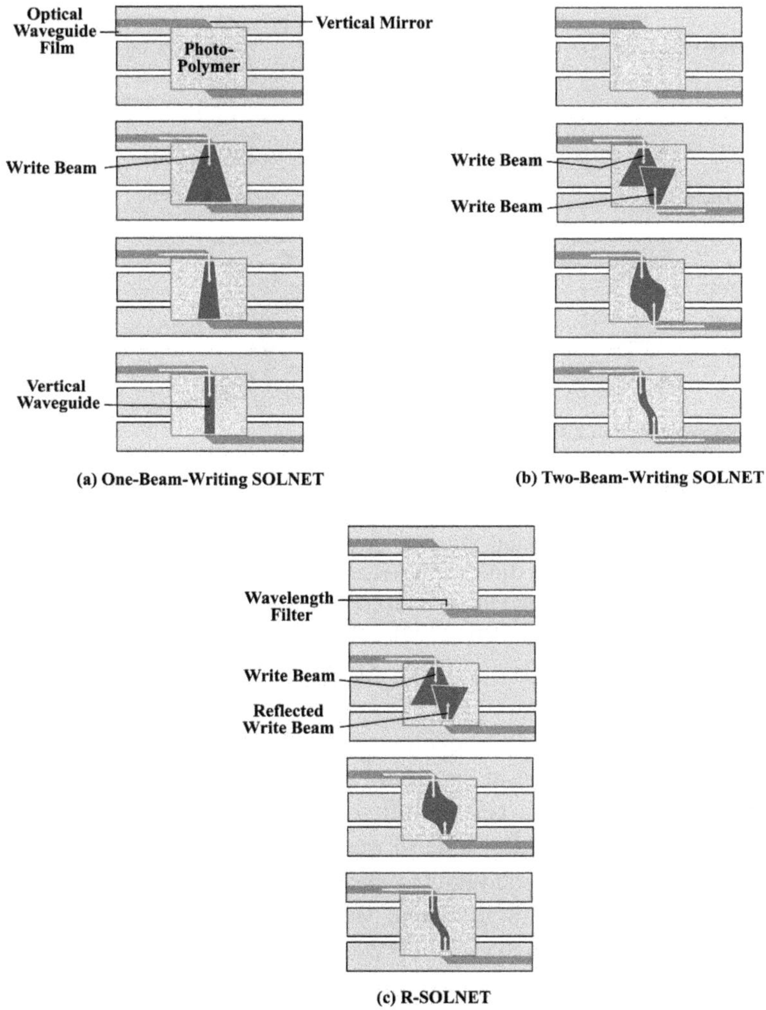

Figure 10.5. Free space optical wiring in optical Z-connections fabricated by (a) one-beam-writing SOLNET, (b) two-beam-writing SOLNET, and (c) R-SOLNET. Reprinted with permission from T. Yoshimura, T. Inoguchi, T. Yamamoto, S. Moriya, Y. Teramoto, Y. Arai, T. Namiki, and K. Asama, "Self-Organized Lightwave Network Based on Waveguide Films for Three-Dimensional Optical Wiring Within Boxes," *J. Lightwave Technol.* **22**, 2091–2100. © 2004 IEEE.

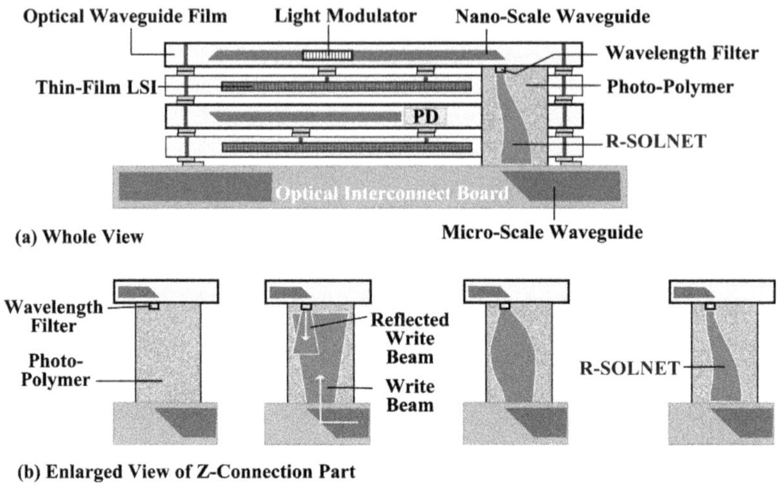

Figure 10.6. Free space optical wiring in optical Z-connection between a micro-scale waveguide and a nano-scale waveguide in a 3-D OE platform. Reprinted with permission from T. Yoshimura and H. Kaburagi, "Self-Organization of Coupling Optical Waveguides by the "Pulling Water" Effect of Write Beam Reflections in Photo-Induced Refractive-Index Increase Media," *Proc. SPIE* **7221**, 722111-1-10 (2009). See also Color Insert.

Another example of waveguide construction in a 3-D free space by R-SOLNET is shown in Fig. 10.6 [10,11]. Films containing nano-scale waveguides with vertical mirrors are stacked together with films containing thin-film LSIs to make a 3-D OE platform. In the films, flakes of light modulators and PDs are embedded for E-O and O-E conversions. Self-aligned vertical waveguides for optical Z-connections are constructed between micro-scale waveguides in an optical interconnect board and the nano-scale optical waviguides in intra-LSI optical circuits by the following process. At the vertical mirror aperture of the nano-scale waveguide, a wavelength filter is deposited. By introducing a write beam from the micro-scale waveguide into the free space filled with photo-polymer via a vertical mirror, R-SOLNET is formed between the micro-scale waveguide and the nano-scale waveguide even when misalignment exists between them.

10.3 Integrated Optical Interconnects within Boxes

With an increase in clock rates and I/O counts of processors, interconnection bottlenecks appear even in circuit-substrate-level interconnects within computers. Optical interconnects are promising to overcome these bottlenecks, providing the following advantages: high frequency, high density, low RC delay, low power dissipation, low noise, open/short acceptability, and so on. From this background, attempts to introduce optical waveguides into circuit substrates have been made [12–17].

"Film/Z-connection" technology for high-density flexible circuit substrates was developed in Fujitsu Computer Packaging Technologies, Inc. (FCPT) [18]. Multi-layer circuit substrates are constructed by stacking films with fine metal patterns and vias. The technology realizes large freedom in substrate structure designs and fabrication processes.

By applying the electrical "Film/Z-connection" to S-FOLM, each device-embedded film element can be made separately, enabling independent process/material optimization without restriction from other film elements. This increases process/material candidates for the film elements in S-FOLM. Both-side processing availability, which is a great feature of the "Film/Z-connection," increases process freedom, for example, making it easy to embed OE devices into films, to reverse the film elements upside-down, and so on. Pre-testing capability for the film element before stacking may improve S-FOLM yield.

In the present section, applications of S-FOLM based on the "Film/Z-connection" technology to optical interconnects within boxes are proposed.

10.3.1 *Optical Interconnects Based on "Film/Z-Connection" Technology*

10.3.1.1 Future image of "within boxes"

Figure 10.7 shows a possible future image of "within boxes" of high-performance computers [19]. E-O conversion and O-E conversion

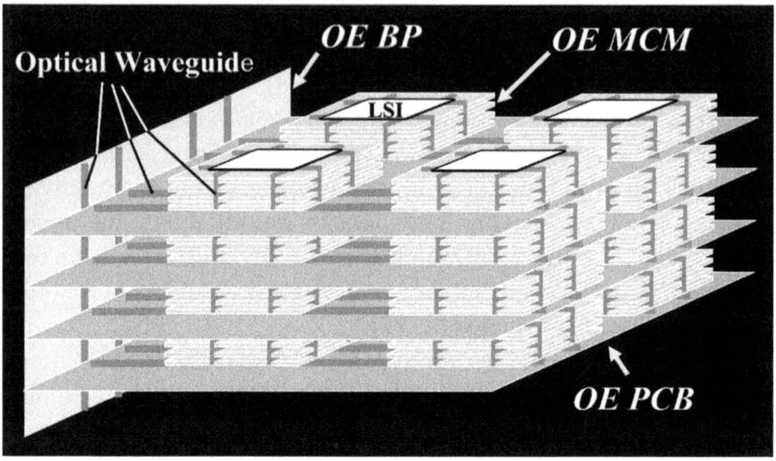

Figure 10.7. Future image of "within boxes" of high-performance computers.

are done at outputs and inputs of LSI, respectively. Optical signals are transmitted from LSI outputs, through optical waveguides of OE MCM, OE PCB and OE BP to LSI inputs.

For extremely high-speed machine cycles, the length of signal propagation during one clock period will get close to size of LSI chip. In this case, to reduce the signal propagation delay between LSI chips, MCM should have a 3D-stack structure as shown in Fig. 10.7, in which the stacked LSI chips are connected through vertical waveguides as well as planar waveguides.

10.3.1.2 Concept of optical interconnects based on "Film/Z-connection" technology

Figure 10.8 shows conceptual illustration of "Film/Z-connection" technology. The upper picture is the original idea for electrical substrates developed in FCPT [18]. First, films with fine electrical lines and pads on both sides as well as vias are made (in the figure electrical lines and vias are not drawn). Then the films are stacked by Z-connection technology to construct multi-layer circuit substrates.

By replacing one or both of the electrical layers by OE films, as shown in Fig. 10.8 (bottom), an OE substrate, in which electrical

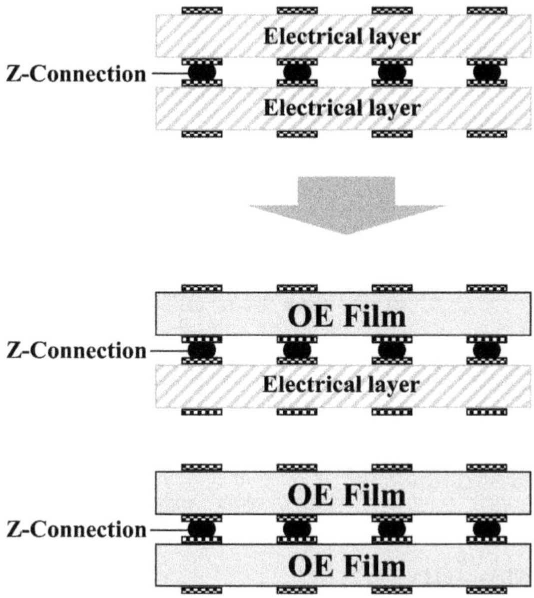

Figure 10.8. Conceptual illustration of Film/Z-connection technology.

and optical interconnects coexist, can be constructed [1, 2, 19–28]. This corresponds to the concept of S-FOLM described in Section 2.2.2. The most significant advantage of the OE substrates with "Film/Z-connection" compared with the conventional "build-up" is large freedom in substrate structures and fabrication processes, enabling us to introduce various functions into substrates by combining OE films. Each type of the OE film can be optimized independently of other types of OE films. Versatility improvement is expected.

Figure 10.9 shows various types of the OE films.

- OE-film-*W*: Ordinary optical waveguide film with metallization.
- OE-film-*D*: OE devices such as VCSEL, PD, light modulator, and optical switch are embedded in a polymer/glass film with metallization.
- OE-film-*DW*: OE devices are embedded in an optical waveguide film with metallization. (V) and (M) indicate that transmitters are VCSELs and light modulators, respectively.

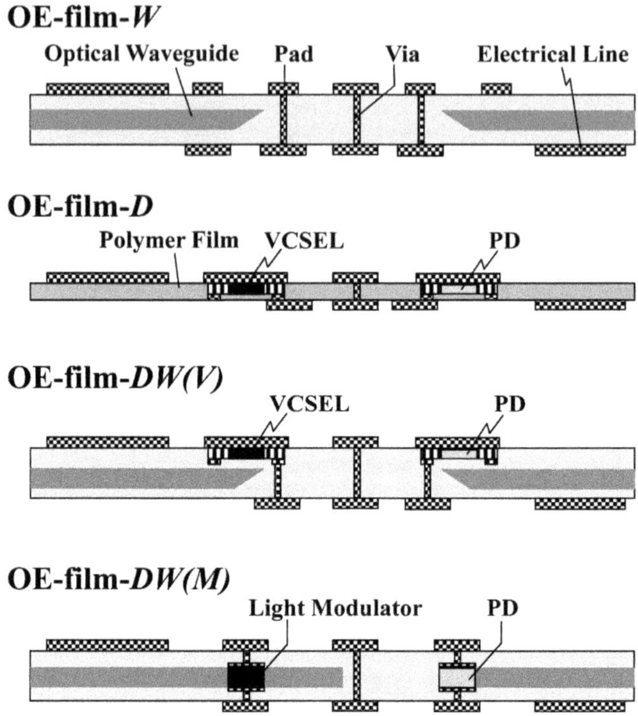

Figure 10.9. Types of OE-films.

It should be noted that the OE films contain electrical lines, pads, and vias, meaning that OE films also have electrical substrate functions. By combining these OE films with electrical layers, various kinds of S-FOLM will be realized according to system requirements. This indicates that using the "Film/Z-connection" technology, all levels of circuit substrates for optical interconnects can be provided with minimum film elements.

10.3.2 *OE Substrates of S-FOLM using "Film/Z-Connection"*

Figure 10.10(a) shows an example of OE PCB and OE BP, which are built from S-FOLM using the "Film/Z-connection" technology. OE PCB is constructed by stacking OE-film-*DW(V)* on electrical board. OE devices and optical waveguides are coupled by 45° mirrors in the optical waveguides. OE BP is constructed by stacking OE-film-*W* on electrical board. OE BP and OE PCB are coupled by 45° mirrors in the optical waveguides. Z-connections have two roles, namely, electrical connections and mechanical supports between OE films and electrical boards. The gap area between films and boards might be filled by dielectric materials if necessary. In this model, OE PCB and OE BP can be OE MCM and OE PCB, respectively. The OE films and the electrical boards are fabricated separately. Therefore, materials and processes for OE films can independently be optimized, not being restricted by the electrical board condition.

Figure 10.10(b) shows an example of OE-film/OE-film stacking, which constructs an OE MCM/ PCB stacked structure. OE MCM consists of OE-film-*DW(V)*, and OE PCB consists of OE-film-*W*. OE MCM and OE PCB are coupled by 45° mirrors. By electrical output signals from an LSI chip, VCSEL in the OE MCM is driven. The emitted light beams from the VCSEL are coupled to the optical waveguide in the OE MCM. The optical signals propagate to a PD in the same OE MCM or go out to OE PCB through back-side connection. Only OE-film-based technology may be able to realize such back-side connection. The PD converts the optical signals to electrical input signals for an LSI chip.

Figure 10.10(c) shows an example of both-side mounting. OE-film-*Ds* are stacked on both sides of OE-film-*W* to construct both-side OE PCB. OE PCBs on the opposite sides can also be connected by both-side OE BP consisting of OE-film-*W*. In this model, OE PCB and OE BP can be OE MCM and OE PCB, respectively. The both-side mounting enables the butterfly-structure, which is important in multi-processor systems. The both-side mounting availability is a significant advantage of the "Film/Z-connection" approach since it is difficult in conventional built-up OE substrates.

Smart pixel, consisting of VCSEL/PD-array chips bonded onto CMOS IC, has received much attention [29, 30]. The OE-film-*D* plays the same role as the array chip. The OE-film-*D* has the following advantages over the VCSEL/PD-array chip: Semiconductor devices exist only at the places where they are necessary in the OE-film-*D*. The remaining area is occupied by polymer film. This is cost effective from the view point of expensive semiconductor material consumption. Furthermore, processability for via/pad/electrical

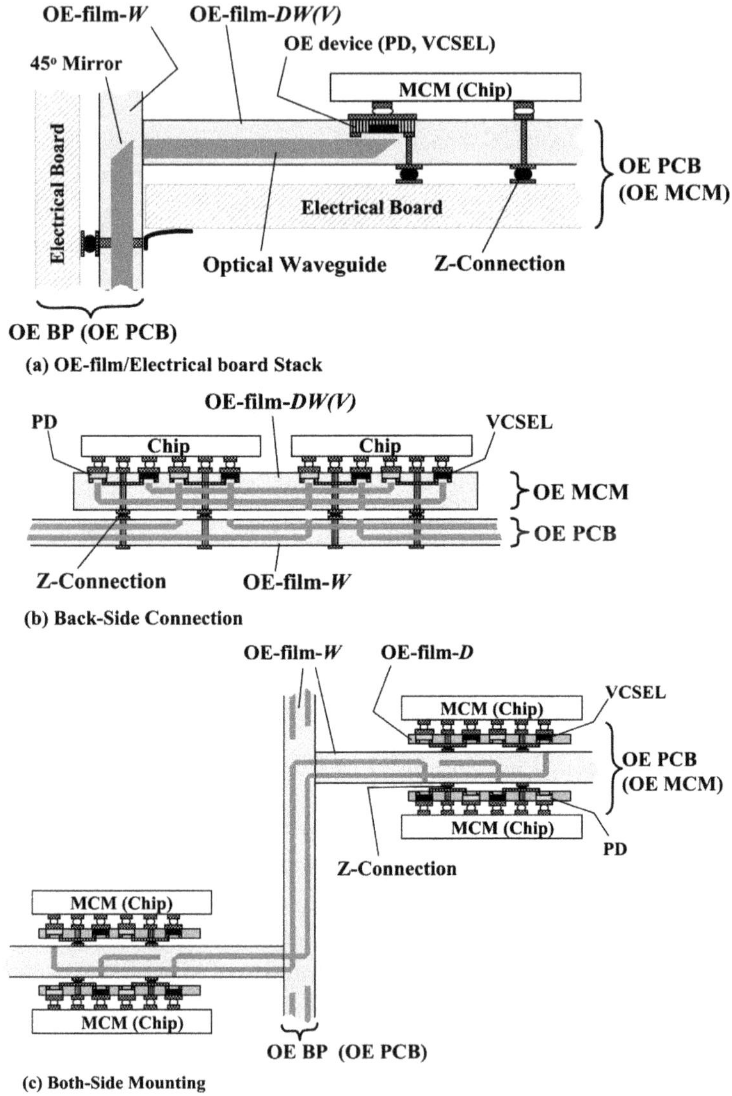
OE-film-W
OE-film-DW(V)
OE device (PD, VCSEL)
45° Mirror
MCM (Chip)
Electrical Board
OE PCB
(OE MCM)
Electrical Board
Optical Waveguide
Z-Connection
OE BP (OE PCB)
(a) OE-film/Electrical board Stack
OE-film-DW(V)
PD
VCSEL
Chip
Chip
OE MCM
OE PCB
Z-Connection
OE-film-W
(b) Back-Side Connection
OE-film-W
OE-film-D
VCSEL
MCM (Chip)
OE PCB
(OE MCM)
MCM (Chip)
PD
Z-Connection
MCM (Chip)
MCM (Chip)
OE BP (OE PCB)
(c) Both-Side Mounting

Figure 10.10.

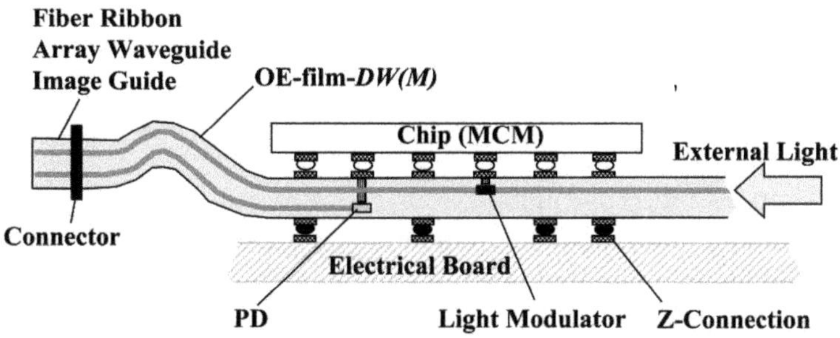

(d) Micro Optical Link Module

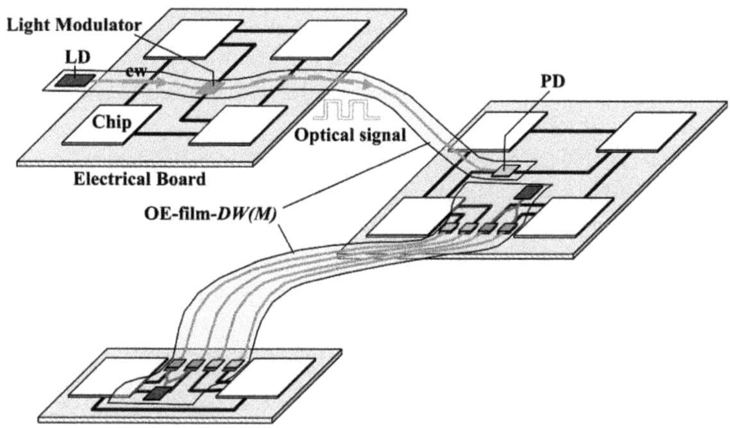

(e) OE Tap Guide

Figure 10.10. Examples of OE substrate structures built from S-FOLM using "Film/Z-connection." Reprinted with permission from T. Yoshimura, J. Roman, Y. Takahashi, M. Lee, B. Chou, S. I. Beilin, W. V. Wang, and M. Inao, "Optoelectronic Scalable Substrates Based on Film/Z-Connection and Its Application to Film Optical Link Module (FOLM)," *Proc. SPIE* **3952**, 202–213 (2000).

line formation is very high. This indicates that OE-film-*D* can contribute to realizing low-cost smart pixels.

Figure 10.10(d) shows an example of micro optical link modules that enable direct jump from LSI chips. In this example, OE-film-*DW(M)* is stacked on an electrical board. Light modulators act as transmitters. External light (cw or pulse train) is modulated by outputs of LSI through the modulators to generate optical signals. If external light of very high power is applied, high-speed and low-heat-generation OE amplifier/driver-less substrate (OE-ADLES) [31–33] described in Section 10.3.5 may be available. OE-film-*DW(V)*, and a combination of OE-film-*D* and OE-film-*W* can also be used for the module. By the micro optical link module, optical signals can directly be extracted from LSIs and transmitted to any levels of units, including to-board and to-shelf. In this model, MCMs can be used instead of LSIs.

Figure 10.10(e) shows examples of OE tap guides using OE-ADLES. OE-film-*DW(M)* is attached to electrodes/pads on an electrical board to connect the electrodes/pads to the light modulators in the OE-film-*DW(M)*. Cw or pulse train lights from LDs are modulated by the electrical signals from the electrodes/pads through the light modulators to generate optical signals. The optical signals are transmitted through optical waveguides to PDs and are converted to electrical signals to be introduced into electrodes/pads of other places. By OE tap guides, optical signals can directly be extracted from electrical substrates and transmitted to any levels of units, and to any places in the same substrate or other substrates.

Figure 10.11 shows possible examples of S-FOLM for WDM optical interconnects. WDM transceiver illustrated in Fig. 10.11(a) can be used in optical backbones, massive parallel optical links, and OE BPs. An OE-film-*DW(V)*, in which VCSELs and PDs for wavelength λ_1 and λ_2 are embedded together with optical waveguides, is attached on an electrical board by an electrical Z-connection. On the OE film, an interface film containing thin-film interface ICs with transceiver circuits for the VCSELs/PDs is attached. The optical waveguides in the OE film are connected to 2-D optical fiber array (or image guide). Power/GND and electrical signals are supplied from the electrical board through electrical via/pad/electrode of the films. Surface normal couplers with wavelength-filter function,

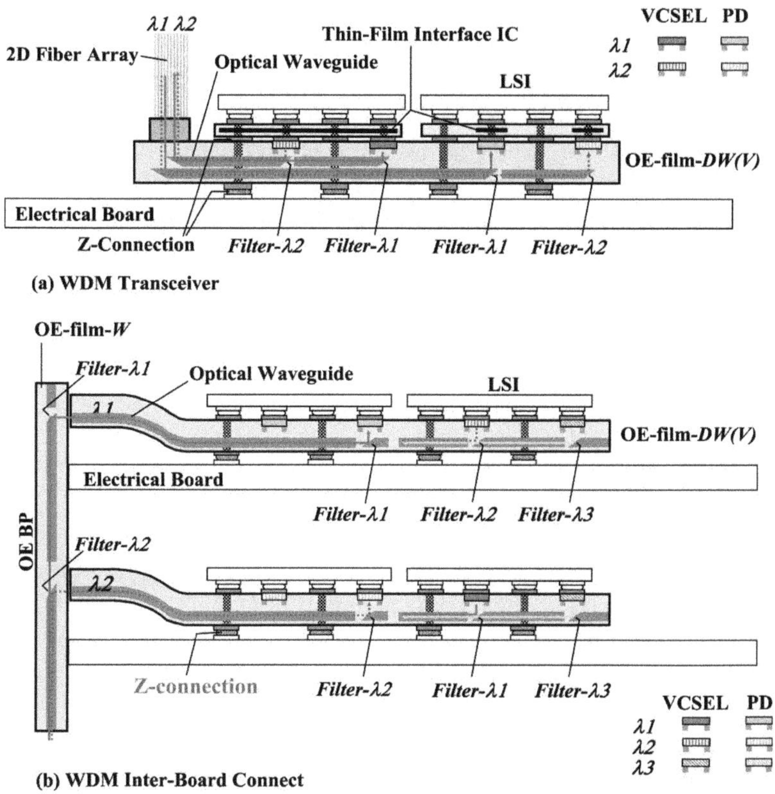

Figure 10.11. Examples of S-FOLM for WDM optical interconnects. Reprinted with permission from T. Yoshimura, J. Roman, Y. Takahashi, M. Lee, B. Chou, S. I. Beilin, W. V. Wang, and M. Inao, "Optoelectronic Scalable Substrates Based on Film/Z-Connection and Its Application to Film Optical Link Module (FOLM)," *Proc. SPIE* **3952**, 202–213 (2000).

which operate as wavelength MUX and DEMUX, achieve optical coupling between the OE devices and optical waveguides. For single wavelength use, the surface normal couplers can be 45° mirrors.

The interface film gives standardized-interface characteristics to the transceiver. Namely, by embedding appropriate thin-film interface ICs into the film, S-FOLM I/O can freely be adjusted to the I/O specs of LSIs to be mounted on the S-FOLM. The interface film can involve many functions such as MUX/DEMUX, error correction, timing controlling, and so on. The thin-film interface IC has through vias or is divided into small pieces for individual pad connection. Another remarkable feature of the WDM transceiver utilizing S-FOLM is flexibility in structures and positions. S-FOLM can be attached at any places with any shapes/sizes on a substrate.

Figure 10.11(b) shows an example of S-FOLM for WDM inter-board connects. An OE-film-*DW(V)*, in which VCSELs and PDs for wavelength λ_1, λ_2, and λ_3 are embedded, is attached on an electrical board by electrical Z-connection, forming an OE board (for simplicity, interface films are neglected in the figure). Surface normal couplers of wavelength filters in OE BP optically connect the OE BP to S-FOLM in the OE board. The couplers act as wavelength MUX/DEMUX.

10.3.3 *Three-Dimensional OE Platforms of S-FOLM using "Film/Z-Connection"*

In 3-D OE platform, 3-D architectures and optical wiring are implemented. Figure 10.12 shows typical examples of the 3-D OE platform for integrated optical interconnects within boxes. The multilayer OE board consists of a plurality of OE films-*W* with vertical mirrors,

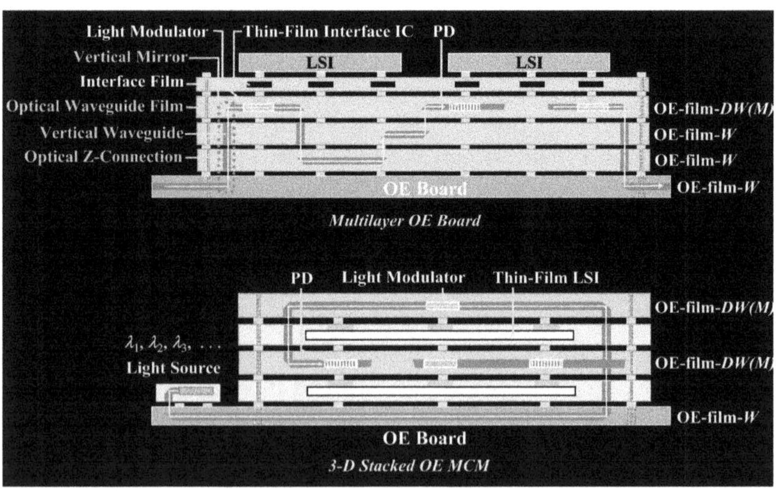

Figure 10.12. Three-dimensional OE platform for integrated optical interconnects within boxes. See also Color Insert.

an OE film-*DW* with embedded light modulators/VCSELs/PDs, and an interface film with thin-film interface ICs. Optical links between the optical waveguide films are carried out by optical Z-connections with micro-lenses or vertical waveguides. The second layer from the top, where thin-film active devices are embedded, is used for E-O and O-E conversions. The third and fourth layers are used for multilayer optical signal routing that avoids wiring interference in high-density interconnect systems. The first film, in which thin-film interface ICs are embedded, is an interface film, on which LSIs are mounted. Electrical signals from output pads of LSIs are converted into optical signals by light modulators or VCSELs. The optical signals are guided to PDs, where, they are converted to electrical signals for input pads of LSIs. The interface approach permits computer designers to build electrical architectures without knowledge of optics.

When we use OE-ADLES, where E-O and O-E conversions are respectively carried out by light modulators directly driven by LSI output and by PDs directly generating LSI input as described in Section 10.3.5, the interface film is not necessary. OE-ADLES is expected to have "low power dissipation" and "high data rate" capability.

In 3-D stacked OE MCM [1, 2, 19–28, 33], light modulator/PD-embedded OE films-*DW(M)* are stacked together with films containing thin-film LSIs. Light sources are placed outside of the MCM. Optical signals, which are generated at an E-O conversion site of the OE film by LSI output, propagate to an O-E conversion site, being converted into electrical signals for LSI input. In some cases, the optical signals are coupled to optical waveguides of the OE board by the back-side connection to be transmitted to other 3-D stacked OE MCMs.

The structure can be regarded as 3-D stacked OE LSI when the stacked thin-film LSIs are divided chips corresponding to blocks of an LSI chip. This structure is a kind of the goal of 3-D OE platform for computers because it realizes "3-D wiring" and "optical wiring" simultaneously. In the 3-D stacked OE MCM, heat generation is a serious problem. So, the implementation of OE-ADLES that uses high-power external light sources of cw or pulse train is mandatory for power dissipation reduction within the 3-D OE platform.

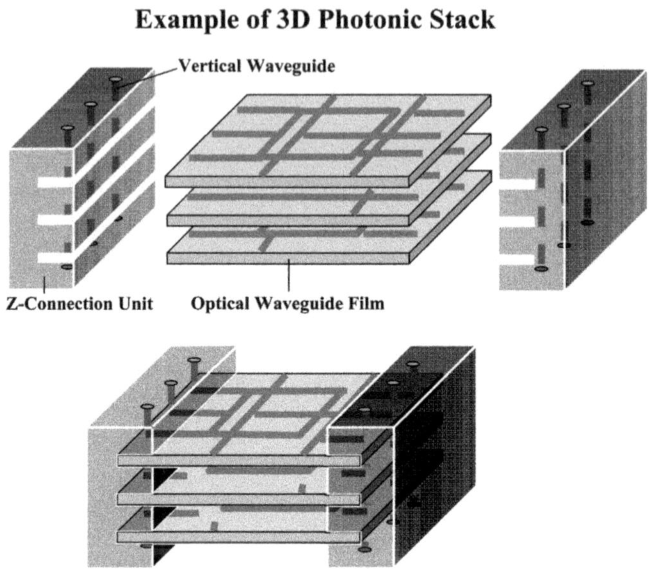

Figure 10.13. An example of 3-D optical circuits consisting of optical waveguide films and Z-connection units for WDM.

By using external light sources with a plurality of wavelengths and by replacing the vertical mirrors with wavelength filters, WDM interconnects can be performed.

Figure 10.13 shows an example of 3-D optical circuit for WDM [22–28]. Optical waveguide films are stacked. They are supported by Z-connection units. In the Z-connection units vertical waveguides are formed, and the optical waveguide films are inserted into gaps in the units.

Here, the important point is the connect point between the optical waveguide in the film and the vertical waveguide in the Z-connection unit. Figure 10.14 shows the examples of the structures for multiplexing and de-multiplexing in WDM optical interconnects [22–28]. 45° wavelength filters are embedded in the optical waveguide films. By embedding semitransparent-mirrors or gratings, 1:N fan-out or N:1 coupling can be achieved. In the case of through, holes are placed instead of wavelength filters, mirrors, and gratings. Figure 10.15 shows examples of signal flows in WDM optical interconnects. At cross points, wavelength filters are formed

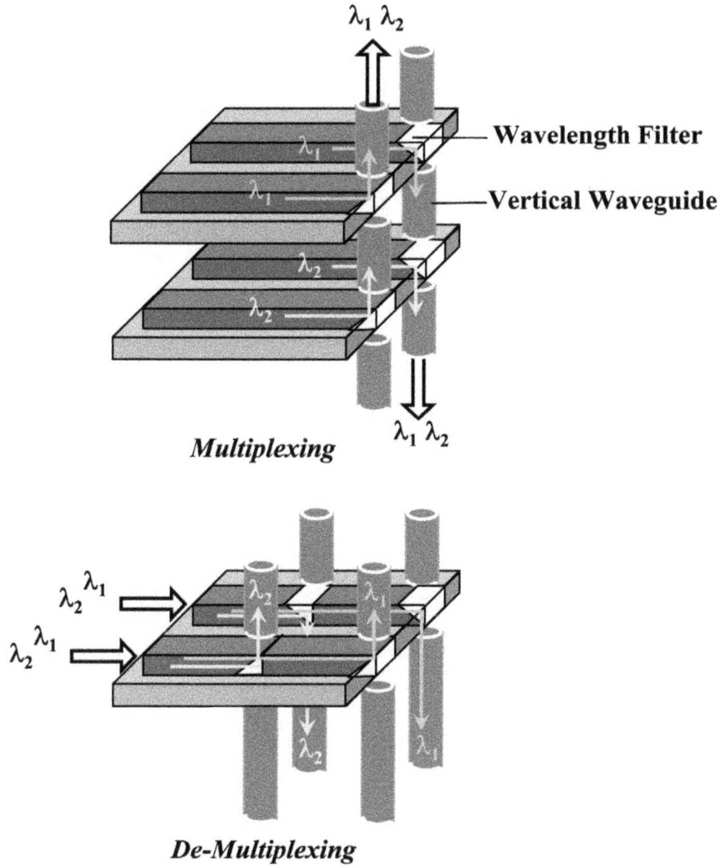

Figure 10.14. Examples of multiplexing and de-multiplexing in WDM optical interconnects.

in the optical waveguide films to achieve wavelength-dependent designated routings.

The Z-connection units can be fabricated by constructing vertical waveguides using build-up process like clad/core (patterning)/clad/..., followed by forming gaps by, for example, laser cut, dicing saw cut. As shown in Fig. 10.16, the Z-connection units can also be fabricated using SOLNET [22–28]. In this case, for the vertical waveguide materials, PRI materials like the DuPont's hologram material are used. By exposing the PRI material to light coming

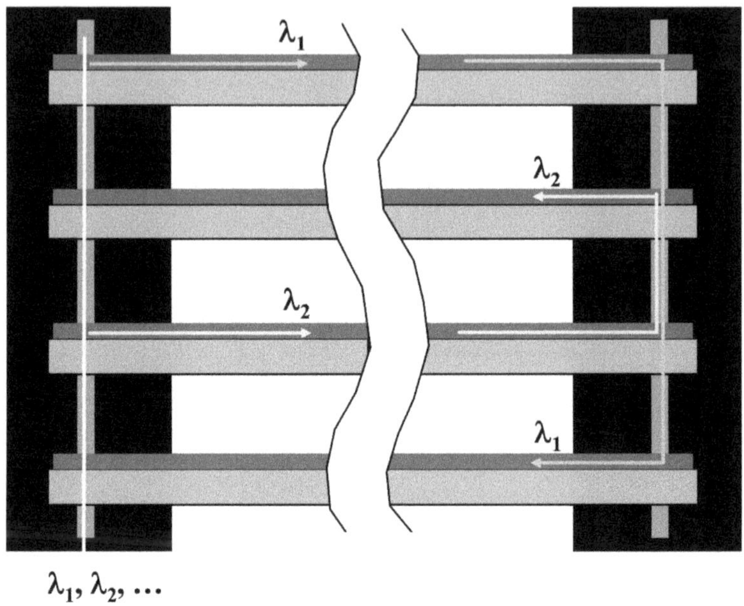

Figure 10.15. Examples of wavelength-dependent signal routings in WDM optical interconnects.

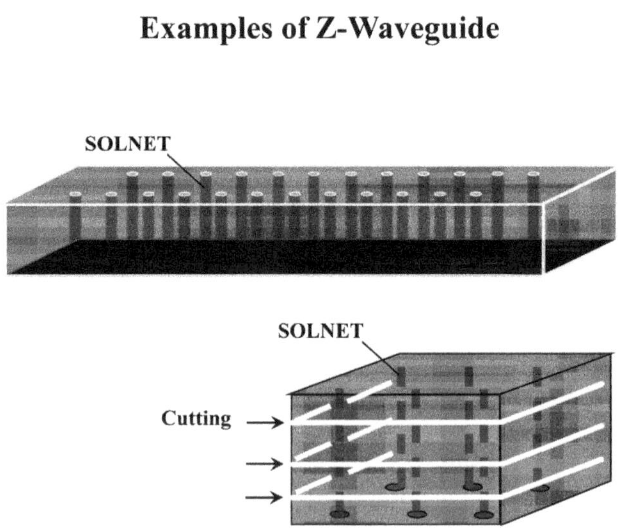

Figure 10.16. Fabrication of Z-connection units using SOLNET.

through mask with a hole array pattern, a vertical waveguide array is automatically constructed along the light propagating direction due to the self-focus effect.

WDM is the final goal for optical interconnects within boxes. By using light modulators, optical switches, and wavelength filters, including tunable filters, flexible optical networks can be constructed in computers with minimized connection line counts as disclosed in [3].

10.3.4 *Fabrication of Optical Waveguide Film/PCB Stacks*

Figure 10.17 shows a process for optical waveguide film preparation and stacking on PCB. Fluorinated polyimide (Hitachi Chemical Co.) was used for waveguide core/clad materials. Refractive indices of clad and core are ~1.52 and ~1.54, respectively. On an Al substrate, under-clad layer (15 μm thick) and core layer (30 μm thick) were successively formed. Multimode waveguide core array of 250 μm pitch were formed by reactive ion etching (RIE) with oxygen/argon mixture. Typical core size is 30 μm × 30 μm. Over-clad layer was

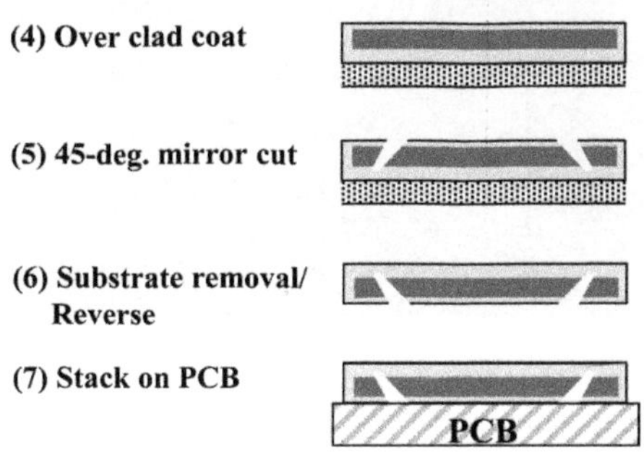

Figure 10.17. Process for optical waveguide film preparation and stacking on PCB.

coated on it. Waveguide film size is 6 inch × 6 inch. Maximum film thickness is ~60 μm.

By UV (THG)-YAG laser cutting, 45° mirrors in the optical waveguides were formed. At this step, the optical waveguide film is still on the Al substrate. To make an angle around 45° for the vertical mirror, the laser incident angle was tilted against the waveguide film surface normal. Depending on applications, we should make outward mirrors or inward mirrors in the optical waveguides. For example, for the application shown in Fig. 10.10(a), inward mirrors are necessary. However, in the "Film/Z-connection" approach, it does not matter which type of mirrors are formed at this step, because we can reverse the film upside-down in the lamination step mentioned below. This is one of the merits of adopting "Film/Z-connection." In the present experiment, since it is easier to make outward mirrors than inward mirrors, we made outward mirrors.

Then the Al substrate was removed to obtain a free-standing waveguide film. The film was reversed upside-down and stacked on PCB with a lamination process. As a result, we obtained an optical waveguide film with inward 45° mirrors on PCB.

Figure 10.18 shows a photograph of a 6′× 6′ optical waveguide film. Figure 10.19(a) shows a cross section of the optical waveguide

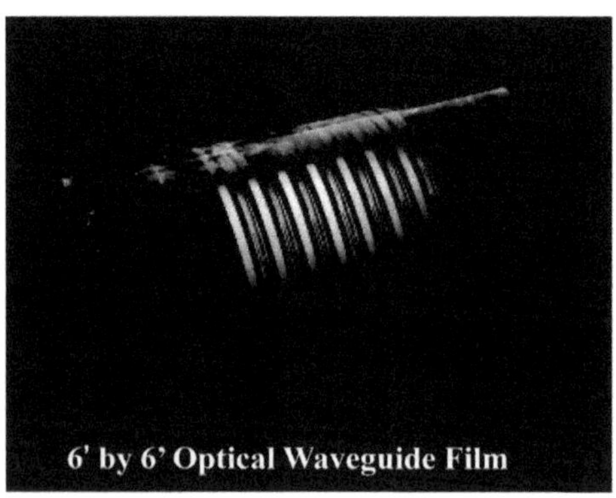

Figure 10.18. Optical waveguide film of 6′× 6′ Reprinted with permission from T. Yoshimura, J. Roman, Y. Takahashi, M. Lee, B. Chou, S. I. Beilin, W. V. Wang, and M. Inao, "Optoelectronic Scalable Substrates Based on Film/Z-Connection and Its Application to Film Optical Link Module (FOLM)," *Proc. SPIE* **3952**, 202–213 (2000).

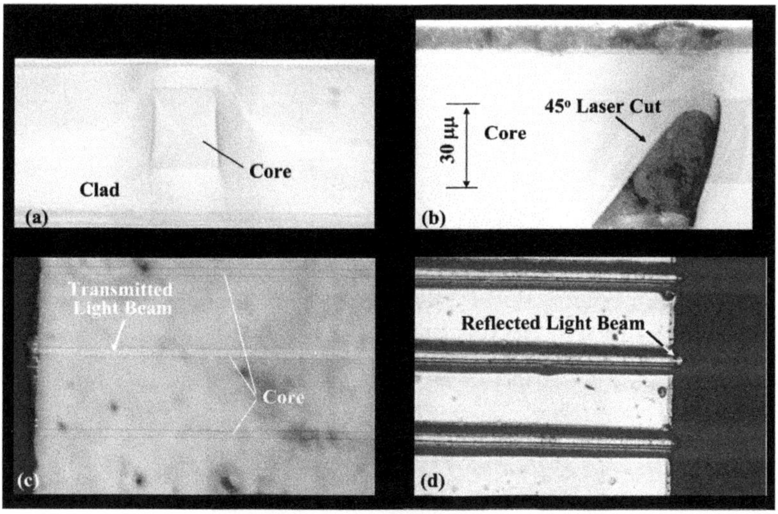

Figure 10.19. (a) Cross section of an optical waveguide, (b) side view of 45° mirror, (c) beam propagation in optical waveguide, and (d) beam reflection from the 45° mirror. Reprinted with permission from T. Yoshimura, J. Roman, Y. Takahashi, M. Lee, B. Chou, S. I. Beilin, W. V. Wang, and M. Inao, "Optoelectronic Scalable Substrates Based on Film/Z-Connection and Its Application to Film Optical Link Module (FOLM)," *Proc. SPIE* **3952**, 202–213 (2000).

film. A slight under-cut is observed. RIE gas pressure reduction from current 40 m Torr will decrease the under-cut. Figure 10.19(b) shows a side view of 45° mirror in the optical waveguide. Actual angle deviates from 45° by a few degrees mainly due to laser beam divergence. Figures 10.19(c) and (d) show beam propagation and reflection from the 45° mirror to a surface normal direction. For the measurement, 640 nm wavelength was used. Excimer laser cutting with high photon energy might be effective for surface quality improvement of the 45° mirrors.

The optical waveguide film and PCB were laminated without peeling at time zero. Figure 10.20(a) shows a sample with optical waveguides attached on Flex PCB. Coexistence of optical and electrical interconnect lines is observed clearly in the substrate. Figure 10.20(b) shows a sample with optical waveguides attached

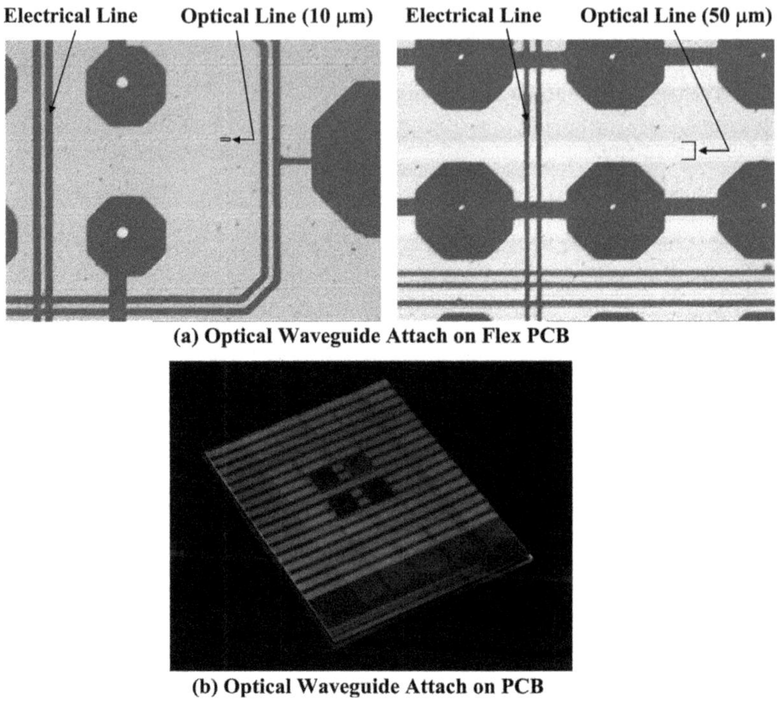

Figure 10.20. Optical waveguide film/PCB stacked samples.

on PCB. These results indicate feasibility of "Film/Z-connection" approach extending to real OE substrate construction.

Figure 10.21 shows a schematic illustration of waveguide routing configurations in an OE transceiver of S-FOLM for 2-D-connection to 2-D optical fiber array. We proposed Type I and Type II. Corresponding to pad positions of chips to be mounted on S-FOLM, surface normal couplers are formed for optical coupling between optical waveguides and OE devices just under the pad locations (OE devices are not shown in the figures). Two types of 2-D-connection parts of 4×4 array are considered. Optical waveguides are routed by corner turning mirrors and S-bending waveguides from the chip-mounting part to the dense 2-D-connection part to an optical fiber. In the present sample, 16×16 pads with 1.2 mm pitch are assumed for chip-mounting parts. Figure 10.22 shows photographs of an optical

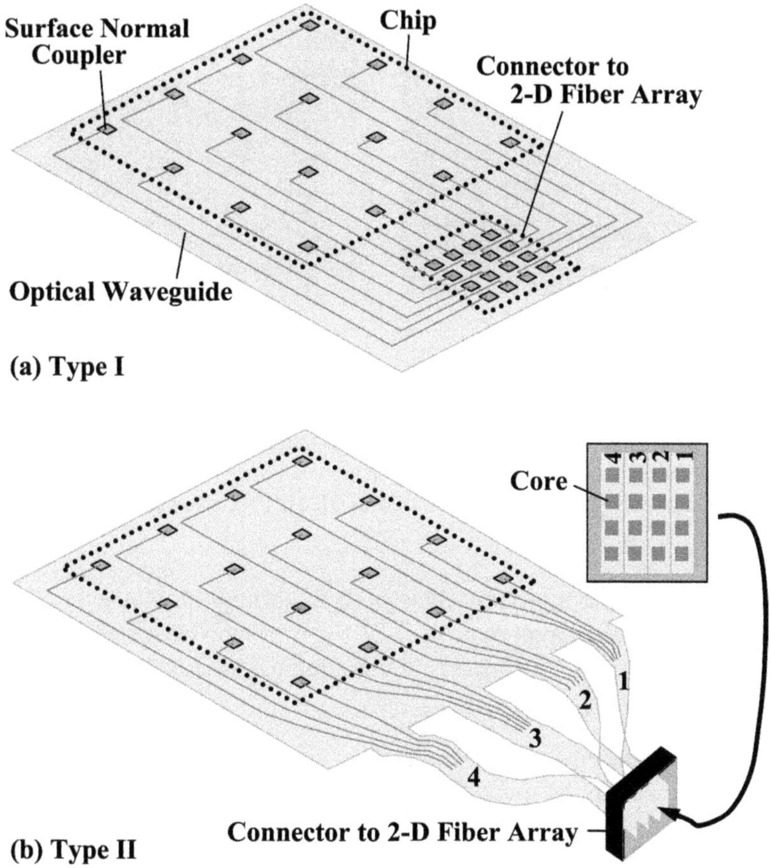

Figure 10.21. Optical waveguide routing configurations for 2D-connection structure in S-FOLM, where an optical waveguide film and a flexible film PCB are stacked. Reprinted with permission from T. Yoshimura, J. Roman, Y. Takahashi, M. Lee, B. Chou, S. I. Beilin, W. V. Wang, and M. Inao, "Optoelectronic Scalable Substrates Based on Film/Z-Connection and Its Application to Film Optical Link Module (FOLM)," *Proc. SPIE* **3952**, 202–213 (2000).

waveguide film. The right-side is a high-magnification photograph of the 2-D-connection part in Type I.

In S-FOLM, surface normal couplers of various directions are needed with high density corresponding to 2-D fiber array pitch and LSI pad pitch. This may be achieved by repeating metal mask

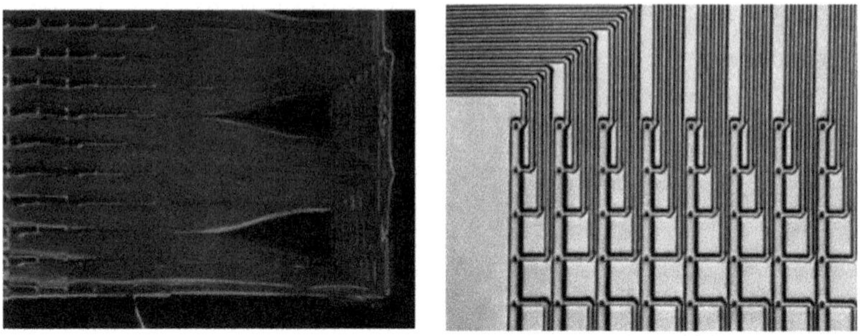

Figure 10.22. Photograph of an optical waveguide film with the 2D-connection structure. Reprinted with permission from T. Yoshimura, J. Roman, Y. Takahashi, M. Lee, B. Chou, S. I. Beilin, W. V. Wang, and M. Inao, "Optoelectronic Scalable Substrates Based on Film/Z-Connection and Its Application to Film Optical Link Module (FOLM)," Proc. SPIE **3952**, 202–213 (2000).

layer patterning steps for beveled wall formation. However, the process results in high cost. To solve the problem, we proposed the following process. The process flow example is schematically shown in Fig. 10.23(a).

(1) A film sample is provided.
(2) On the sample, metal mask layer is formed with apertures at all the sites, where walls will be made, regardless of the wall directions.
(3) Putting cover mask 1 on the sample having apertures at the sites where vertical walls should be formed. The sample is exposed vertically to RIE beam or laser beam to form vertical walls.
(4) Putting cover mask 2 having apertures at the sites where beveled walls of direction 1 should be formed. Oblique exposure of direction 1 is performed.
(5) Putting cover mask 3, oblique exposure of direction 2 is performed.
(6) Beveled walls with three different directions are fabricated.

We call the proposed process as Moving Neon Etching (MNE) and Moving Neon Ablation (MNA) for RIE use and UV laser ablation use, respectively. The process enables wall fabrication with arbitrary

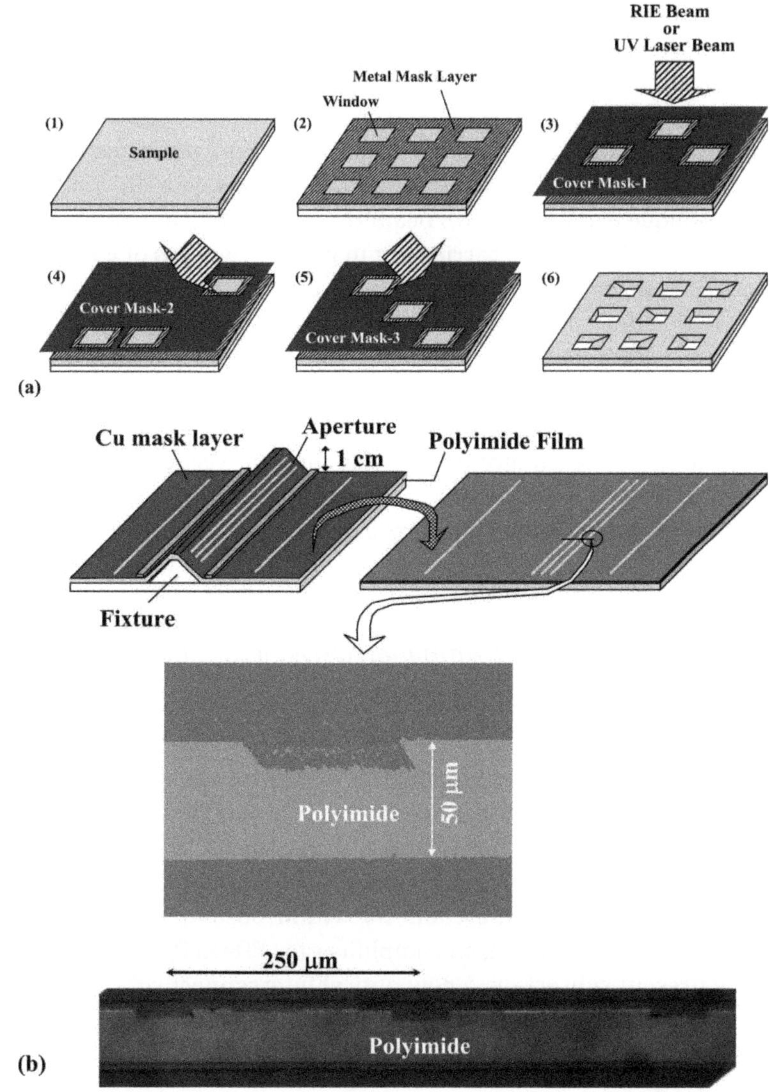

Figure 10.23. (a) MNA and MNE process flow, and (b) wall formation of three different angles with 250 μm pitch. Reprinted with permission from T. Yoshimura, J. Roman, P. Reilly, Y. Takahashi, M. Lee, B. Chou, S. I. Beilin, W. V. Wang, and M. Inao, "Proposal of Film Optical Link Module (FOLM) and Beveled Cut Process for the High Density Waveguide Couplers," *Proc. 1999 IEEE LEOS Annual Meeting Conference*, San Francisco, California, 27–28 (1999).

directions/locations by only one-step formation of a fine mask layer with assistance of cover masks.

In Fig. 10.23(b), an MNE example is shown. On a 50 μm-thick Kapton polyimide film surface, 200 nm-thick Cu mask was deposited with windows of 250 μm pitch. The window sizes range from 30 μm to 100 μm. The film was set on a stainless fixture with 45° slope to apply oblique exposure of RIE beam that was first demonstrated by Chen et al. [34]. RIE was carried out in pure oxygen gas of 2 Pa with 450 W rf power. The etch rate was ~5 μm/h. A beveled structure was obtained as shown in the middle photograph in Fig. 10.23(b). Three-step RIE for oblique angles of −45°, 0°, and +45° was carried out by changing cover masks step by step. In Fig. 10.23(b) (bottom), 250 μm-pitch walls of three different directions can be seen. The result indicates feasibility of MNE for forming vertical mirrors in optical waveguides with arbitrary angles at arbitrary locations with one fine mask formation step, enabling S-FOLM compatible coupler density.

10.3.5 *Optoelectronic Amplifier/Driver-Less Substrate (OE-ADLES)*

Although the optical interconnect has advantages to overcome the problems of electrical interconnects, there is a large barrier that prevents optics from getting into boxes. It is ascribed to the optics excess shown in Fig. 10.24: space, components and packaging, and delay and skew in the E-O and O-E conversion. So far, the demonstrated optical interconnects within boxes use drivers for VCSELs in E-O conversion, and amplifiers for PDs in O-E conversion. In the present subsection, a new optical interconnect substrate, OE amplifier/driver-less substrate (OE-ADLES) [19, 20, 31–33], which will remove the optics excess, is presented.

10.3.5.1 Concept of OE-ADLES

In OE-ADLES, as shown in Fig. 10.24, E-O and O-E conversion functions by light modulators and PDs are completely buried in an "Optical Layer" by integrating EO materials and semiconductors together with optical waveguides. Optical interconnection is

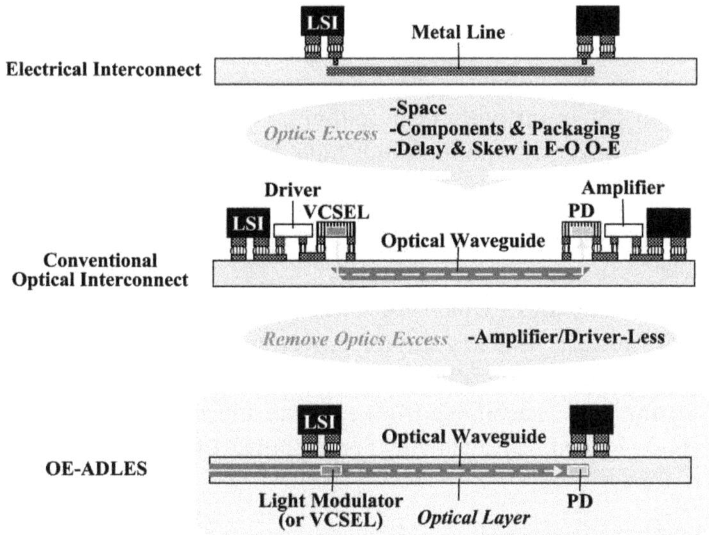

Figure 10.24. Comparison of optical interconnects based on OE-ADLES with electrical interconnects and conventional optical interconnects.

performed by directly connecting I/O pads or electrodes of LSIs to the electrodes of light modulators and PDs. In some cases, VCSELs with a high quantum efficiency might be able to be used instead of the light modulators.

Figure 10.25 shows OE-ADLES for OE MCM. Strong light beams (cw or pulse train) are introduced from outside into an OE MCM via an OE PCB. The external light beams are directly modulated by LSI

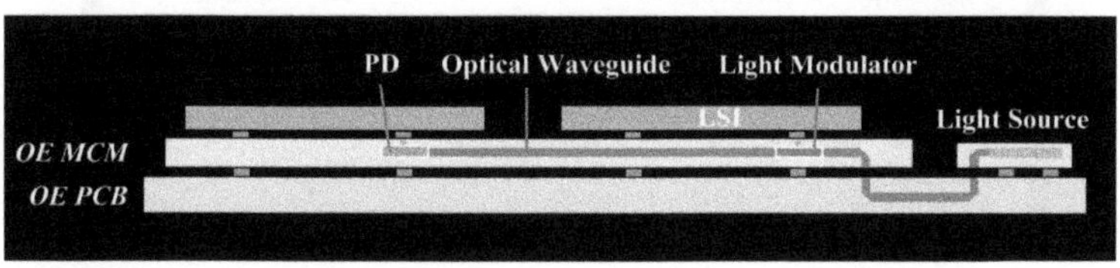

Figure 10.25. OE-ADLES for OE MCM. Reprinted with permission from T. Yoshimura, Y. Suzuki, N. Shimoda, T. Kofudo, K. Okada, Y. Arai, and K. Asama, "Three-Dimensional Chip-Scale Optical Interconnects and Switches with Self-Organized Wiring Based on Device-Embedded Waveguide Films and Molecular Nanotechnologies," *Proc. SPIE* **6126**, from Photonics West 2006, San Jose, California, 612609-1-15 (2006).

output through the light modulators without drivers. The generated optical signals are transmitted through optical waveguides to the PDs, where the optical signals are converted back to electrical signals of LSI input without amplifiers. OE-ADLES is expected to have "low power dissipation" and "high data rate" capability due to the following reasons:

- Light sources are put outside, reducing power dissipation within the interconnect systems.
- Light modulators, which are voltage-driven devices, are used for transmitters, reducing driving power and rising/falling time for optical signal generation.
- Strong light is available from external sources (high-power LD, mode-locked LD), achieving fast response of PDs.

10.3.5.2 Estimation of power dissipation and delay in OE-ADLES

In order to estimate power dissipation and delay in OE-ADLES, models shown in Fig. 10.26 are considered. For electrical substrates,

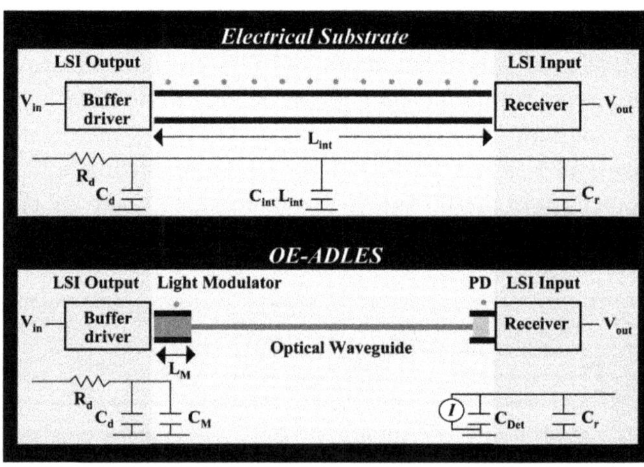

Figure 10.26. Models for an electrical substrate and OE-ADLES. Reprinted with permission from T. Yoshimura, Y. Suzuki, N. Shimoda, T. Kofudo, K. Okada, Y. Arai, and K. Asama, "Three-Dimensional Chip-Scale Optical Interconnects and Switches with Self-Organized Wiring Based on Device-Embedded Waveguide Films and Molecular Nanotechnologies," *Proc. SPIE* **6126**, from Photonics West 2006, San Jose, California, 612609-1-15 (2006).

data transmissions are treated as charge/discharge process of capacitances of a buffer driver, a transmission line, and a receiver through a gate resistance of the buffer driver. For OE-ADLES, on the other hand, data transmissions are treated as charge/discharge process of capacitances of a buffer driver and a light modulator through a gate resistance of the buffer driver, and charge/discharge process of capacitances of a PD and a receiver by currents generated in the PD.

In Table 10.1, expressions for the delay and power dissipation are summarized as well as parameters for calculations. Attenu-

Table 10.1. Expressions and parameters for calculations of delay and power dissipation

Electrical	Delay	$\tau = R_d(C_{int} L_{int} + C_d + C_r)$
Subtract	Power dissipation	$P = f(C_{int}L_{int} + C_d + C_r)V_{out}^2 \;\; C_{int} = \varepsilon_0\varepsilon_i w/d$
OE-ADLES	Delay	$\tau = R_d(C_M + C_d) + V_{out}(C_{Det} + C_r)/(\eta_c P_{Det}) + \tau_P$
	Power dessipation	$P = f[(C_M + C_d)V_{in}^2 + (C_{Det} + C_r)V_{out}^2] \;\; C_M = \varepsilon_0\varepsilon_i A_M/d$

		Electrical substrate	*OE-ADELS*	
			PLZT	Polymer MQDs (Calc.)
Dielectric constant	ε_i	3	300	4
Refractive index	n		2.48	2
EO coefficient	R, r		$R = 5.7 \times 10^{-16}$ m^2/V^2	$r = 1.35 \times 10^{-9}$ m/V

f	Data rate	d	Gap between electrodes: 1 μm
V_{in}	Power supply voltage: 3 V	A_M	Light modulator area: 16 × 16 μm^2
V_{out}	Received detector voltage: 3 V	C_M	Light modulator capacitance
C_{int}	Transmission line capacitance per cm	C_{Det}	Photodetector capacitance: 0.01 pF
R_d	Driver gate resistance: 50 Ω	η_c	Responsibility of photodetector: 0.8 A/W
C_d	Driver capacitance: 0.02 pF	τ_p	Photodetector internal propagation delay: 25 ps
C_r	Receiver capacitance: 0.02 pF	P_{Det}	Incident power to photodetector
L_{int}	Transmission line length	λ	Wavelength: 1.3 μm
w	Electrode width: 4 μm		

Source: Reprinted with permission from T. Yoshimura, Y. Suzuki, N. Shimoda, T. Kofudo, K. Okada, Y. Arai, and K. Asama, "Three-Dimensional Chip-Scale Optical Interconnects and Switches with Self-Organized Wiring Based on Device-Embedded Waveguide Films and Molecular Nanotechnologies," *Proc. SPIE* **6126**, from Photonics West 2006, San Jose, California, 612609 - 1 -15 (2006).

ation along the transmission lines is neglected. External power dissipation at light sources is not included. WPD optical switch described in Sections 9.1.3 and 9.1.5 is assumed as the light modulator for the transmitter.

In Fig. 10.27, delay and power dissipation of an electrical substrate and OE-ADLES are compared. PLZT and a polymer MQD are assumed as EO materials for WPD optical switches. Here, the polymer MQD is an artificial material consisting of polymer wires

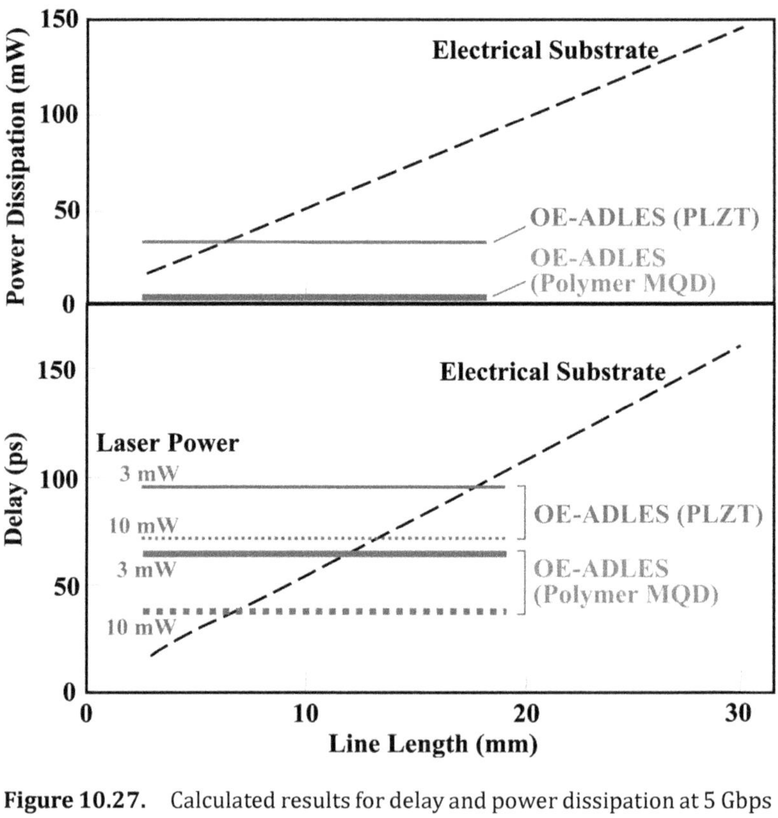

Figure 10.27. Calculated results for delay and power dissipation at 5 Gbps in an electrical substrate and OE-ADLES. Reprinted with permission from T. Yoshimura, Y. Suzuki, N. Shimoda, T. Kofudo, K. Okada, Y. Arai, and K. Asama, "Three-Dimensional Chip-Scale Optical Interconnects and Switches with Self-Organized Wiring Based on Device-Embedded Waveguide Films and Molecular Nanotechnologies," *Proc. SPIE* **6126**, from Photonics West 2006, San Jose, California, 612609-1-15 (2006).

with MQDs, whose EO coefficient was calculated by using the molecular orbital method. The detail is given in Section 12.2.

For delay, OE-ADLES with the polymer MQD has an advantage over the electrical substrate in a line length range longer than 1.2 cm for light power of 3 mW. With increasing the light power introduced from external light sources, the response becomes faster, moving the break-even point toward shorter line length of mm ranges. For power dissipation, drastic reduction can be seen in OE-ADLES with the polymer MQD. The break-even point is located at line length of sub mm ranges. The break-even point in OE-ADLES with PLZT is around 6 mm. In OE-ADLES with PLZT, compared with OE-ADLES with the polymer MQD, the break-even point is located at longer line length regions for both delay and power dissipation. This comes from the large dielectric constant of PLZT.

The superiority of optics in the calculated delay and power dissipation is attributed to the capacitance reduction. Namely, in the electrical substrate a large amount of charge is required to charge up the electrical transmission line due to the large line capacitance, while in OE-ADLES, only a small amount of charge is required to charge up the electrodes of the modulator and PD due to their small capacitances.

Key issue in OE-ADLES is EO material availability with a large EO coefficient and a small dielectric constant like the polymer MQD fabricated by the molecular layer deposition (MLD). Size reduction of the light modulators and PDs is another important issue. When they are miniaturized by one order of magnitude for each dimension, that is, the device thickness is reduced by one order of magnitude and the device area by two orders of magnitude, their capacitances as well as the PD internal propagation delay will be reduced by one order of magnitude in principle. This will move the break-even points to sub mm ranges for both delay and power dissipation.

10.4 Optical Switching Systems

High-speed/large-scale optical switching systems will be required for rapid growth of network traffic. Currently, as shown in Fig. 10.28, large-scale optical switching systems are actively developed using

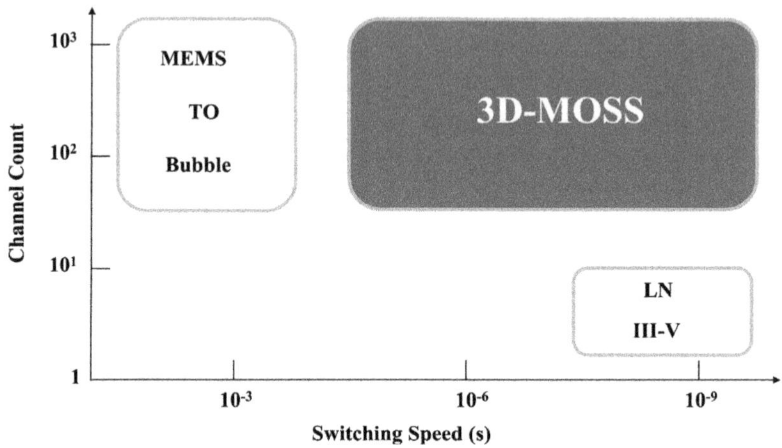

Figure 10.28. Road map of optical switching systems.

the micro electro-mechanical system (MEMS), the thermo-optics (TO), and the bubble reflection. These elements, however, have switching speed limit around ms. Meanwhile, waveguide-type optical switches using EO materials such as lithium niobate (LN), III-V semiconductor, PLZT, and organic materials exhibit switching time of ns~ps order. Unfortunately, large-scale integration has not been accomplished using the EO materials.

In the present section, a new architecture "Three-Dimensional Micro Optical Switching System (3D-MOSS)" [8, 35–39] with large-scale integration of EO waveguide optical switches is reviewed. 3D-MOSS is based on S-FOLM consisting of film stacks with embedded thin-film OE devices and optical waveguides. It will enable a large-scale switching (channel count over several tenth) with high-speed (switching speed below μs).

10.4.1 *Optical Switching Systems with 3-D Architecture*

Figure 10.29 shows schematic illustration of 3D-MOSS application images. For optical switching in fiber communications, 3D-MOSSs are mounted on OE PCB, to which input and output optical fibers are connected. For reconfigurable optical interconnects, 3-D stacked OE MCMs and 3D-MOSSs are mounted on OE PCB. It might be possible to mount conventional LSIs or 3-D stacked LSIs instead of the 3-D

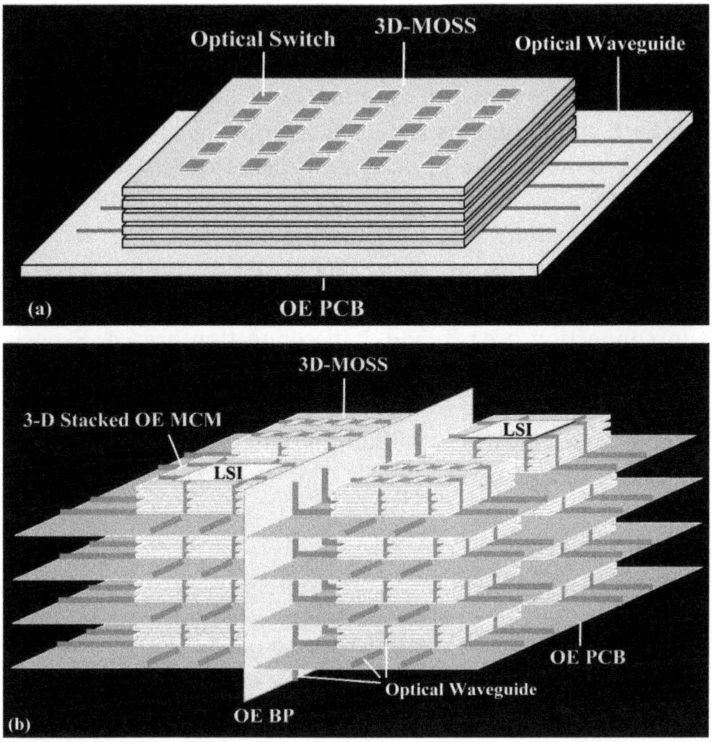

Figure 10.29. 3D-MOSS (a) for optical switching and (b) for reconfigurable optical interconnects.

stacked OE MCMs. Optical signals in the OE PCB are switched in 3D-MOSS to enable dynamic routing between LSIs. In the case of multiprocessor systems, miniaturized X-BAR switch configurations may be available on a board.

Optical switches in 3D-MOSS can be tunable wavelength filters or wavelength converters, indicating that 3D-MOSS are also regarded as a WDM switching system.

10.4.2 *Concept of the 3-D Micro Optical Switching System (3D-MOSS)*

In this subsection, the concept of 3D-MOSS is explained for a preliminary model of Banyan network with 32 $\times$ 32 channels.

10.4.2.1 Structures

10.4.2.1.1 Network structure Figure 10.30 shows a conventional 32 × 32 Banyan network that consists of five optical switch arrays. Each array contains sixteen 2 × 2 optical switches. Adjacent arrays are connected by optical waveguides in "connection area." Adjacent optical switch arrays i and $i + 1$ are connected by optical wiring in connection area i. In the present case, four waveguide connection areas exist. In connection area 1, an optical waveguide experiences 15 waveguide crossing points, 7 in connection area 2, 3 in connection area 3, and 1 in connection area 4. Especially in connection area 1, considerably long optical waveguides are necessary. Therefore, the connection area 1 is critical.

To construct 3D-MOSS, the Banyan network is divided into four sub-network blocks of BL 1, 2, 3, and 4 as shown in Fig. 10.30.

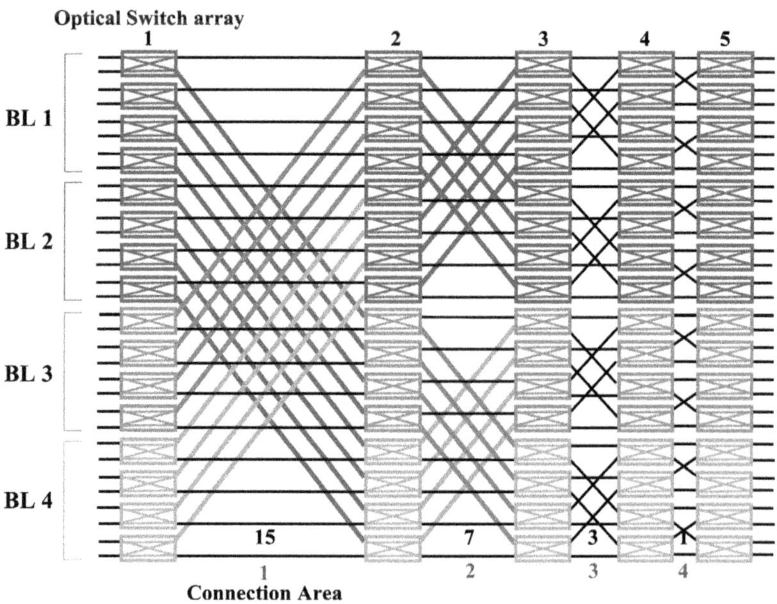

Figure 10.30. A 32 × 32 Banyan network that consists of five optical switch arrays. Reprinted with permission from T. Yoshimura, S. Tsukada, S. Kawakami, M. Ninomiya, Y. Arai, H. Kurokawa, and K. Asama, "Three-Dimensional Micro-Optical Switching System Architecture Using Slab-Waveguide-Based Micro-Optical Switches," *Opt. Eng.* **42**, 439–446 (2003).

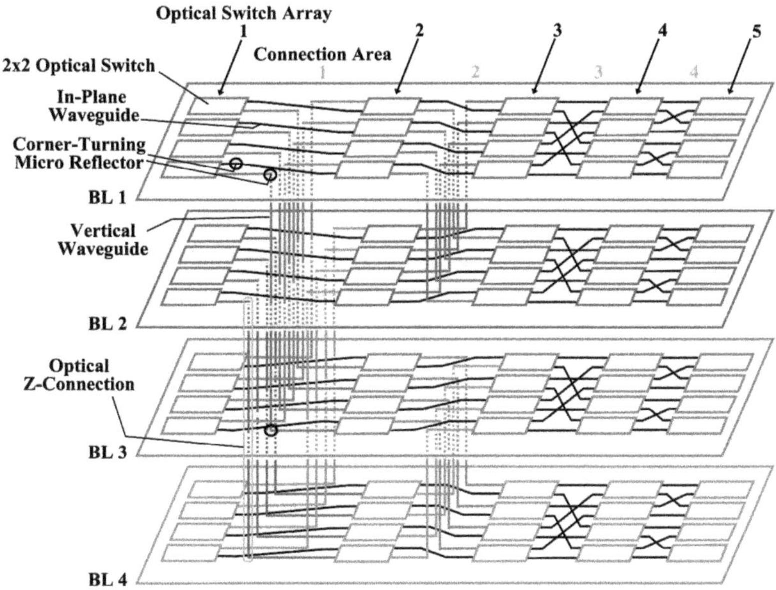

Figure 10.31. Network configuration of 3D-MOSS for 32 × 32 Banyan network.

Inter- and intra-block connections are represented by thick and thin lines, respectively. As shown in Fig. 10.31, the blocks are separated and stacked to each other in the order of BL 1, 2, 3, and 4. The long optical waveguides in connection area 1 are replaced with very-short-distance vertical light paths, that is, optical Z-connections, connecting BL 1 to 3 and BL 2 to 4. The optical Z-connections consist of corner-turning micro reflectors and vertical waveguides. The long optical waveguides in connection area 2 are also replaced with optical Z-connections between BL 1 and 2, and BL 3 and 4. Thus, 3D-MOSS drastically reduces waveguide crossing point count and wiring distance. 3D-MOSS also contributes to size reduction and scalability.

The advantages of 3D-MOSS become remarkable with increasing the channel count as described in Section 10.4.3. For 1024 × 1024 Banyan network, an optical waveguide in the connection area 1 experiences 511 crossing points in conventional planar configurations, which may induce terrible signal degradation. 3D-MOSS with eight sub-network blocks will reduce the crossing point count from 511 to 0, improving optical signal quality. Reduction by a factor of more than one order of magnitude may be expected in wiring distance and system size.

Connection areas consisting of in-plane optical connections with in-plane waveguides and corner-turning micro reflectors are denoted by "in-plane connection areas", and those consisting of vertical optical connections with optical Z-connections for inter-plane links by "vertical connection areas." Connection areas 3 and 4 are in-plane connection areas, and connection areas 1 and 2 are vertical connection areas.

The degree of three-dimensional characteristics of the networks is represented by D_{3D}, which indicates the number of vertical connection areas in a 3D-MOSS. For example, in a conventional planar structure $D_{3D} = 0$. For 3D-MOSS with two layers, in which optical wiring in connection area 1 is carried out by vertical optical connections, $D_{3D} = 1$. With increasing layer counts, D_{3D} increases. In the 3D-MOSS shown in Fig. 10.31, $D_{3D} = 2$.

Figure 10.32 shows a possible 3D-MOSS structure for 32 × 32 Banyan network. On an OE PCB, OE films for BL 1, 2, 3, and 4, in which optical switch flakes and corner-turning micro reflectors such as filters/mirrors are embedded with optical waveguides, are stacked. In each block, an interface film, in which thin-film ICs are embedded, is inserted for driving switches. The output terminals of the ICs are vertically connected to the electrodes of the optical switches by, for example, metal diffusion technique. The situation

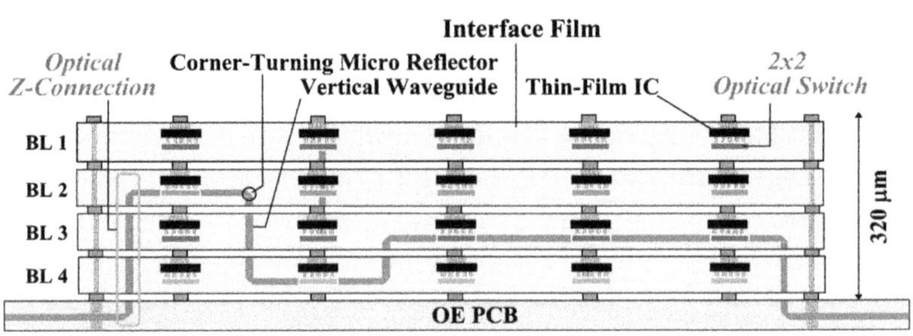

Figure 10.32. Possible film-based structure of 3D-MOSS for 32 × 32 Banyan network.

is illustrated in Fig. 9.3. The optical waveguides in the OE PCB are connected to optical waveguides in the OE films by optical Z-connections similarly to inter-OE-film connections. When input light signals introduced into 3D-MOSS from the OE PCB are led to BL 2 through an optical Z-connection, the signals are switched in BL 2 to be led into BL 4, and are again switched, followed by switching in BL 3, and, finally, are output into the OE PCB.

In the conventional 32 × 32 Banyan network, 176 waveguide crossing points exist in connection area 1. In 3D-MOSS, the crossing point count can be reduced to 0 in principle. 3D-MOSS thickness is estimated to be ~320 μm for four OE films and four interface films, assuming that the films are 30 μm thick and the gap space per a block is 10 μm.

A network configuration example is shown in Fig. 10.33 for BL 2. The 3-D optical circuit backbone is assumed to consist of optical waveguides having 4 × 4 μm^2 core cross sections. The block size is 600 μm wide and 2250 μm long. The optical switch size is 60 × 390 μm^2, assuming the 390 μm-long 2 × 2 WPD optical

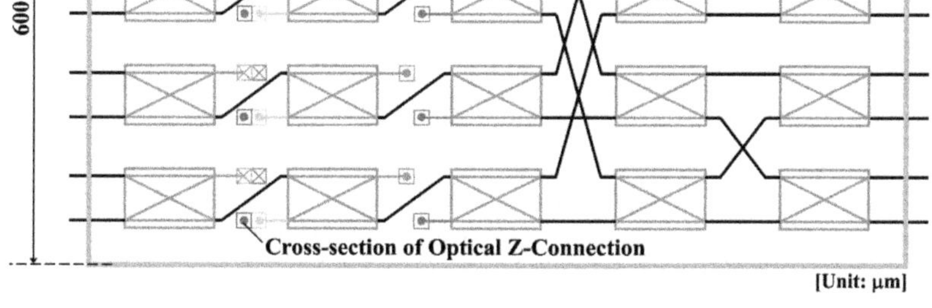

Figure 10.33. Network Configuration in BL 2. Reprinted with permission from T. Yoshimura, S. Tsukada, S. Kawakami, M. Ninomiya, Y. Arai, H. Kurokawa, and K. Asama, "Three-Dimensional Micro-Optical Switching System Architecture Using Slab-Waveguide-Based Micro-Optical Switches," *Opt. Eng.* **42**, 439–446 (2003).

Table 10.2. Switching characteristics of 2 × 2 WPD optical switch with 4 μm-thick SPCD for input 1 to output 1 and input 1 to output 2

	Output 1	Output 2
For Input 1 to Output 1	0.9	0.06
For Input 1 to Output 2	0.036	0.88

Note: The values simulated by BPM represent light power normalized by the input light power.

switch shown in Fig. 9.9. Although in the model of the optical switch the SPCD slab waveguide thickness is 400 nm, in the present model the thickness is assumed 4 μm in order to match it to core thickness of the 3-D optical circuit backbone. In this case, switching voltage is increased from 11 V to110 V. The switching characteristics simulated by BPM are summarized in Table 10.2, in which the light power is normalized by the input light power. The gap area between the switches is 60×60 μm^2, indicating that at most $3600/(4 \times 4) = 225$ Z-connections can path through the area in the case that cross section of an optical Z-connection is 4×4 μm^2. In the present model, only 4 or 2 optical Z-connections pass through the gap area.

10.4.2.1.2 Optical Z-connections Optical Z-connection is performed using corner turning micro reflectors and light guiding devices. For the former, we have two choices — vertical mirrors or vertical filters — and for the latter, three choices — vertical waveguide, micro lens, or SOLNET.

Figure 10.34 shows BPM/FDTD coupled simulation for an optical Z-connection. 4 μm-thick waveguides are in upper and lower layers. Gap between the waveguides of the adjacent layers is 16 μm. The optical waveguides are connected by two corner-turning micro reflectors and a vertical waveguide of 4 μm wide. The micro reflectors can be metal mirrors, dielectric filters, gratings, or photonic crystals. In the present work, metal mirrors with refractive index of 0.4 for real part and −5 for imaginary part is assumed for the micro reflectors. It can be seen from a bird's-eye view for the propagating light intensity in Fig. 10.34 that a light beam smoothly propagates from the upper layer to the lower layer through optical Z-connection with little mode disturbance. Calculated loss of the

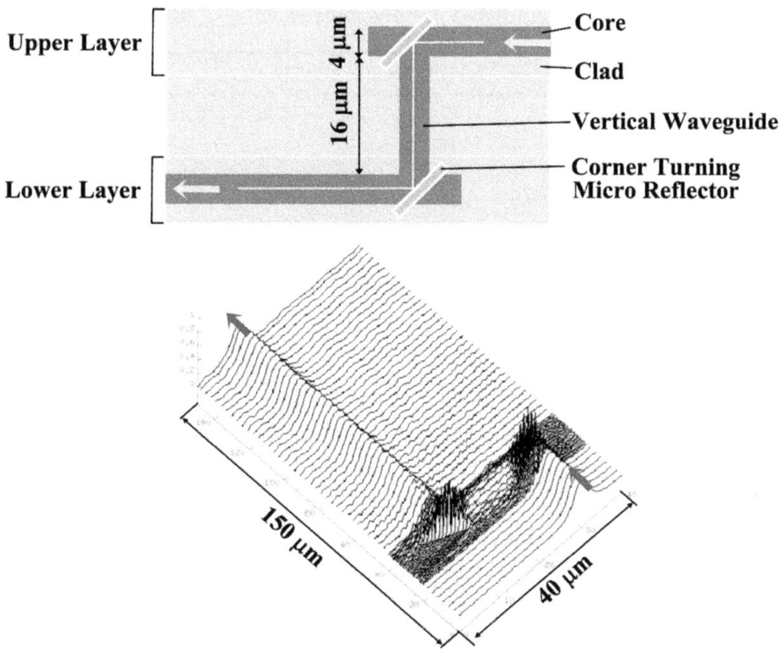

Figure 10.34. BPM/FDTD coupled simulation for optical Z-connection. Reprinted with permission from T. Yoshimura, M. Ojima, Y. Arai, and K. Asama, "Three-Dimensional Self-Organized Micro Optoelectronic Systems for Board-Level Reconfigurable Optical Interconnects — Performance Modeling and Simulation," *IEEE J. Select. Topics Quantum Electron.* **9**, 492–511. © 2003 IEEE.

optical Z-connection (γ_Z) is 0.97 dB. The loss is basically induced at the mirror reflection parts.

10.4.2.2 Performance

Table 10.3 summarizes simulated performance of 3D-MOSS for the 32 × 32 Banyan network. System size is 2250 µm long, 600 µm wide, and 320 µm high. Operation voltage is 110 V.

Switching speed and power consumption can be estimated using the charge-discharge model described in Section 9.1.2. In the case that switch driver on-resistance (R_{on}) is 100 Ω and EO material dielectric constant is $4\varepsilon_0$, capacitance of the capacitor (C) and

Table 10.3. Simulated performance of 32 × 32 3D-MOSS

	Length 2250 μm	Width 600 μm	Height 320 μm
Operation voltage	110 V		
Switching speed	1.8 ps		
Power consumption at switching rate of 1.5×10^7/s	510 mW		
Maximum insertion loss	14 dB		

switching time (τ) are calculated by Eqs. (9.9) and (9.8) to be 1.1 fF and 0.11 ps, respectively, for a prism-shaped capacitor of 128 μm^2 area with an electrode distance of 4 μm. The number of the prism-shaped electrodes in one switching route is 16 for four-step cascade, implying that total capacitance is 18 fF and the switching time is 1.8 ps.

Using the same capacitor model with applied voltage of 110 V, energy consumption for one charge-discharge cycle is revealed to be $CV^2 = 13.3$ pJ for one prism-shaped capacitor. In a 2 × 2 WPD optical switch, 32 capacitors are contained. In a film block, 20 WPD optical switches are contained. Then, total number of the capacitors in a 32 × 32 3D-MOSS is 32 × 20 × 4 = 2560. Therefore, the energy consumption in a 32 × 32 3D-MOSS is 34 nJ. When switching operation is carried out at a switching rate of 1.5×10^7/s for ~100 ns-long packet switching, power consumption is calculated to be 510 mW, which corresponds to power consumption per unit area of 37 W/cm^2. The value is close to acceptable heat generation of 30 W/cm^2 that was reported by K. Takahashi [40] for four-layer-stacked LSI of 1 cm^2 area with 10 μm^{ϕ} cupper thermal vias of 100 μm pitch and an active thermal sink of 0.3K/W.

For maximum insertion loss calculation, optical switch loss, corner turning loss, waveguide-switch coupling loss, and waveguide propagation loss are considered. Estimated loss and quantity of loss-inducing parts for the maximum-loss-inducing route is listed in Table 10.4. The loss of corner turning and waveguide-switch coupling is for the case of 1 μm misalignment. For waveguide-switch alignment, it is not difficult to make photolithographic patterns with an accuracy of 1 μm by using conventional aligners or steppers. For film-stacking alignment, it was reported that high-accuracy bonding

Table 10.4. Estimated loss and quantity of loss-inducing parts for insertion loss evaluation of 32 × 32 3D-MOSS

Parts	Loss (dB)	Quantity
Optical switch	0.63 dB	5
Corner turning with 1 μm misalignment	0.43 dB	12
Waveguide-switch coupling	0.36 dB	10
Optical waveguide	0.3 dB/cm	0.9 cm

equipment enables layer stacking with a positional accuracy of 1 μm for 1 cm^2 chips [40]. Maximum total loss of the 3D-MOSS is calculated by summing all the losses listed in Table 10.4, namely, $0.63 \times 5 + 0.43 \times 12 + 0.36 \times 10 + 0.3 \times 0.9$, to be 12 dB.

These results indicate that 32 × 32 3D-MOSS is viable from view points of channel count, switching speed, thermal management, and power budget, implying that larger-scale 3D-MOSS, say, 1024 × 1024 3D-MOSS, that can be constructed by building up sixty-four 32 × 32 3D-MOSS units, may also viable by inserting one or two optical amplifier arrays. Detailed consideration on optimized structure is described in Section 10.4.3.

10.4.2.3 Material/cost saving heterogeneous integration

Figure 10.35 shows an example of SORT applied to heterogeneous integration in 3D-MOSS. First, optical switch array, micro lens array, and vertical mirror array are prepared on different device growth wafers. Next, the device arrays are picked up onto supporting substrates from device growth wafers. Then, some of the optical switch flakes are partially transferred from the supporting substrate onto a final substrate at one time. Similarly, some of the micro lens flakes and vertical mirror flakes are transferred onto the final substrate at one time successively.

SORT is effective for reducing material consumption. For example, in the case shown in Fig. 10.35, three optical switch flakes are distributed on one final substrate. Total flake count on the supporting substrate is 36. Therefore, the supporting substrate can provide the optical switch flakes to 12 final substrates, realizing drastic material and process effort saving, resulting in

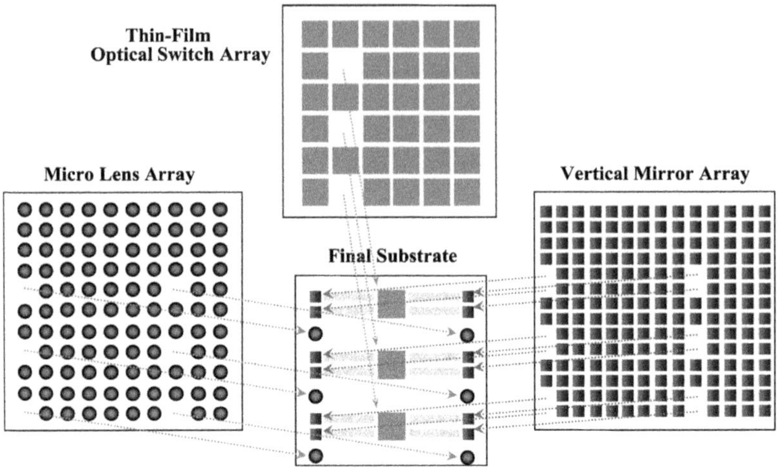

Figure 10.35. SORT concept for 3D-MOSS fabrication. Reprinted with permission from T. Yoshimura, S. Tsukada, S. Kawakami, Y. Arai, H. Kurokawa, and K. Asama, "3D Micro Optical Switching System (3D-MOSS) Architecture," *Proc. SPIE* **4653**, 62–70 (2002).

cost reduction. SORT is followed by embedding and metallization processes to make OE films following the steps of PL-Pack with SORT described in Section 8.2.

10.4.2.4 Implementation of self-organized 3-D optical circuits

In 3D-MOSS, channel count is quite large, resulting in a large number of optical coupling points. This forces us to align a lot of OE devices and layers precisely with enormous efforts. Furthermore, vertical waveguides are necessary for inter-layer connections. In order to achieve a drastic cost reduction of 3D-MOSS, implementation of SOLNET is proposed for construction of self-organized 3-D optical circuits.

Figure 10.36 shows fabrication procedure of self-organized 3-D optical circuits. First, OE device flakes with optical waveguides are arranged in a designed configuration in optical waveguide films to make OE films. The OE films are stacked to construct a 3-D structure. The OE device flakes within film planes are initially positioned so that some excitation will achieve self-alignment. The

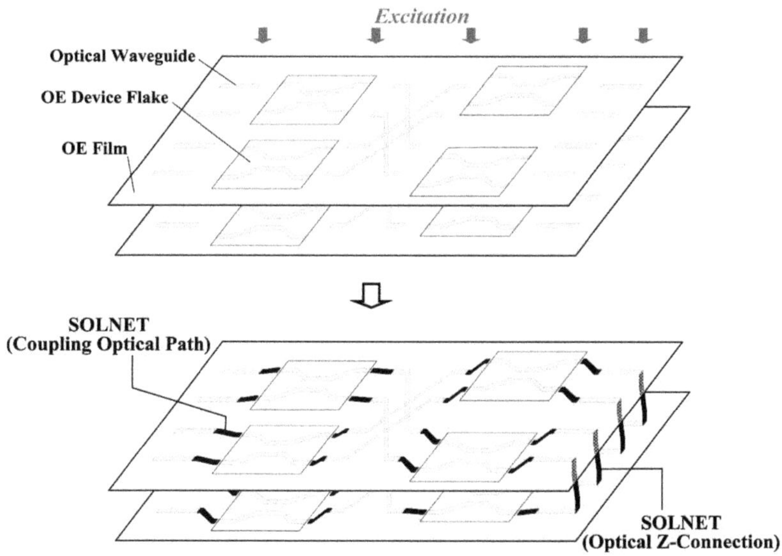

Figure 10.36. Fabrication procedure of self-organized 3-D optical circuits. Reprinted with permission from T. Yoshimura, M. Ojima, Y. Arai, and K. Asama, "Three-Dimensional Self-Organized Micro Optoelectronic Systems for Board-Level Reconfigurable Optical Interconnects — Performance Modeling and Simulation," *IEEE J. Select. Topics Quantum Electron.* **9**, 492–511. © 2003 IEEE.

OE films are also arranged so that the excitation will form optical Z-connections. Next, by introducing the excitation to the 3-D structure from outside, a self-aligned 3-D optical circuit is automatically formed within the structure, including coupling optical paths and optical Z-connections.

SOLNET is a promising candidate for the self-aligned 3-D optical circuits due to its "self-organized" characteristics. By SOLNET implementation, even if some misalignment of the optical waveguides and the stacked films exist, automatic optical couplings can be achieved, which will reduce fabrication cost, especially cost for assembly and optical Z-connection formation as described in Sections 10.1 and 10.2.

For SOLNET implementation, write beam insertion is a serious concern. For example, because 3D-MOSS has a large number of connection parts and some connections are terminated by two

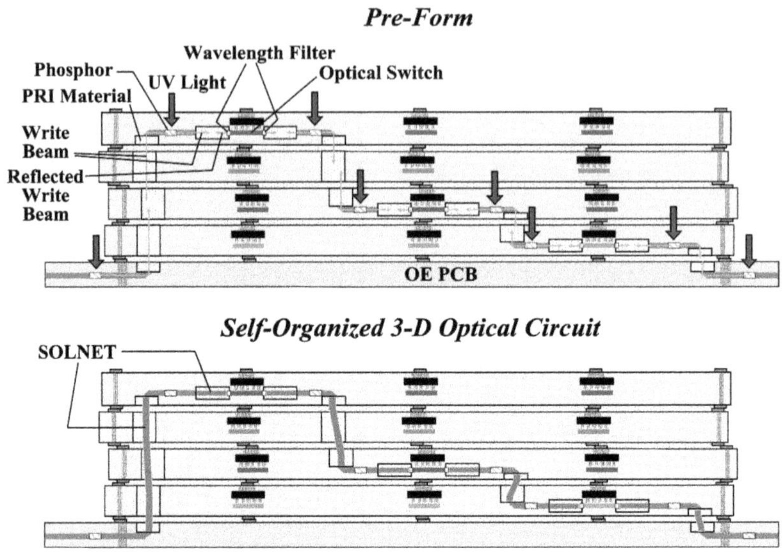

Figure 10.37. A possible fabrication process of self-organized 3-D optical circuits. Reprinted with permission from T. Yoshimura, M. Ojima, Y. Arai, and K. Asama, "Three-Dimensional Self-Organized Micro Optoelectronic Systems for Board-Level Reconfigurable Optical Interconnects — Performance Modeling and Simulation," *IEEE J. Select. Topics Quantum Electron.* **9**, 492–511. © 2003 IEEE. See also Color Insert.

optical switches, which make it impossible to introduce write beams from outside. R-SOLNET and P-SOLNET solve the problem as shown in Fig. 10.37.

First, OE films are made by transferring optical switch flakes onto final substrates using SORT described in Chapter 8. Before the transfer, wavelength filters, which selectively reflect write beams of wavelengths typically between 0.35 μm and 0.8 μm and pass signal beams of 0.4–1.5 μm, are deposited just on core surfaces at the edge of optical switch flakes using chemical vapor deposition (CVD), ALD, or MLD with photolithography. Next, a pre-form is constructed by stacking the OE films. For write beam generation, phosphor is partially doped in optical waveguides. PRI materials are inserted in parts where SOLNET is necessary.

Then, by illuminating the phosphor with UV light from outside of 3D-MOSS, write beams generated in the phosphor-doped regions propagate in optical waveguides to contribute to P-SOLNET formation. For vertical waveguide construction, the write beams are radiated into the PRI materials from upper and lower optical waveguides through corner turning micro reflectors to form SOLNET. For connections between an in-plane waveguide and an optical waveguide within an optical switch, a write beam introduced into the PRI material from the in-plane waveguide is reflected by a wavelength filter on the waveguide core edge of the optical switch flake. The write beam from the in-plane waveguide and the reflected write beam from the optical switch flake are attracted each other to construct R-SOLNET. Thus, a self-organized 3-D optical circuit is constructed. UV light from outside corresponds to the "excitation" in Fig. 10.36.

10.4.2.5 Impact of HIC waveguide implementation into 3D-MOSS

The 3D-MOSS model evaluated so far consists of the 3-D optical circuit backbone with optical waveguides having a 4 × 4 μm^2 core cross section. Consequently, the electrode gap of the WPD optical switch is 4 μm in the model. The gap of 4 μm is one order of magnitude larger comparing with the gap in the model shown in Fig. 9.9, where the gap is 400 nm. The increase in the electrode gap from 400 nm to 4 μm raises the switching voltage from 11 V to 110 V.

If HIC waveguides having 400 × 400 nm^2 core cross sections are used for the 3-D optical circuit backbone in the 3D-MOSS, the 400 nm-electrode-gap WPD optical switch shown in Fig. 9.9 can be used, enabling switching voltage reduction to 11 V. Low-voltage switching is favorable for low-power and high-speed operation.

The HIC-waveguide-based optical circuits might possibly be fabricated, for example, by the PRI sol-gel materials described in Sections 5.4 and 6.4.

10.4.3 *3D-MOSS for 1024 × 1024 Banyan Network*

In this section, to demonstrate a solution for "miniaturization of high-speed massive optical switching systems," a detail performance analysis of the 3D-MOSS for 1024 × 1024 Banyan network is presented. Basic concepts obtained in the Banyan network are expected to be applicable to other advanced network configurations.

10.4.3.1 Structural model

Table 10.5 summarizes structural parameters for the 1024 × 1024 3D-MOSS model. The 3-D optical circuit backbone consists of optical waveguides with core cross section of 4 × 4 μm^2, core refractive index n_{Core} of 1.52, and clad refractive index n_{Clad} of 1.50. In this case, two modes, that is, the ground mode of $m = 0$ and the first excited mode of $m = 1$, exist theoretically for wavelength of 1.3 μm. It is known, however, that optical waveguides having a few modes actually behave as single mode waveguides due to weak confinement for the higher modes. Then, the optical waveguides in the 3D-MOSS is regarded as a single mode network in practice. Therefore, in the present work, calculations of coupling efficiency or insertion loss are carried out for the ground mode light beams.

Figure 10.38 shows top-view illustrations of Banyan network models for a conventional planar structure and a 3D-MOSS structure. Channel count is N_{C}, meaning $N_{\text{C}} \times N_{\text{C}}$ switching is considered.

Table 10.5. Structural parameters for 1024 × 1024 3D-MOSS model

Channel count: N_C		1024
Layer count: N_L		16
Layer thickness/gap: T_L / T_G		46 μm/4 μm
Degree of 3D characteristics: $D_{3\text{D}}$		4
Wavelength: λ		1.3 μm
Optical waveguide		
Width: w		4 μm
Refractive index	Clad: n_{Clad}	1.50
	Core: n_{Core}	1.52
Pitch: P_{WG}		20 μm
2 × 2 Optical switch		
Size: length L_{SW}, Width: W_{SW}		1190 μm, 100 μm
Step count: S_{SW}		10
Pitch in an array: P_{SW}		200 μm

Source: Reprinted with permission from T. Yoshimura, M. Ojima, Y. Arai, and K. Asama, "Three-Dimensional Self-Organized Micro Optoelectronic Systems for Board-Level Reconfigurable Optical Interconnects — Performance Modeling and Simulation," *IEEE J. Select. Topics Quantum Electron.* **9**, 492–511. © 2003 IEEE.

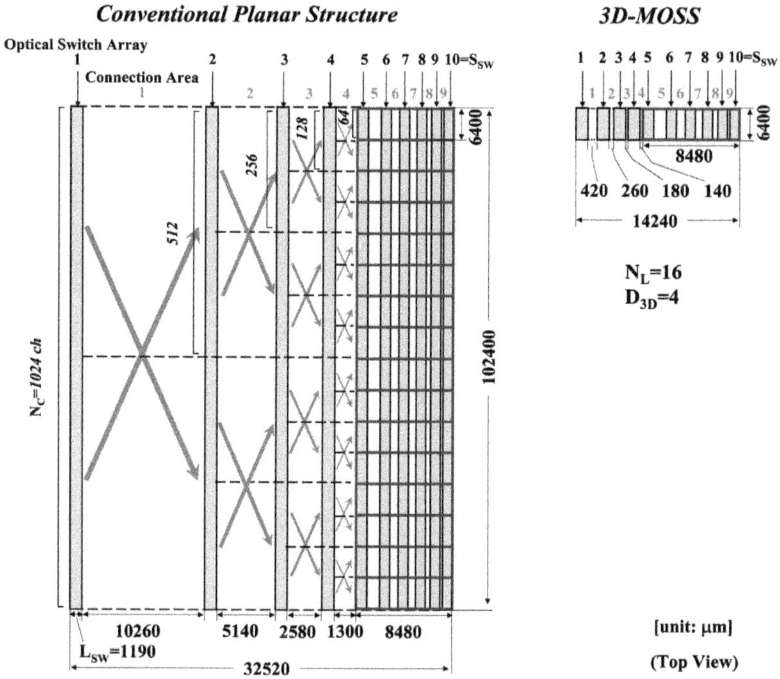

Figure 10.38. Top view of 1024 × 1024 Banyan network models for a conventional planar structure and a 3D-MOSS structure. Reprinted with permission from T. Yoshimura, M. Ojima, Y. Arai, and K. Asama, "Three-Dimensional Self-Organized Micro Optoelectronic Systems for Board-Level Reconfigurable Optical Interconnects — Performance Modeling and Simulation," *IEEE J. Select. Topics in Quantum Electron.* **9**, 492–511. © 2003 IEEE.

In the present model, N_C is 1024. The network consists of 2 × 2 optical switch arrays, which are connected by optical waveguides in connection areas. The optical switch is the WPD optical switch shown in Fig. 9.19. In the model in Fig. 9.19, the PLZT thickness is 400 nm. In the present model, however, in order to match the PLZT thickness to core thickness of the 3-D optical circuit backbone, the PLZT thickness is assumed 4 μm. In this case, switching voltage becomes 12 V. One array contains $N_C/2$ 2 × 2 optical switches with length L_{SW} of 1190 μm and width W_{SW} of 100 μm. The length of the array equals individual switch length L_{SW}. The optical switch array count is S_{SW}, which is 10 in the present model.

In 3D-MOSS, the network is divided into N_L blocks, which are stacked to make a N_L-layer 3-D structure. As mentioned in Section 10.4.2, the degree of 3-D characteristics of networks, D_{3D}, indicates the number of vertical connection areas in a 3D-MOSS. The general expression for D_{3D} is as follows:

$$D_{3D} = \log_2 N_L \tag{10.1}$$

This gives $D_{3D} = 4$ for the present model with $N_L = 16$.

It is found from Fig. 10.38 that the area of the network is drastically reduced in 3D-MOSS, compared with the conventional planar structure. The dimensions inserted in Fig. 10.38, as well as in Figs. 10.39–10.41, are calculated using the general expressions (10.2–10.6) described later.

Figure 10.39 shows a side-view illustration of 3D-MOSS. Since D_{3D} is the number of vertical connection areas, optical wiring is carried out by vertical optical connections in connection areas 1 to D_{3D}, and by in-plane optical connections in connection areas D_{3D} +1 to S_{SW} −1. In the model shown in Fig. 10.39, $D_{3D} = 4$; then, vertical optical connections are in connection areas 1–4, and in-plane optical connections in connection areas 5–9. The region

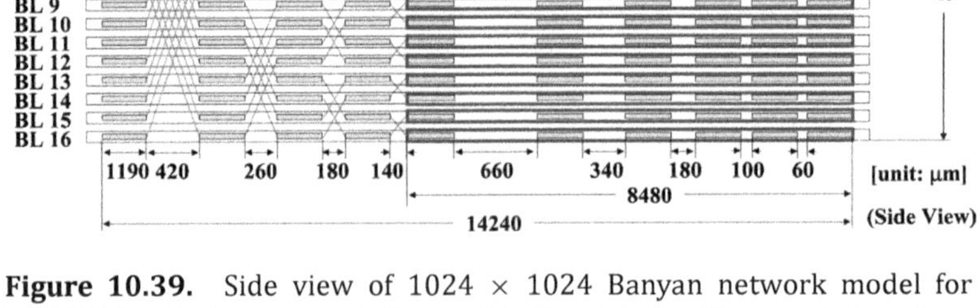

Figure 10.39. Side view of 1024 × 1024 Banyan network model for 3D-MOSS with layer count $N_L = 16$. Reprinted with permission from T. Yoshimura, M. Ojima, Y. Arai, and K. Asama, "Three-Dimensional Self-Organized Micro Optoelectronic Systems for Board-Level Reconfigurable Optical Interconnects — Performance Modeling and Simulation," *IEEE J. Select. Topics Quantum Electron.* **9**, 492–511. © 2003 IEEE.

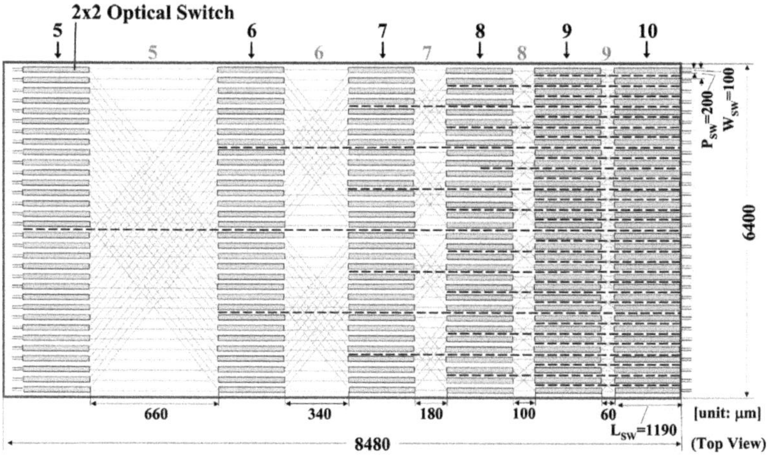

Figure 10.40. Top view of 64 × 64 Banyan network with in-pane optical connections. Reprinted with permission from T. Yoshimura, M. Ojima, Y. Arai, and K. Asama, "Three-Dimensional Self-Organized Micro Optoelectronic Systems for Board-Level Reconfigurable Optical Interconnects — Performance Modeling and Simulation," *IEEE J. Select. Topics Quantum Electron.* **9**, 492–511. © 2003 IEEE.

that includes connection areas 5–9 corresponds to 64 × 64 Banyan network with a conventional planar structure shown in Fig. 10.40. T_L and T_G are layer thickness and gap thickness, respectively. In the present model, T_L is 46 μm and T_G is 4 μm. Thickness per one layer is 50 μm.

As an example of in-plane connection area, detail of connection area 5 is shown in Fig. 10.41(a). Connections between optical switch arrays are constructed by crank-shaped waveguides consisting of a pair of corner-turning micro reflectors. P_{SW} and P_{WG} are the pitch of the individual optical switches and pitch of optical waveguides, respectively. Inter-switch connection line count is expressed by $N_C/2^5$ for connection area 5. Then, with adding P_{WG} for buffer spaces, the length of connection area 5 is given by $(N_C/2^5 + 1)P_{WG}$. The width is given by $((N_C/2)/2^{D_{3D}})P_{SW}$. Thus, the expressions for connection area length and width of in-plane connection area i can be generalized as follows:

$$L(IP)_i = (N_C/2^i + 1)P_{WG} \tag{10.2}$$

$$W(IP) = \left((N_C/2)/2^{D_{3D}}\right) P_{SW} \tag{10.3}$$

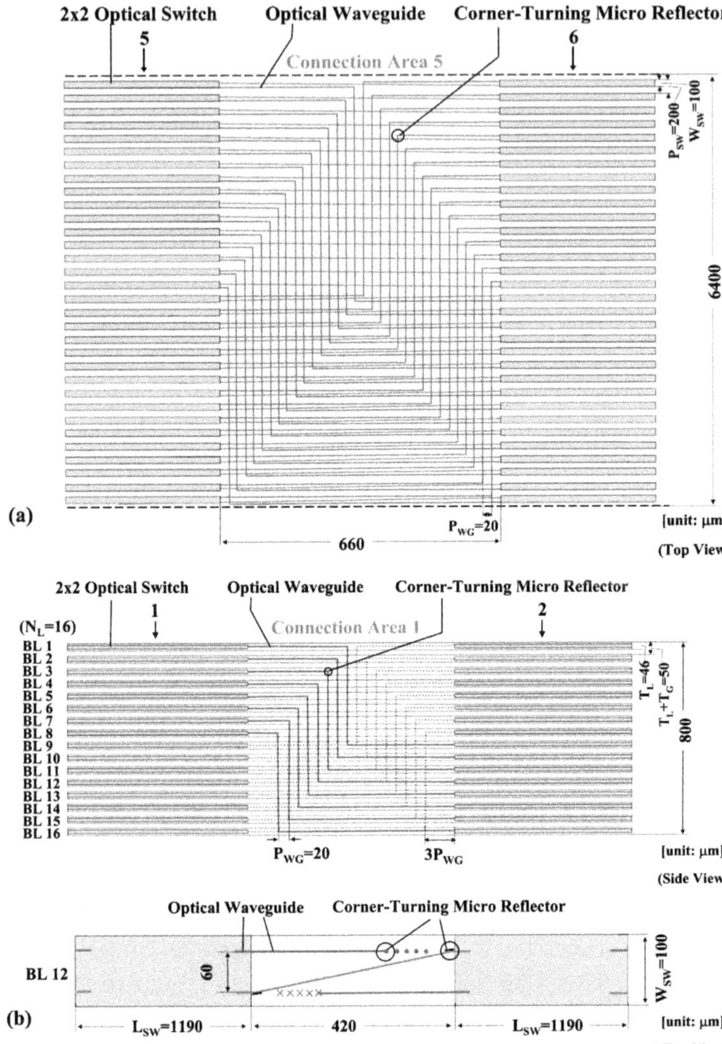

Figure 10.41. Detail of assumed connection patterns. (a) Top view of in-plane optical connection of connection area 5, and (b) side view of vertical optical connection of connection area 1 and top view of BL 12. Reprinted with permission from T. Yoshimura, M. Ojima, Y. Arai, and K. Asama, "Three-Dimensional Self-Organized Micro Optoelectronic Systems for Board-Level Reconfigurable Optical Interconnects — Performance Modeling and Simulation," *IEEE J. Select. Topics Quantum Electron.* **9**, 492–511. © 2003 IEEE.

In the present model, $P_{SW} = 200$ μm and $P_{WG} = 20$ μm. By substituting the parameters into Eqs. (10.2) and (10.3), the size of connection area 5 can be calculated to be 660 μm long and 6400 μm wide. In similar ways, the dimensions for $i = 6 \sim 9$ can be calculated as shown in Figs. 10.39 and 10.40. P_{WG} of 20 μm is wide enough for parallel wiring with length of 1 cm to keep cross talk less than –30 dB for the assumed 4 μm-wide waveguides and 1.3 μm wavelength.

As an example of vertical connection area, detail side view of connection area 1 and a top view of a part of the connection area in BL 12 are shown in Fig. 10.41(b). Inter-switch connection line count in a cross section is expressed as $N_L/2^{1-1}$ for connection area 1. Then, adding buffer spaces of $3P_{WG}$ for corner-turning micro reflectors at front and tail ends of the connection area, the length of connection area 1 is calculated to be $(N_L/2^{1-1} + 5)P_{WG}$. The width is the same as in the in-plane optical connection case. The height is $N_L(T_L + G_L)$. Thus, the expressions for connection area length $L(V)$, width $W(V)$, and height $H(V)$ of vertical connection area i can be generalized as follows:

$$L(V)_i = (N_L/2^{i-1} + 5)P_{WG} \tag{10.4}$$

$$W(V) = \left((N_C/2)/2^{D_{3D}}\right) P_{SW} \tag{10.5}$$

$$H(V) = N_L(T_L + G_L) \tag{10.6}$$

Parameter substitution into Eqs. (10.4–10.6) gives the size of connection area 1 of 420 μm long, 6400 μm wide, and 800 μm high. In similar ways, the dimensions for $i = 2 \sim 4$ can be calculated as shown in Figs. 10.38 and 10.39.

10.4.3.2 Simulated performance

Figure 10.42 shows a schematic illustration of a stacked structure model for simulation of performance in 3-D MOSS, where only a part of the model is drawn for simplicity. OE films are put with a pitch of $T_L + T_G$. In OE films, 2 × 2 WPD optical switches of length L_{SW} and width W_{SW} are arranged with a pitch of P_{SW}. Intra-layer wiring is carried out by in-plane waveguides with a pitch of P_{WG} while inter-layer wiring by optical Z-connections. Major losses caused in 3D-MOSS are indicated by "γ" in the figure. γ_P is propagation loss of optical waveguides, γ_{CR} loss induced at a right-angle

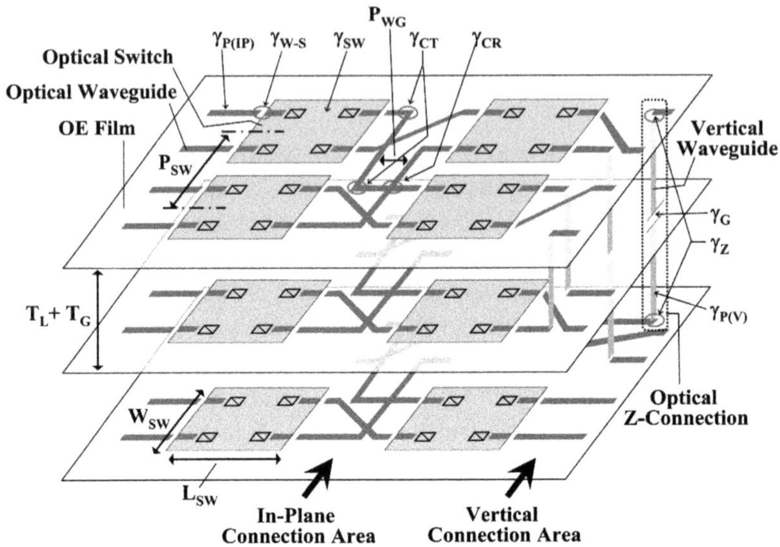

Figure 10.42. A schematic illustration of a stacked structure model for simulation of performance in 3D-MOSS. Reprinted with permission from T. Yoshimura, Y. Arai, H. Kurokawa, and K. Asama, "Predicted Insertion Loss Reductions Achieved by Implementing Three-Dimensional Micro Optical Network in Chip-Scale Optical Interconnects," *IEEE Photon. Technol. Lett.* **16**, 647–649. © 2003 IEEE.

waveguide-waveguide cross point, γ_Z loss of an optical Z-connection, γ_{CT} loss of a pair of corner turnings with two mirrors, γ_G loss at a layer gap, γ_{SW} loss within a 2 × 2 optical switch, and γ_{W-S} waveguide-switch coupling loss.

Simulated system performance of 1024 × 1024 3D-MOSS is summarized in Table 10.6 as well as those of conventional planar structures for comparison. The process to derive the performance is described below.

10.4.3.3 Size and insertion loss

For 3D-MOSS with S_{SW} and D_{3D}, connection areas 1 to D_{3D} have vertical optical connections, while connection areas $D_{3D} + 1$ to S_{SW} – 1 have in-plane optical connections. The connection area length for the in-plane connection areas and the vertical connection areas

Table 10.6. Summary of simulated system performance of 1024 × 1024 3D-MOSS with $D_{3D} = 4$. Size and insertion loss of a conventional in-plane structure are inserted for comparison

Size	Length	Width	Height
Conventional planar structure	3.252 cm	10.24 cm	0.005 cm
3D-MOSS ($D_{3D} = 4$)	1.424 cm	0.64 cm	0.08 cm
Insertion loss			
Conventional planar structure	43 dB		
3D-MOSS ($D_{3D} = 4$) Without misalignment	29 dB		
With 25% misalignment	73 dB		
With 25% misalignment using SOLNET	32 dB		
Electrical characteristics			
Operation voltage: V	12 V		
Switching speed: τ	456 ps		
Power Consumption at a Switching Rate: P at F	36 W/cm^2 at 1×10^7/s		

Source: Reprinted with permission from T. Yoshimura, M. Ojima, Y. Arai, and K. Asama, "Three-Dimensional Self-Organized Micro Optoelectronic Systems for Board-Level Reconfigurable Optical Interconnects — Performance Modeling and Simulation," *IEEE J. Select. Topics Quantum Electron.* **9**, 492–511.

are respectively expressed by Eqs. (10.2) and (10.4) and the width by Eqs. (10.3) and (10.5). Therefore, total 3D-MOSS length $L_{3D-MOSS}$, width $W_{3D-MOSS}$, and height $H_{3D-MOSS}$, including switch array length, are given by

$$L_{3D-MOSS} = L_{SW}S_{SW} + \left(\sum_{i=1}^{D_{3D}} (N_L/2^{i-1} + 5)\right) P_{WG} + \left(\sum_{i=D_{3D}+1}^{S_{SW}-1} (N_C/2^i + 1)\right) P_{WG} \quad (10.7)$$

$$W_{3D-MOSS} = \left((N_C/2)/2^{D_{3D}}\right) P_{SW} \quad (10.8)$$

$$H_{3D-MOSS} = N_L(T_L + G_L) \quad (10.9)$$

Major losses in 3D-MOSS are listed in Table 10.7. Propagation loss γ_P of optical waveguides is assumed 0.5 dB/cm, which is typical in fluorinated polyimide. Loss induced at a right-angle waveguide-waveguide cross point γ_{CR} is estimated by the BPM simulation to be 0.0088 dB. Loss of an optical Z-connection γ_Z of 0.97 dB was

Table 10.7. Major losses in 3D-MOSS

Propagation in optical waveguide: γ_P		0.5 dB/cm
Waveguide-waveguide cross point: γ_{CR}		0.0088 dB
Pair of corner turnings (2 micro reflectors): γ_{CT}		0.97 dB
Optical Z-connection: γ_Z		0.97 dB
Layer gap : γ_G	With no misalignment	0.0088 dB
	With 25% misalignment	0.89 dB
	With 25% misalignment using SOLNET	0.14 dB
2×2 Optical switch: γ_{SW}		1.7 dB
Waveguide-switch coupling: γ_{W-S}	With no misalignment	0 dB
	With 25% misalignment	0.83 dB
	With 25% misalignment using SOLNET	0.14 dB

Source: Reprinted with permission from T. Yoshimura, M. Ojima, Y. Arai, and K. Asama, "Three-Dimensional Self-Organized Micro Optoelectronic Systems for Board-Level Reconfigurable Optical Interconnects — Performance Modeling and Simulation," *IEEE J. Select. Topics Quantum Electron.* **9**, 492–511. © 2003 IEEE.

calculated in Section 10.4.2. Since a pair of corner turnings with two mirrors has the same structure as the optical Z-connection, the loss γ_{CT} can also be 0.97 dB. A cross point of a vertical waveguide and a layer gap of 4 μm can be regarded as a kind of right-angle waveguide-waveguide cross point. So, loss at a layer gap γ_G can be 0.0088 dB. BPM calculation reveals that γ_G with 25% misalignment is 0.89 dB, and loss for SOLNET coupling with 25% misalignment is 0.14 dB. Therefore, γ_G with 25% misalignment using SOLNET is 0.14 dB, neglecting the SOLNET length dependence of γ_G. Loss within a 2 × 2 optical switch γ_{SW} was estimated to be 1.7 dB in Section 9.1.3. Waveguide-switch coupling loss γ_{W-S} is basically 0 dB for no misalignment. For the case with 25% misalignment, the BPM calculation reveals that γ_{W-S} is 0.83 dB. For SOLNET coupling with 25% misalignment, similarly to the case of γ_G, γ_{W-S} is 0.14 dB. In the present simulation, boundary reflection losses are neglected, assuming AR coating on the edge surfaces of individual parts.

General expressions for eight kinds of losses in 3D-MOSS are derived in terms of the losses listed in Table 10.7. The results are summarized in Table 10.8. In-plane optical connections include in-plane propagation, waveguide-waveguide cross points, pairs of corner turnings, optical switches, and waveguide-switch couplings. Vertical optical connections include vertical propagation, optical

Table 10.8. General expressions for losses in 3D-MOSS

In-plane optical connections		
In-plane propagation		
$\left[\left(\sum_{i=1}^{D_{3D}} (N_L/2^{i-1}+5)\right) P_{WG} + \left(\sum_{i=D_{3D}+1}^{S_{SW}-1} (N_C/2^i+1)\right) P_{WG} + \left(\sum_{i=D_{3D}+1}^{S_{SW}-1} (N_C/2)/2^i\right) P_{SW}\right] \gamma_P$		
Waveguide-waveguide cross point	Pair of corner turnings	
$\left[\sum_{i=D_{3D}+1}^{S_{SW}-1} N_C/2^i\right] \gamma_{CR}$	$[S_{SW}-1-D_{3D}]\,\gamma_{CT}$	
Optical switch	Waveguide-Switch Coupling	
$[S_{SW}]\,\gamma_{SW}$	$[2S_{SW}]\,\gamma_{W-S}$	
Vertical optical connections		
Vertical propagation		
$[2N_L(T_L+T_G)]\,\gamma_P$		
Optical Z-connection	Layer gap	
$[2+D_{3D}]\,\gamma_Z$	$[2N_L]\,\gamma_G$	$[2+D_{3D}]\,\gamma_G$ for SOLNET

Source: Reprinted with permission from T. Yoshimura, M. Ojima, Y. Arai, and K. Asama, "Three-Dimensional Self-Organized Micro Optoelectronic Systems for Board-Level Reconfigurable Optical Interconnects — Performance Modeling and Simulation," *IEEE J. Select. Topics Quantum Electron.* **9**, 492–511. © 2003 IEEE.

Z-connections, and layer gaps. For in-plane propagation, the first and the second terms are caused from horizontal line length given by the second and the third terms of Eq. (10.7). The third term is caused from line length perpendicular to the horizontal direction. Since lines in vertical connection areas have no contribution to the line length perpendicular to the horizontal direction basically, the summation is carried out only for the in-plane connection areas. For waveguide-waveguide cross points, as shown in Fig. 10.41(a), an optical waveguide has a cross-point count of $N_C/2^i$ in in-plane connection area i. In vertical connection areas, no waveguide-waveguide cross points exist. Thus, the summation for the total loss calculation is carried out for the in-plane connection areas. Total loss due to a pair of corner turnings is determined by the number of pair of corner turnings that equals in-plane connection area count since optical signals experience one pair of corner turnings in one in-plane connection area. Total losses within optical switches and at waveguide-switch coupling points are respectively induced by S_{SW} optical switches and two S_{SW} waveguide-switch coupling points. For vertical propagation, as can be seen in Figs. 10.39 and 10.41(b), the total loss is determined by optical path length that corresponds to two times 3D-MOSS thickness. Since optical signals experience one optical Z-connection in one vertical connection area as shown in Fig. 10.41(b), total loss caused by optical Z-connections, with adding two optical Z-connections for input and output, is calculated to be $[2 + D_{3D}]\gamma_Z$. Total loss due to layer gaps is determined by two times the layer gap count contained in a 3D-MOSS. When SOLNET is implemented in vertical optical connections, total loss due to layer gaps is replaced with a product of total Z-connection count times γ_G for SOLNET.

By substituting the parameters in Table 10.5 into Eqs. (10.7), (10.8), and (10.9), length, width, and height of 3D-MOSS can be respectively calculated to be 1.424 cm, 0.64 cm, and 0.08 cm, as shown in Table 10.6. For conventional planar structures, length is 3.252 cm, width 10.24 cm, and height 0.005 cm. Therefore, it is found that drastic occupation area reduction of 3/100 is achieved in 3D-MOSS with $D_{3D} = 4$.

Insertion losses of 3D-MOSS, shown in Table 10.6, can be obtained by summation of eight losses listed in Table 10.8. While insertion loss of the conventional planar structure is 43 dB, that of the 3D-MOSS is 29 dB, indicating that 14 dB loss reduction is achieved in 3D-MOSS. In Fig. 10.43, occupation areas and insertion losses of 3D-MOSS are shown as a function of D_{3D}. $D_{3D} = 0$ implies conventional structure, and $D_{3D} = 9$ implies that all the connection areas consist of vertical optical connections. It is found that the occupation area decreases drastically with D_{3D}. Insertion loss, on the other hand, has a peak in the region of $D_{3D} = 4{\sim}6$. For small D_{3D}, main contribution to the loss is waveguide-waveguide cross points while layer gaps for large D_{3D}. The results indicate that size and loss of 3D-MOSS can be controlled by selecting D_{3D}.

Misalignment issue is a serious concern in 3D-MOSS. As shown in Table 10.7, misalignment between adjacent layers causes layer gap loss γ_G of 0.89 dB. At a waveguide-switch coupling, which is equivalent to a coupling between 4 μm wide waveguides, 25% misalignment induces loss γ_{W-S} of 0.83 dB. By substituting these loss values into the summation of the eight losses shown in Table 10.8, insertion loss of 3D-MOSS with 25% misalignment is estimated to be 73 dB as shown in Table 10.6. In this case, in order to transmit optical signals with reasonable power, threefold optical

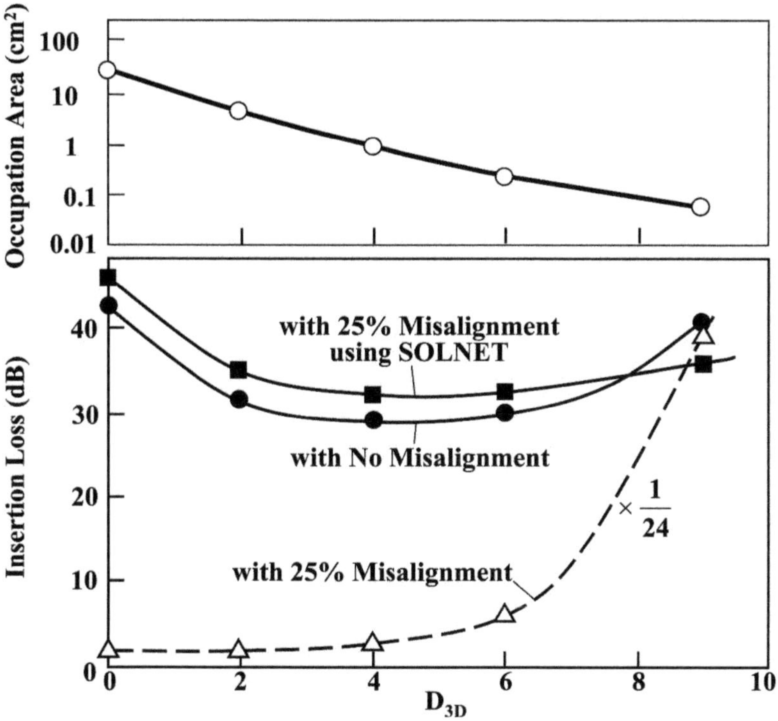

Figure 10.43. D_{3D} dependence of occupation areas and insertion losses of 3D-MOSS. Reprinted with permission from T. Yoshimura, Y. Arai, H. Kurokawa, and K. Asama, "Predicted Insertion Loss Reductions Achieved by Implementing Three-Dimensional Micro Optical Network in Chip-Scale Optical Interconnects," *IEEE Photon. Technol. Lett.* **16**, 647–649. © 2003 IEEE.

amplification might be required in 3D-MOSS by inserting amplifier layers. As can be seen in Fig. 10. 43, SOLNET implementation is effective to reduce the insertion loss. By using γ_G and γ_{W-S} of 0.14 dB listed in Table 10.7 for 25% misalignment using SOLNET, the insertion loss of the SOLNET-implemented 3D-MOSS is calculated to be 32 dB, which is close to the loss in 3D-MOSS without misalignment.

It was reported that high-accuracy bonding equipment enables layer stacking with a positional accuracy of 1 μm for 1 cm^2 chips [40]. Therefore, for 3D-MOSS of 1.424 cm × 0.64 cm, it is consistent to assume maximum positional misalignment of $\sim$1 μm that corresponds to 25% misalignment for optical circuit backbones consisting of optical waveguides with $4 \times 4\ \mu m^2$ core cross section.

The insertion loss of 32 dB for 1024 × 1024 SOLNET-implemented 3D-MOSS is still high, which requires an optical amplifier layer insertion or 10 mW-order input light signals. For further simplification of the system, each loss listed in Table 10.7 should be decreased by optimizing individual structures and materials.

10.4.3.3.1 Electrical characteristics Switching speed and power consumption can be estimated based on the charge-discharge model for the prism-shaped capacitor in the waveguide prism deflector. The WPD optical switch model used in the present calculation is the same as the model shown in Fig. 9.19 except that the PLZT thickness is increased from 400 nm to 4 μm. Then, the capacitance C, switching speed τ, and power consumption Power that are given by Eqs. (9.9), (9.8), and (9.10) for one prism-shaped capacitor are respectively calculated to be 0.57 pF and 57 ps for driver on-resistivity $R_{ON} = 100\ \Omega$, and 41 mW for $F = 5\times 10^8$ 1/s and V =12 V.

When the number of the waveguide prism deflectors in a WPD optical switch is N_{Prism}, switching speed τ for R_{ON} is given by

$$\tau = N_{Prism}\ R_{ON}C \tag{10.10}$$

For power consumption, total WPD optical switch count in a 3D-MOSS is $(N_C/2)S_{SW}$, giving $N_{Prism}(N_C/2)S_{SW}$ for total count of the prism-shaped capacitors contained in a 3D-MOSS. Then, total capacitance arising from the deflectors is given by $C\ N_{Prism}(N_C/2)S_{SW}$. As a result, power consumption at a switching rate of F in a 3D-MOSS is expressed as follows:

$$\text{Power} = F\ C\ V^2\ N_{Prism}\ (N_C/2)S_{SW} \tag{10.11}$$

By substituting $C = 0.57$ pF, $V = 12$ V, $N_{Prism} = 8$, $N_C = 1024$, and $S_{SW} = 10$ into (10.10) and (10.11), $\tau = 456$ ps for $R_{ON} = 100\ \Omega$, and Power = 33 W for $F = 1 \times 10^7$ 1/s are obtained. The power consumption corresponds to power density of 36 W/cm^2 since the area of the 3D-MOSS model is 1.42×0.64 cm^2. This value is comparable to acceptable heat generation of 30 W/cm^2 reported

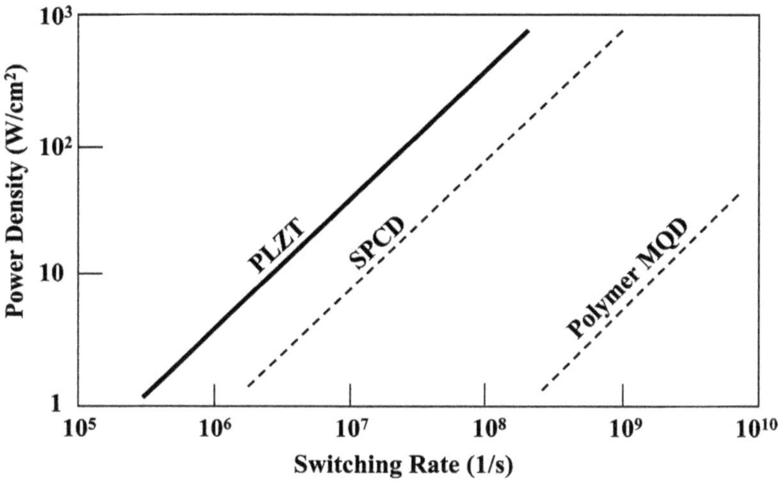

Figure 10.44. Dependence of power density generated in 3D-MOSS on switching rates for various EO materials of WPD optical switch. Reprinted with permission from T. Yoshimura, M. Ojima, Y. Arai, and K. Asama, "Three-Dimensional Self-Organized Micro Optoelectronic Systems for Board-Level Reconfigurable Optical Interconnects — Performance Modeling and Simulation," *IEEE J. Select. Topics Quantum Electron.* **9**, 492–511. © 2003 IEEE.

by K. Takahashi [40] for stacked LSI of 1 cm^2 area. These results are included in Table 10.6.

Figure 10.44 shows switching rate dependence of power density generated in 3D-MOSS for various EO materials of WPD optical switch. Power density increases with switching rate. Possible switching rate is determined by a balance of heat generation and releasing rates in 3D-MOSS. For a given heat releasing rate, the lower the heat generation rate becomes, the higher the available switching rate becomes as far as switching speed of WPD optical switch τ is sufficiently fast.

The power density versus switching rate relationship varies depending on EO materials. In PLZT, switching rate is limited below 1×10^7 1/s when the limit of the heat releasing capability is 30 W/cm^2. In SPCD organic crystals, which has EO coefficient of 430 pm/V, expected dielectric constant of $\sim 4\varepsilon_0$ and refractive index of 1.55 as mentioned in Section 9.1.2, the line move to the right-hand side, indicating that higher switching rates are expected than in PLZT. Drastic switching rate improvement is expected in polymer MQDs [41–44], which has a potentiality of EO coefficient of $\sim$1000 pm/V, dielectric constant of $\sim 4\varepsilon_0$ and refractive index of $\sim$2, resulting in a possibility of switching rate close to 10^{10} 1/s.

These results of performance simulation indicate that 1024 $\times$ 1024 3D-MOSS is viable for a switching system in board-level reconfigurable optical interconnects as well as for massive optical packet switching systems from viewpoints of size, insertion loss, switching speed, and thermal management.

10.4.3.4 Impact of HIC waveguide implementation into 3D-MOSS

When HIC waveguides having a 400 $\times$ 400 nm^2 core cross section are used for the 3-D optical circuit backbone in the 3D-MOSS, as described in Section 10.4.2, the PLZT thickness, namely, the electrode gap in the WPD optical switches can be reduced to 400 nm. In this case, switching voltage is reduced to 1.2 V to achieve low-voltage switching, which is favorable for low-power and high-speed operation. The HIC-waveguide-based optical circuits might be fabricated by the PRI sol-gel materials described in Sections 5.4 and 6.4.

References

1. T. Yoshimura, J. Roman, Y. Takahashi, W. V. Wang, M. Inao, T. Ishitsuka, K. Tsukamoto, K. Motoyoshi, and W. Sotoyama, "Self-Organizing Waveguide Coupling Method "SOLNET" and Its Application to Film Optical Circuit Substrates," *Proc. 50th Electron. Components Technol. Conf. (ECTC)*, 962–969 (2000).
2. T. Yoshimura, J. Roman, Y. Takahashi, W. V. Wang, M. Inao, T. Ishituka, K. Tsukamoto, K. Motoyoshi, and W. Sotoyama, "Self-Organizing Lightwave Network (SOLNET) and Its Application to Film Optical Circuit Substrates," *IEEE Trans. Comp., Packag. Technol.* **24**, 500–509 (2001).
3. T. Yoshimura, W. Sotoyama, K. Motoyoshi, T. Ishitsuka, K. Tsukamoto, S. Tatsuura, H. Soda, and T. Yamamoto, "Method of Producing Optical

Waveguide System, Optical Device and Optical Coupler Employing the Same, Optical Network and Optical Circuit Board," U.S. Patent 6,081,632 (2000).

4. T. Yoshimura, A. Hori, Y. Yoshida, Y. Arai, H. Kurokawa, T. Namiki, and K. Asama, "Coupling Efficiencies in Reflective Self-Organized Lightwave Network (R-SOLNET) Simulated by the Beam Propagation Method," *IEEE Photon. Technol. Lett.* **17**, 1653–1655 (2005).
5. H. Kaburagi and T. Yoshimura, "Simulation of Micro/Nanometer-Scale Self-Organized Lightwave Network (SOLNET) Using the Finite Difference Time Domain Method," *Opt. Commun.* **281**, 4019–4022 (2008).
6. T. Yoshimura and H. Kaburagi, "Self-Organization of Optical Waveguides Between Misaligned Devices Induced by Write-Beam Reflection," *Appl. Phys. Express* **1**, 062007-1-3 (2008).
7. T. Yoshimura, T. Inoguchi, T. Yamamoto, S. Moriya, Y. Teramoto, Y. Arai, T. Namiki, and K. Asama, "Self-Organized Lightwave Network Based on Waveguide Films for Three-Dimensional Optical Wiring Within Boxes," *J. Lightwave Technol.* **22**, 2091–2100 (2004).
8. T. Yoshimura, M. Ojima, Y. Arai, and K. Asama, "Three-Dimensional Self-Organized Micro Optoelectronic Systems for Board-Level Reconfigurable Optical Interconnects — Performance Modeling and Simulation," *IEEE J. Select. Topics Quantum Electron.* **9**, 492–511 (2003).
9. T. Yoshimura, M. Miyazaki, Y. Miyamoto, N. Shimoda, A. Hori, and K. Asama, "Three-Dimensional Optical Circuits Consisting of Waveguide Films and Optical Z-Connections," *J. Lightwave Technol.* **24**, 4345–4352 (2006).
10. T. Yoshimura, K. Wakabayashi, and S. Ono, "Analysis of Reflective Self-Organized Lightwave Network (R-SOLNET) for Z-Connections in Three-Dimensional Optical Circuits by the Finite Difference Time Domain Method," *IEEE J. Select. Topics Quantum Electron.* **17**, 566–570 (2011).
11. T. Yoshimura and H. Kaburagi, "Self-Organization of Coupling Optical Waveguides by the 'Pulling Water' Effect of Write Beam Reflections in Photo-Induced Refractive-Index Increase Media," *Proc. SPIE* **7221**, from Photonics West 2009, San Jose, California, 722111-1-10 (2009).
12. Y.S. Liu, "Lighting the Way in Computer Design," *Circuits Devices*, **January**, 23–31 (1998).
13. A. Glebov, M. G. Lee, D. Kudzuma, J. Roman, M. Peters, L. Huang, and S. Zhou, "Integrated Waveguide Microoptic Elements for 3D Routing

in Board-Level Optical Interconnects," *Proc. SPIE* **6126**, 61260L-1-10 (2006).

14. M. S. Bakir, A. L. Glebov, M. G. Lee, P. A. Kohl, and J. D. Meindl, "Mechanically Flexible Chip-to-Substrate Optical Interconnections Using Optical Pillars," *IEEE Trans. Adv. Packag.* **31**, 143–153 (2008).
15. T. Matsumoto, Y. Noguchi, Y. Kudoh, and M. Koyanagi, "Polyimide Optical Waveguide with Multi-Fan-Out for Multi-Chip Module Application," *Jpn. J. Appl. Phys.* **36**, 1903–1906 (1997).
16. S. Koike, F. Shimokawa, T. Matsuura, and H. Takahara, "New Optical and Electrical Packaging Techniques Using Optical Waveguides for Optoelectronic Multichip Modules," *IEEE Trans. Comp. Hybrids. Manuf. Technol.* **19**, 124–130 (1996).
17. T. Mikawa, M. Kinoshita, K. Hiruma, T. Ishitsuka, M. Okabe, S. Hiramatsu, H. Furuyama, T. Matsui, K. Kumai, O. Ibaragi, and M. Bonkohara, "Implementation of Active Interposer for High-Speed and Low-Cost Chip Level Optical Interconnects," *IEEE J. Select. Topics Quantum Electron.* **9**, 452–459 (2003).
18. B. Chou, S. Beilin, H. Jiang, D. Kudzuma, M. Lee, M. McCormack, T. Massingill, M. Peters, J. Roman, Y. Takahashi, and W. V. Wang, "Multilayer High Density Flex Technology," *Proc. 49th Electron. Components Technol. Conf. (ECTC),* 1181–1189 (1999).
19. T. Yoshimura, J. Roman, Y. Takahashi, M. Lee, B. Chou, S. Beilin, W. Wang, and M. Inao, "Optoelectronic Scalable Substrates Based on Film/Z-Connection and Its Application to Film Optical Link Module (FOLM)," *Proc. SPIE* **3952**, from Photonics West 2000, San Jose, California, 202–213 (2000).
20. T. Yoshimura, J. Roman, Y. Takahashi, M. Lee, B. Chou, S. I. Beilin, W. V. Wang, and M. Inao, "Proposal of Optoelectronic Substrate with Film/Z-Connection Based on OE-Film," *Proc. 3rd IEMT/IMC Symp.*, 140–145 (1999).
21. T. Yoshimura, T. Inoguchi, T. Yamamoto, S. Moriya, Y. Teramoto, Y. Arai, T. Namiki, and K. Asama, "Self-Organized Lightwave Network Based on Waveguide Films for Three-Dimensional Optical Wiring Within Boxes," *J. Lightwave Technol.* **22**, 2091–2100 (2004).
22. T. Yoshimura, Y. Takahashi, M. Inao, M. Lee, W. Chou, S. Beilin, W.-C. Wang, J. Roman, T. Massingill, "Systems Based on Opto-Electronic Substrates with Electrical and Optical Interconnections and Methods for Making," U.S. Patent 6,343,171 B1 (2002).
23. T. Yoshimura, Y. Takahashi, M. Inao, M. Lee, W. Chou, S. Beilin, W-C. Wang, J. Roman, T. Massingill, "Optoelectronic Substrates with Electrical and

Optical Interconnections and Method for Making," U.S. Patent 6,611,635 B1 (2003).

24. T. Yoshimura, Y. Takahashi, J. Roman, M. T. McCormack, S. Beilin, W.-C. Wang, M. Inao, "Optical Coupling Structures and the Fabrication Processes," U.S. Patent 6,684,007 B2 (2004).
25. T. Yoshimura, Y. Takahashi, M. Inao, M. Lee, W. Chou, S. Beilin, W.-C. Wang, J. Roman, T. Massingill, "Three-Dimensional Opto-Electronic Modules with Electrical and Optical Interconnections and Methods for Making," U.S. Patent 6,690,845 B1 (2004).
26. T. Yoshimura, Y. Takahashi, K. Kuwabara, S. Beilin, M. Peters, W.-C. Wang, M. Inao, "Optical Reflective Structures and Method for Making," U.S. Patent 6,706,546 B2 (2004).
27. T. Yoshimura, Y. Takahashi, J. Roman, M. T. McCormack, S. Beilin, W.-C. Wang, M. Inao, "Single and Multilayer Waveguides and Fabrication Process," U.S. Patent 6,785,447 B2 (2004).
28. T. Yoshimura, Y. Takahashi, M. Inao, M. Lee, W. Chou, S. Beilin, W.-C. Wang, J. Roman, T. Massingill, "Multi-Layer Opto-Electronic Substrates with Electrical and Optical Interconnections and Methods for Making," U.S. Patent 6,845,184 B1 (2005).
29. N. M. Jokerst, M. A. Brooke, J. Laskar, D. S. Wills, A. S. Brown, O. Vendier, S. Bond, J. Cross, M. Vrazel, M. Thomas, M. Lee, S. Jung, Y. J. Joo, and J. J. Chang, "Smart Photonics: Optoelectronics Integrated Onto Si CMOS Circuits," *Proc. 1999 IEEE LEOS Annu. Meeting Conf.*, San Francisco, California, 423–424 (1999).
30. Y. Liu, "Smart Pixel Module Development for Free Space Optical Interconnect," *Optics in Computing '98* (OC '98), *Proc. SPIE* **3490**, from Optics in Computing 1998 (OC 98), Bruges, Belgium, 528–531 (1998).
31. T. Yoshimura, J. Roman, Y. Takahashi, S. Beilin, W. Wang, and M. Inao, "Optoelectronic Amplifier/Driver-Less Substrate <OE-ADLES> for Polymer-Waveguide-Based Board-Level Interconnection — Calculation of Delay and Power Dissipation," 4th International Conference on Organic Nonlinear Optics (ICONO'4), Extended Abstracts, Chitose, Japan, 278 (1998).
32. T. Yoshimura, J. Roman, Y. Takahashi, S. I. Beilin, W. V. Wang, and M. Inao, "Optoelectronic Amplifier/Driver-Less Substrate <OE-ADLES> for Polymer-Waveguide-Based Board-Level Interconnection — Calculation of Delay and Power Dissipation," *Nonlinear Opt.* **22**, 453–456 (1999).
33. T. Yoshimura, Y. Suzuki, N. Shimoda, T. Kofudo, K. Okada, Y. Arai, and K. Asama, "Three-Dimensional Chip-Scale Optical Interconnects

and Switches with Self-Organized Wiring Based on Device-Embedded Waveguide Films and Molecular Nanotechnologies," *Proc. SPIE* **6126**, from Photonics West 2006, San Jose, California, 612609-1-15 (2006).

34. Y. Lin, L. Lin, C. Choi, B. Bihara, and R. Chen, "Optoelectronic Integration of Polymer Waveguide Array and Metal–Semiconductor–Metal Photodetector Through Micromirror Couplers," *IEEE Photon. Technol. Lett.* **13**, 355–357 (2001).
35. T. Yoshimura, Y. Arai, H. Kurokawa, and K. Asama, "Predicted Insertion Loss Reductions Achieved by Implementing Three-Dimensional Micro Optical Network in Chip-Scale Optical Interconnects," *IEEE Photon. Technol. Lett.* **16**, 647–649 (2003).
36. T. Yoshimura, S. Tsukada, S. Kawakami, Y. Arai, H. Kurokawa, and K. Asama, "3D Micro Optical Switching System (3D-MOSS) — Packaging Design," *2001 IEEE/LEOS Annu. Meeting Conf. Proc.*, San Diego, California, 499–500 (2001).
37. T. Yoshimura, S. Tsukada, S. Kawakami, M. Ninomiya, Y. Arai, H. Kurokawa, and K. Asama, "Three-Dimensional Micro-Optical Switching System Architecture Using Slab-Waveguide-Based Micro-Optical Switches," *Opt. Eng.* **42**, 439–446 (2003).
38. T. Yoshimura, S. Tsukada, S. Kawakami, Y. Arai, H. Kurokawa, and K. Asama, "3D Micro Optical Switching System (3D-MOSS) Architecture," *Proc. SPIE* **4653**, from Photonics West 2002, San Jose, California, 62–70 (2002).
39. T. Yoshimura, Y. Arai, "3D Optoelectronic Micro System," U.S. Patent 2006/0261432 A1 (2006).
40. K. Takahashi, "Research and Development on Ultra-High-Density 3-Dimensional LSI-Chip-Stack Packaging Technologies," *3rd Annual Meeting on Electronics System Integration Technologies Digest* (edited and published by the Electronic System Integration Technology Research Department, Association of Super-Advanced Electronics Technologies (ASET)), 43–94 (2002).
41. T. Yoshimura, "Enhancing Second-Order Nonlinear Optical Properties by Controlling the Wave Function in One-Dimensional Conjugated Molecules," *Phys. Rev.* **B 40**, 6292–6298 (1989).
42. T. Yoshimura, S. Tatsuura, W. Sotoyama, A. Matsuura, and T. Hayano, "Quantum wire and Dot Formation by Chemical Vapor Deposition and Molecular Layer Deposition of One-Dimensional Conjugated Polymer," *Appl. Phys. Lett.* **60**, 268–270 (1992).

43. T. Yoshimura, A. Oshima D. Kim and Y. Morita "Quantum Dot Formation in Polymer Wires by Three-Molecule Molecular Layer Deposition (MLD) and Applications to Electro-Optic/Photovoltaic Devices" *ECS Trans.* **25** (4) "Atomic Layer Deposition Applications 5," from the Vienna, Austria Meeting, 15–25 (2009).
44. T. Yoshimura, *Thin-Film Organic Photonics: Molecular Layer Deposition and Applications*, CRC/Taylor & Francis, Boca Raton, Florida, (2011).

Chapter 11

Integrated Solar Energy Conversion Systems

The integrated film systems shown in Fig. 11.1 will provide various types of systems with ubiquitous functions, including solar energy conversion systems, wearable computers, system glasses, thin-film speakers, and so on [1]. The integrated film systems, whose structures are similar to those of the OE films for optical interconnects within boxes and 3D-MOSS, are one of the important applications of core technologies of optical electronics.

In this chapter, the concept of the film-based integrated solar energy conversion systems and the related devices such as the integrated photonic/electronic/chemical system (IPECS) and the light beam collecting films are proposed [1–10]

11.1 Concept

Solar energy conversion systems have received much attention because of their capability to solve the problems of environment and energy shortage. In conventional energy conversion systems, semiconductors such as Si are placed all over the solar energy

Optical Electronics: Self-Organized Integration and Applications
Tetsuzo Yoshimura

ISBN 978-981-4310-90-1 (Hardcover), 978-981-4364-08-9 (eBook)
www.panstanford.com

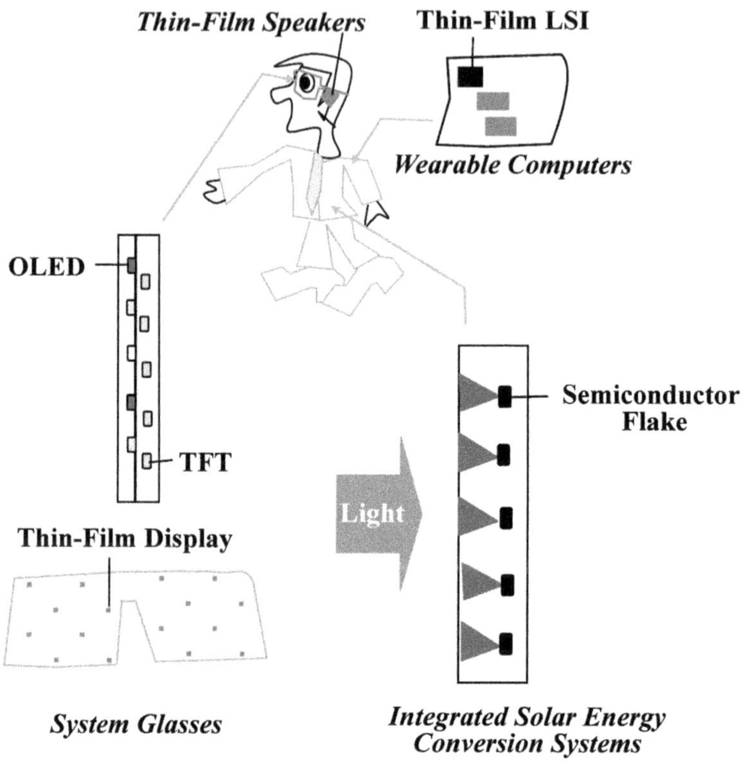

Figure 11.1. Integrated film systems.

conversion modules; so it is foreseen that the cost of the systems will increase by the shortage of semiconductor materials. In systems with Fresnel lens/mirror concentrators, semiconductor materials can be saved because solar beams are focused on small semiconductor cells. However, the system size becomes large, and furthermore, they need solar-beam tracking mechanisms to keep the focusing point of the solar beams on the small-size semiconductor cells.

To reduce the consumption of semiconductor materials and to provide wide-angle light beam collecting capability to the systems, and at the same time, to make systems flexible and compact, we proposed the film-based integrated solar energy conversion systems with optical waveguides [1–10].

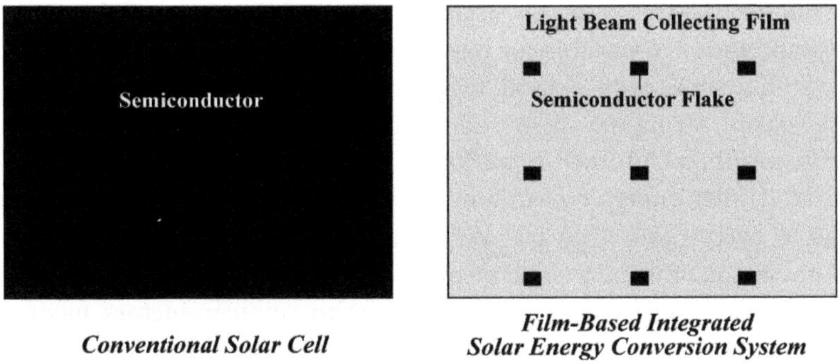

Figure 11.2. Schematic illustrations of (a) the conventional solar cell and (b) the proposed film-based integrated solar energy conversion system.

Figure 11.2 shows schematic illustrations of the conventional solar cell and the proposed film-based integrated solar energy conversion system. In the proposed system, semiconductor flakes are placed partially in a light beam collecting film by the heterogeneous integration process such as PL-Pack with SORT [11].

The advantages and possible applications of the film-based integrated solar energy conversion systems are summarized in Table 11.1. The proposed structures can reduce semiconductor material consumption. If necessary, expensive III-V compounds, for

Table 11.1. Advantages and possible applications of the film-based integrated solar energy conversion systems

Advantages
• Reduction of semiconductor/ITO consumption, cost, size, and weight
• Flexibility
• Wide-angle light beam collection capability
• Function availability by implementation of optical circuits
○ Demultiplexing for delivering appropriate wavelength light beams to semiconductors having matched sensitivity spectra
• Function availability by embedding thin-film batteries and ICs
○ Charge storage and current regulation capability
• Preferable for integration of the waveguide-type photo-voltaic devices
Applications
• Curtains, sunshades/umbrellas, sheets, clothes, etc.
• Put on buildings, walls, fences, roads, cars, robots, computers, desks, cellular phones, hand-carrying equipment, consumer electronics, etc.

example, GaAs, could be used for the semiconductor. The III-V compounds are suitable for high-efficiency solar energy conversion systems since they exhibit higher electron mobility than Si. The proposed structures also reduce transparent electrode material consumption. Another remarkable feature of the film-based integrated solar energy conversion systems is flexibility, which enables us to carry them with us everywhere, and put them everywhere. Conversion efficiency could be improved by delivering collected light beams of different wavelength regions to semiconductors having sensitivity in matched wavelength regions using wavelength filters embedded in the optical circuits of the light beam collecting films. It might be possible to embed thin-film batteries and ICs in the films. The film-based systems have additional feature that the waveguide-type photo-voltaic devices described in Section 11.4 are easily integrated into the optical waveguides in the light beam collecting films.

Due to the flexible film structures, as illustrated in Fig. 11.3, the film-based integrated solar energy conversion systems are expected to be used as curtains, sunshades/umbrellas, and sheets with

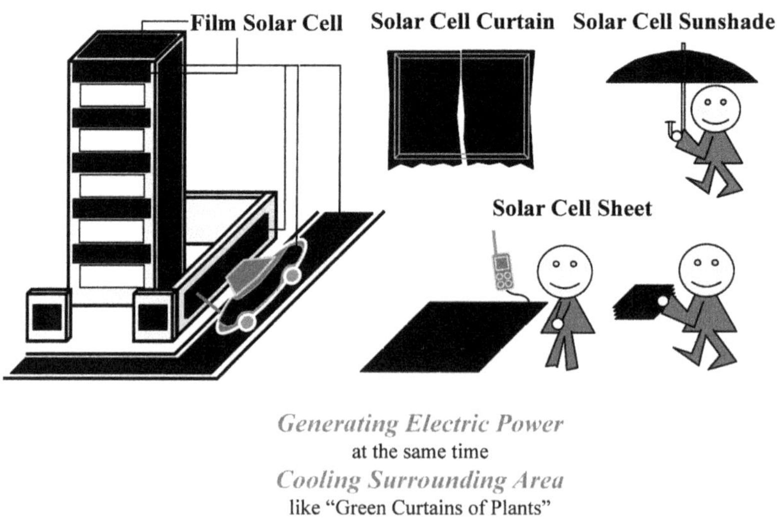

Figure 11.3. Schematic illustrations of applications of the film-based integrated solar energy conversion system.

electric power generation capabilities. The film-based systems can also be put on buildings, walls, fences, roads, cars, robots, etc. Since the solar energy conversion systems have an effect to suppress heat generation arising from solar beam exposure, they act as the "green curtains of plants" to cool down the temperature of surrounding areas, including temperature of the city.

The film-based integrated solar energy conversion system is regarded as a kind of the integrated IPECS. The goal of IPECS is schematically shown in Fig. 11.4(a). IPECS has a similar structure as the integrated optical interconnects shown in Fig. 11.4(b). IPECS contains micro fluidic circuits (MFCs) and semiconductors, realizing thin-film photo-voltaic devices and photo-synthesis devices to generate electrical power and chemical products. The MFCs add chemical reactions to the self-organized 3-D OE platform, providing energy/bio OE systems with 3-D optical circuits.

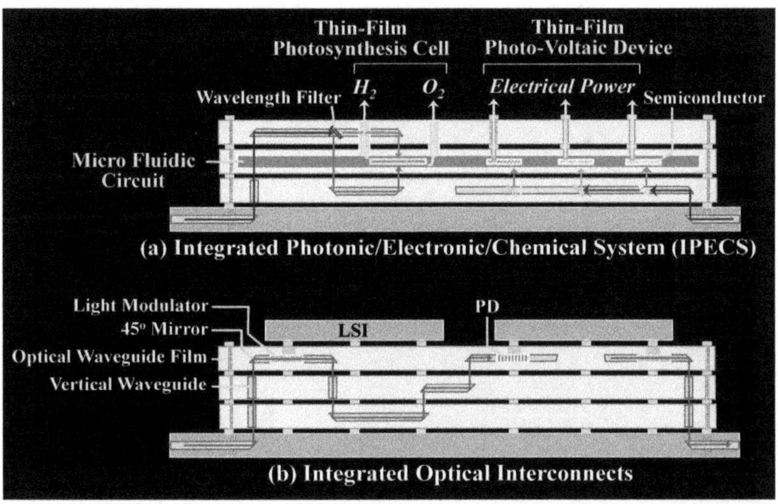

Figure 11.4. (a) Integrated photonic/electronic/chemical system (IPECS) and (b) integrated optical interconnects. Reprinted with permission from T. Yoshimura, "Proposed Applications of 3-D Optical Interconnect Technologies to Integrated Chemical Systems," 2007 Digest of the LEOS Summer Topical Meetings [Bio-Inspired Sensors and Application/Imprinting on Photonic Integrated Circuits], Portland, Oregon, 129–130. © 2007 IEEE.

11.2 Light Beam Collecting Films

11.2.1 *Structures*

We proposed three types of light beam collecting films: tapered vertical waveguide type, tapered vertical/horizontal waveguide type, and multi-layer waveguide type.

Figure 11.5 shows an example of the integrated solar energy conversion system with a tapered vertical waveguide-type light beam collecting film. Semiconductor flakes for thin-film photovoltaic or photo-synthesis devices are embedded in the film. The incident light beams received by the tapered vertical waveguides are guided to the semiconductor flakes, whose areas are smaller than the areas of the light beam collecting aperture of the vertical waveguide. This enables the reduction of semiconductor consumption, namely, cost for solar energy conversion. In this example, the material consumption can be reduced to ~1/9 of the consumption in the conventional solar cells.

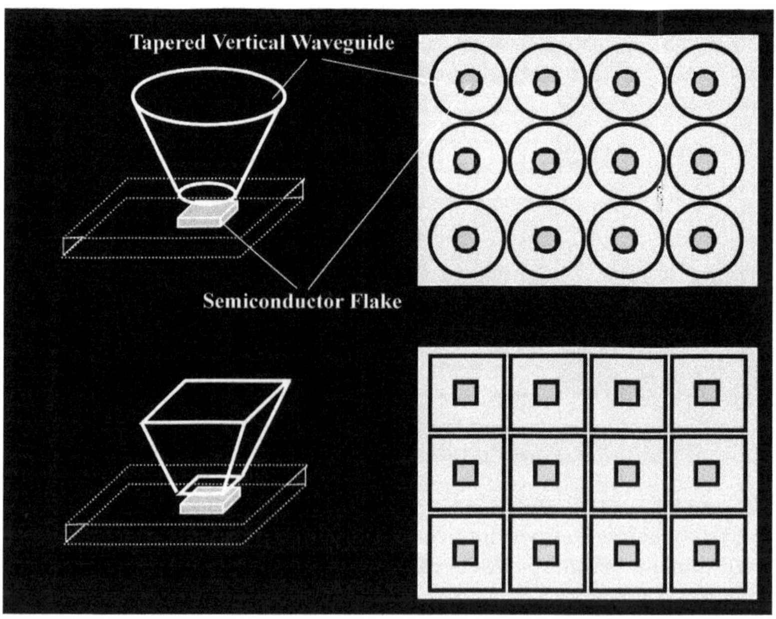

Figure 11.5. Integrated solar energy conversion system with a tapered vertical waveguide-type light beam collecting film.

Figure 11.6 shows an example of the system with a tapered vertical/horizontal waveguide-type light beam collecting film. A film with tapered vertical waveguides and a film with horizontal waveguides having vertical mirrors are stacked. The incident light beams are introduced into the horizontal waveguides through the vertical mirrors and guided to semiconductor flakes embedded in the horizontal waveguide film. This enables further reduction of semiconductor consumption. In this example, the material consumption can be reduced to $\sim(1/9) \times (1/16) = \sim 1/144$ of the

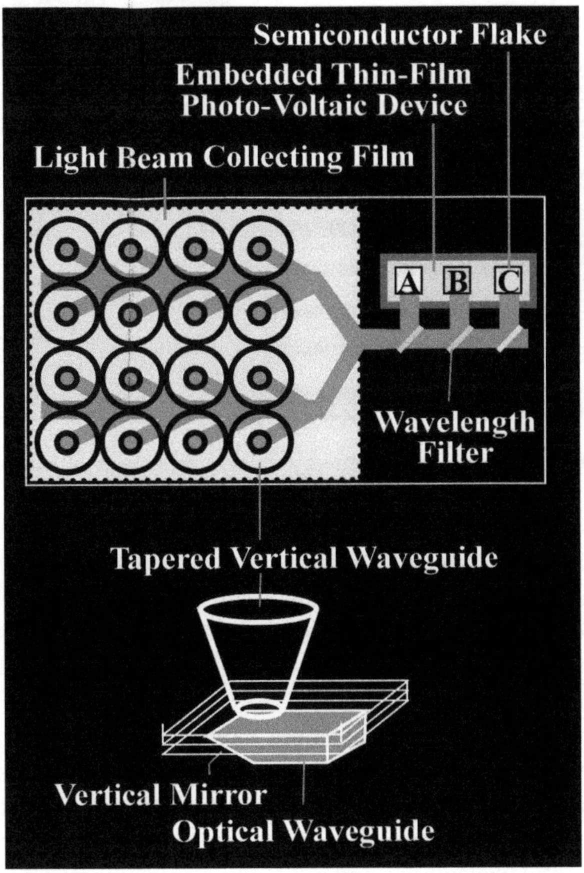

Figure 11.6. Integrated solar energy conversion system with a tapered vertical/horizontal waveguide-type light beam collecting film.

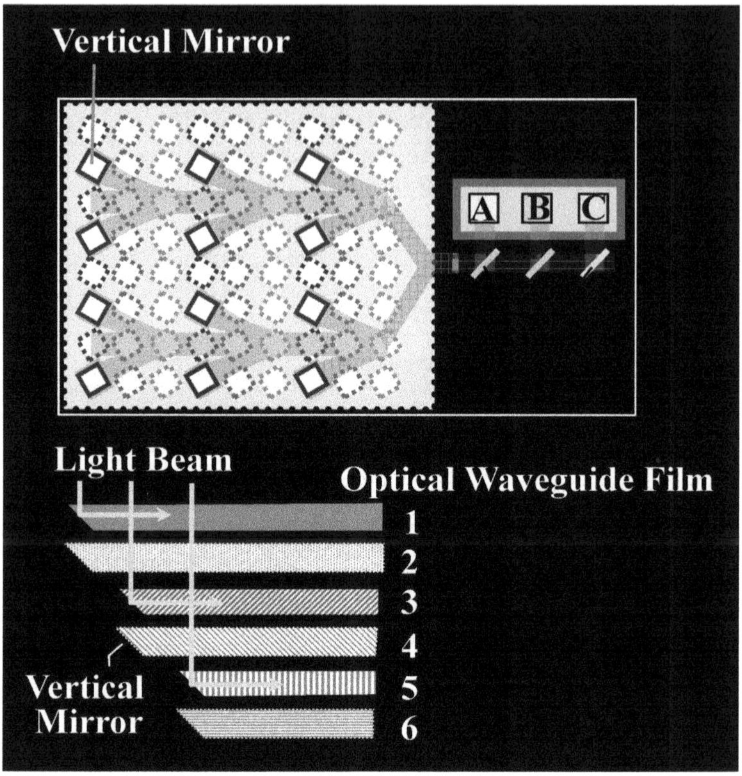

Figure 11.7. Integrated solar energy conversion system with a multi-layer waveguide-type light beam collecting film.

consumption in conventional solar cells. By embedding wavelength filters into the optical waveguide films, solar beams can branch to semiconductors having appropriate matched spectral responses. For example, UV-blue beams to ZnO, visible beams to Si, and infrared beams to InP.

Figure 11.7 shows an example of the system with a multi-layer waveguide-type light beam collecting film. Many films having horizontal waveguides with vertical mirrors are stacked. The incident light beams are introduced through the vertical mirrors into the horizontal waveguides and guided to semiconductor flakes.

Figure 11.8 shows an example of coloring of the integrated solar energy conversion systems by back lights. In conventional solar cells,

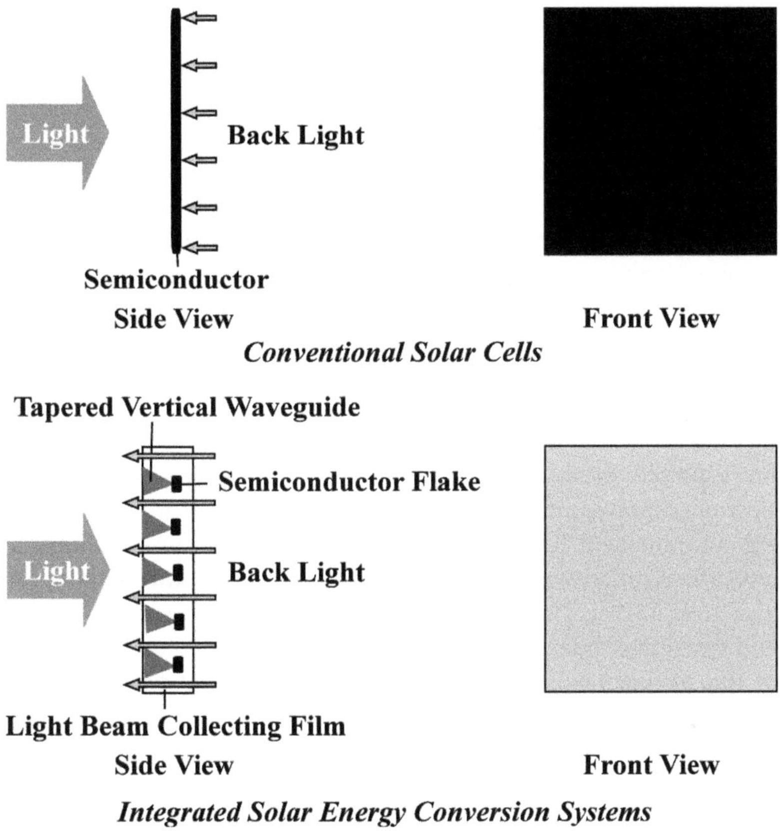

Figure 11.8. Coloring of the integrated solar energy conversion systems by back lights.

since semiconductors are placed on the whole substrate, the cells look black. To avoid the monochromatic characteristics of solar cells, in some dye-sensitized solar cells, colored appearance is available by suppressing light absorption in some wavelength regions. This causes the reduction of energy conversion efficiency. In the case of the integrated solar energy conversion systems, on the other hand, back light can pass through the light beam collecting film. By adjusting the color of the back lighting, favorite coloration of the film may be achievable, which improves appearance of solar cells.

11.2.2 *Design*

In this subsection, the results on the design of the light beam collecting films and their light beam collecting ability estimated by the BPM and the FDTD method are described [8].

The tapered vertical waveguide-type light beam collecting film and the tapered vertical/horizontal waveguide-type light beam collecting film are basically regarded as a similar structure except for the merging waveguides. So, the simulation was carried out for the tapered vertical/horizontal waveguide-type light beam collecting film and the multi-layer waveguide-type light beam collecting film.

11.2.2.1 Simulation procedure

An example of a simulation model for the tapered vertical/horizontal waveguide-type light beam collecting film is shown in Fig. 11.9 Incident light beams from a metal mirror are introduced into a horizontal waveguide via a tapered vertical waveguide and a vertical mirror. The refractive index of optical waveguides is 1.6, and refractive index of background is 1.0 assuming air. The width of the horizontal waveguide is 4 μm. AR coating with 100 nm thickness and refractive index of 1.3 is applied to the top and the bottom of the optical waveguide. Incident light beams with a large

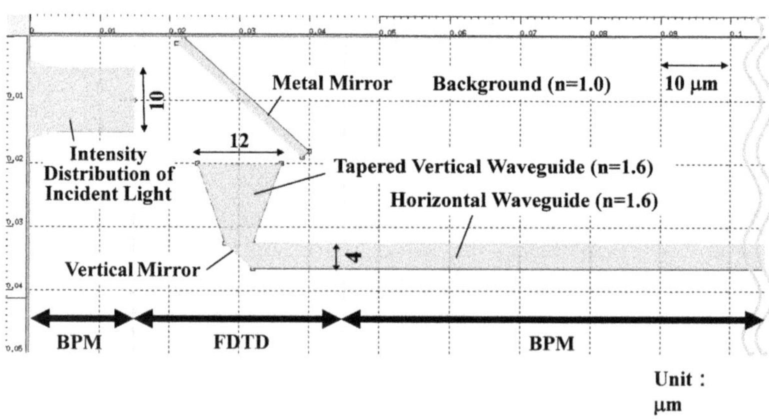

Figure 11.9. An example of a simulation model for the tapered vertical/horizontal waveguide-type light beam collecting film.

width are preferable to reproduce the condition of plane waves of sunlight. However, when the width is broadened, the area of the FDTD calculation must be large, making computation time extremely long. So in the present work, the simulations were performed using fundamental guided beams in an optical waveguide with a width of 10 μm for incident light beams. Incident angles of the light beams are adjusted by the metal mirror. The wavelengths of incident light beams are 400, 600, and 800 nm. The length of the model from the left boundary to the right boundary is 200 μm. Light intensity at the output (the right boundary of the model) is calculated to estimate light beam collecting efficiency for various models.

11.2.2.2 Tapered vertical/horizontal waveguide-type light beam collecting films

In Fig. 11.10, a model and the propagating light intensity for a waveguide-type light beam collecting film without the tapered vertical waveguide is shown. The angle of the vertical mirror is 45°. Although considerable leakage is observed at the vertical mirror a part of the incident light beam is guided to the output. The light beam collecting efficiency was 62%.

Figure 11.11 shows models and the propagating light intensity for the tapered vertical waveguide with high-refractive-index upper

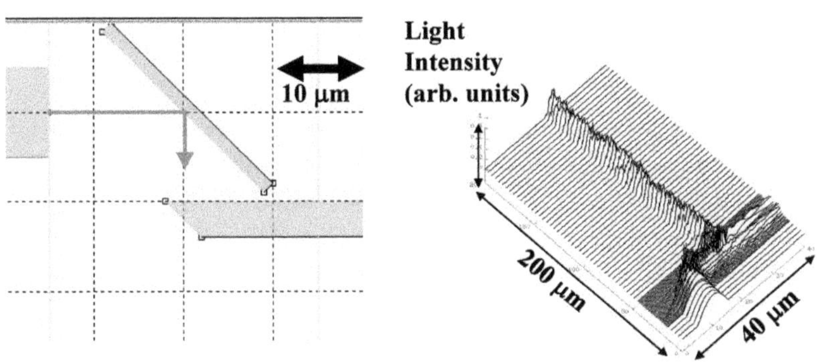

Figure 11.10. A model and the propagating light intensity for a waveguide-type light beam collecting film without tapered vertical waveguides.

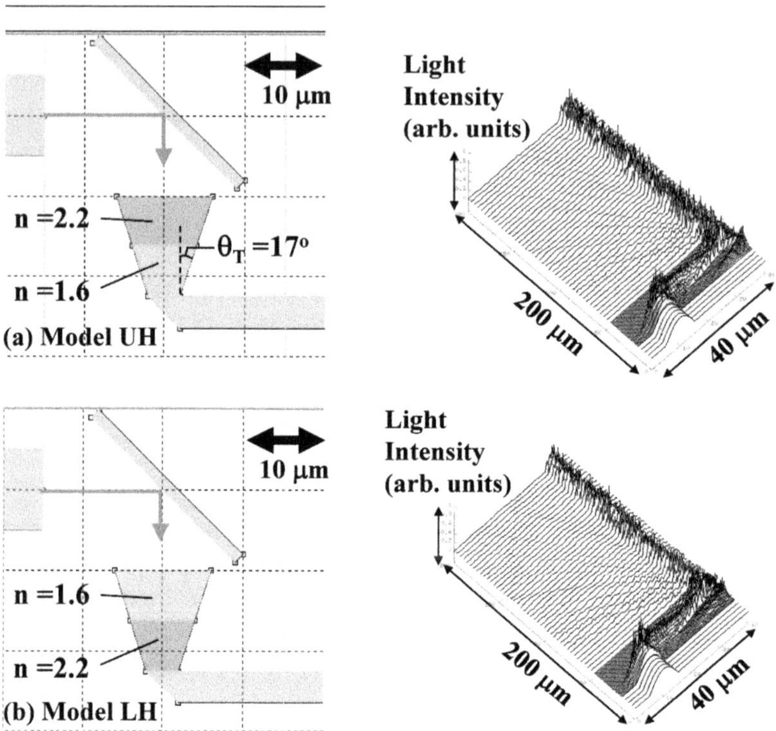

Figure 11.11. Models and the propagating light intensity for the tapered vertical/horizontal waveguide-type light beam collecting film. (a) The upper part of the tapered vertical waveguide has high refractive index (Model UH). (b) The lower part of the vertical waveguide has high refractive index (Model LH).

part (Model UH) and with high-refractive-index lower part (Model LH). The tapered angle of the tapered vertical waveguide is $\theta_T = 17°$. While the light beam collecting efficiency was 61% in Model UH, the efficiency increased up to 77% in Model LH.

This can be explained as follows. Since the light beam spot size becomes small with propagation of light beams to the lower region of the tapered vertical waveguide, stronger light beam confinement is necessary in the lower region. By putting the high-refractive-index material in the lower regions, strong light beam confinement can be achieved effectively, resulting in the enhancement of the light beam collecting efficiency.

Since the light beam collecting aperture of the tapered vertical waveguides is 12 μm and that of the vertical mirror is 4 μm, light beam collection with a factor of 3 is achieved in this two-dimensional model. For three-dimensional model, light beam collection with a factor of $3 \times 3 = 9$ is expected.

11.2.2.3 Multi-layer waveguide-type light beam collecting films

A model for the multi-layer waveguide-type light beam collecting film is shown in Fig. 11.12. It consists of stacked waveguide films [8].

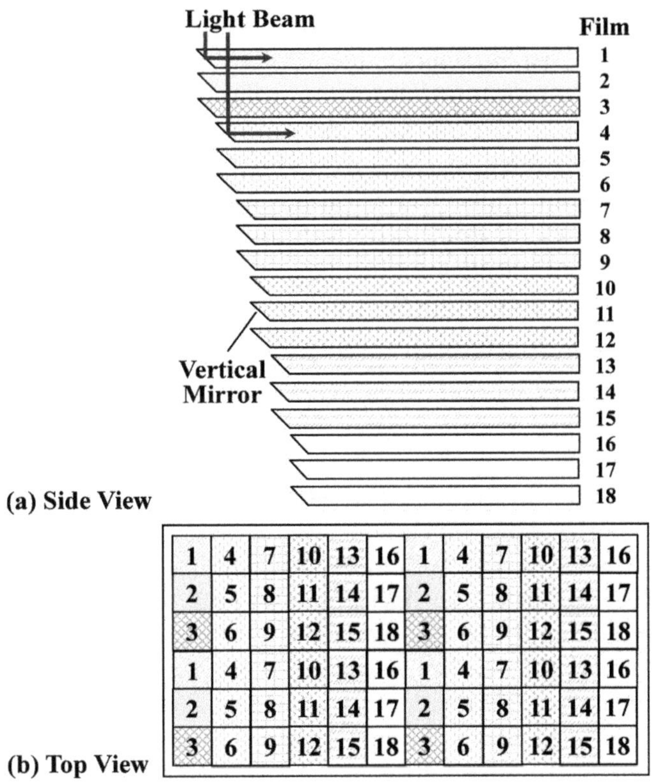

Figure 11.12. A model for the multi-layer waveguide-type light beam collecting film. Reprinted with permission from R. Shioya and T. Yoshimura, *J. Renewable Sustainable Energy* **1**, 033106 1-15 (2009). Copyright 2009, American Institute of Physics.

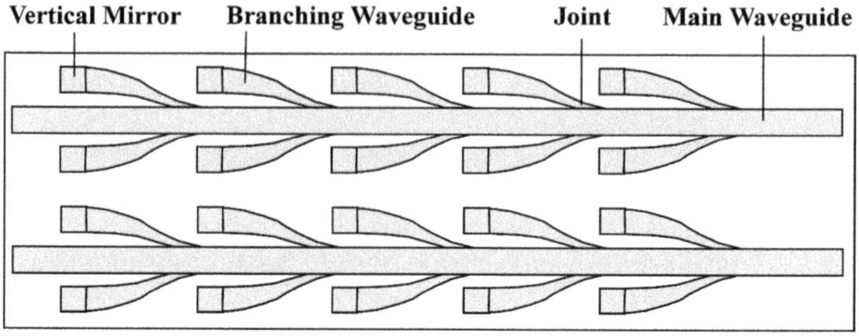

Figure 11.13. An example of optical circuits of merging waveguides.

In this example, as shown in the side view, 18 optical waveguide films are stacked. Numbers in the top view indicate vertical mirror apertures, namely, the light beam collecting apertures, of the film denoted by the numbers. An example of optical circuits in a film is illustrated in Fig. 11.13. Merging waveguides for light beam collection from the vertical mirrors through the branching waveguides are integrated. By arranging the films regularly, the surface can be occupied by the light beam collecting apertures in high density, achieving aperture occupation of 100% theoretically.

11.2.2.3.1 Effects of incident angles on collecting efficiency for single-layer Optical Waveguide Films Figure 11.14 shows models with 45° mirrors and the propagating light intensity for incident angles of 0°, 45°, and -45°. The wavelength of incident light beams is 600 nm. Here, in the model of -45°, the waveguide direction is flipped horizontally, which effectively reproduces a configuration for −45° incidence with the same incident light beam direction as for the 45° incidence. For incident angle of 0°, the light beams are introduced into the optical waveguide through the 45° mirror and then guided in the optical waveguide smoothly, indicating that the fundamental mode is dominant. For incident angles of 45° and -45°, a lot of leakage of light beams passing through the 45° mirror is observed. At the same time considerably higher modes are excited in the guided beams.

Figure 11.15 shows models with 30° mirrors and the propagating light intensity. For incident angle of 0°, the light beams guided in the optical waveguide contain considerably higher modes. For

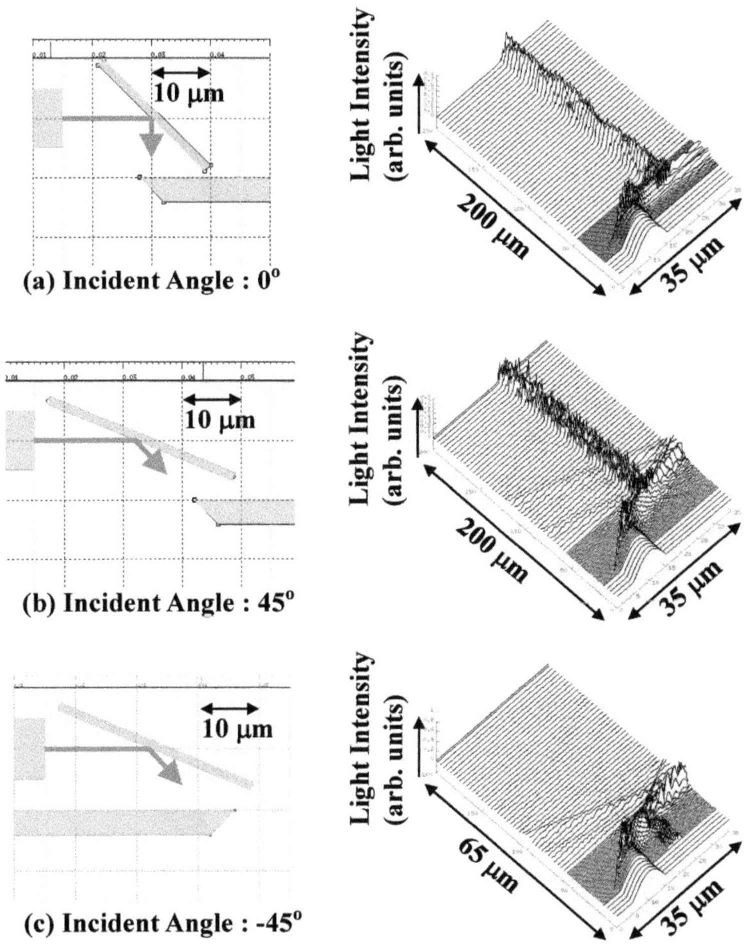

Figure 11.14. Models with 45° mirrors and the propagating light intensity. Reprinted with permission from R. Shioya and T. Yoshimura, *J. Renewable Sustainable Energy* **1**, 033106 1-15 (2009). Copyright 2009, American Institute of Physics.

incident angle of 45°, on the other hand, the guided beams propagate smoothly with small amount of higher modes. Figure 11.16 shows models having 60° mirrors and the propagating light intensity. The guided beams contain a lot of higher modes for the both incident angles of 0° and 45°.

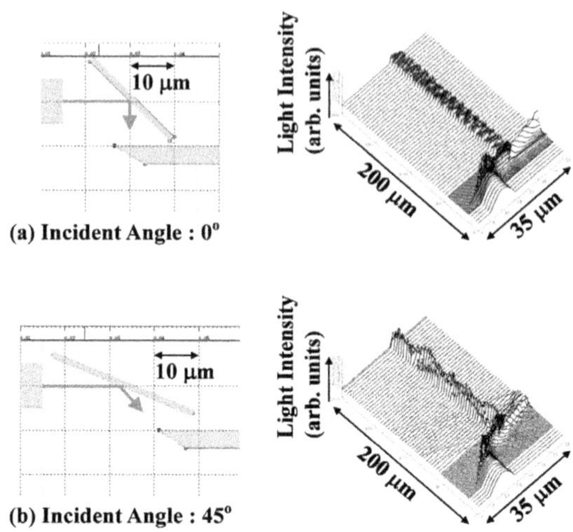

Figure 11.15. Models with 30° mirrors and the propagating light intensity. Reprinted with permission from R. Shioya and T. Yoshimura, *J. Renewable Sustainable Energy* **1**, 033106 1-15 (2009). Copyright 2009, American Institute of Physics.

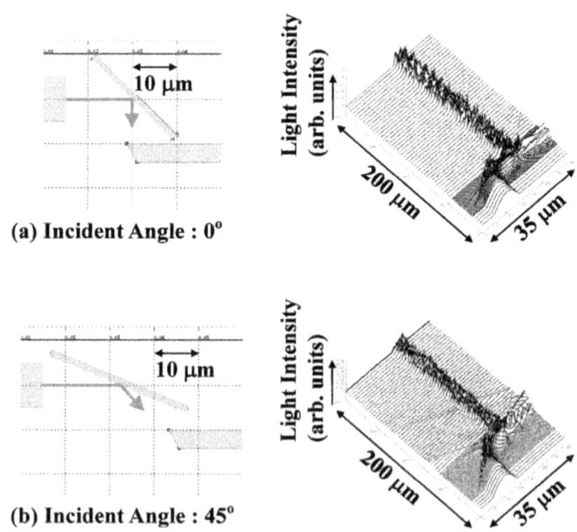

Figure 11.16. Models with 60° mirrors and the propagating light intensity. Reprinted with permission from R. Shioya and T. Yoshimura, *J. Renewable Sustainable Energy* **1**, 033106 1-15 (2009). Copyright 2009, American Institute of Physics.

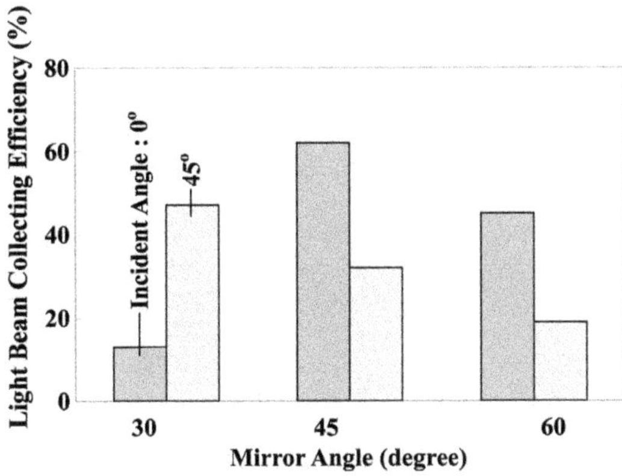

Figure 11.17. Light beam collecting efficiency in models with 30°, 45°, and 60° mirrors for incident angles of 0° and 45°. Reprinted with permission from R. Shioya and T. Yoshimura, *J. Renewable Sustainable Energy* **1**, 033106 1-15 (2009). Copyright 2009, American Institute of Physics.

The light beam collecting efficiency is summarized in Fig. 11.17 for incident angles of 0° and 45°. It is found that for incident angle of 0°, the 45° mirror exhibits the largest efficiency of 62%, and for incident angle of 45°, the 30° mirror exhibits the largest efficiency of 47%. The 60° mirror does not exhibit large efficiency for both incident angles of 45° and 0°. From these results, it is concluded that optical waveguides with 45° mirrors and optical waveguides with 30° mirrors are respectively useful for incident light beams with vertical direction and for those with tilted directions.

11.2.2.3.2 Effects of layer counts on collecting efficiency Figure 11.18 shows a model of a three-layer waveguide structure with 45° mirrors and the propagating light intensity for vertical incident light beams with wavelength of 600 nm. The distance between optical waveguides is 1 μm. It is found that the incident light beams which go through the first layer, are received by the second layer and the third layer to increase the light beam collecting efficiency. As shown in Table 11.2 light beam collecting efficiency increases from 62% to 87% with increasing the layer counts from one to three.

11.2.2.3.3 Collecting efficiency in multi-layer waveguides with 45° and 30° mirrors As mentioned above, optical waveguides with 45°

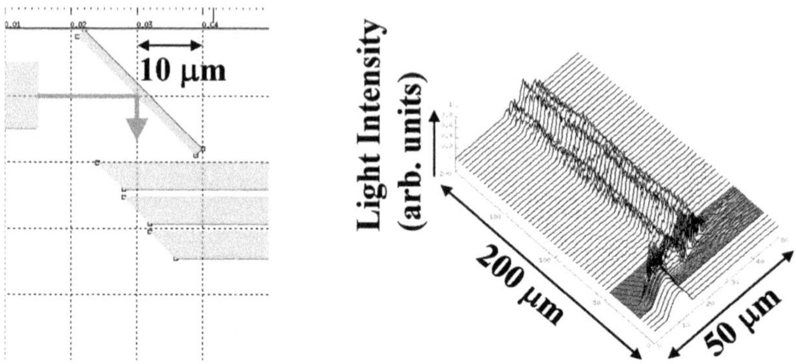

Figure 11.18. Model of a three-layer waveguide structure with 45° mirrors and the propagating light intensity for incident angle of 0°. Reprinted with permission from R. Shioya and T. Yoshimura, *J. Renewable Sustainable Energy* **1**, 033106 1-15 (2009). Copyright 2009, American Institute of Physics.

mirrors and optical waveguides with 30° mirrors are respectively preferable to collect light beams with vertical incident angles and tilted incident angles. Based on the results, we proposed light beam collecting films consisting of multi-layer optical waveguides with 45° and 30° mirrors, denoted by 45°/30°/30°. An example of a three-layer structure is illustrated in Fig. 11.19. The first layer is an optical waveguide with 45° mirror, and the second layer and the third layer are optical waveguides with 30° mirror. The two optical waveguides with 30° mirror are arranged with opposite directions each other. The first layer is for incident light beams with vertical directions. The second layer is for beams from the

Table 11.2. Light beam collecting efficiency for layer counts of one, two, and three

Layer count	Light beam collecting efficiency (%)
1	62
2	83
3	87

Source: Reprinted with permission from R. Shioya and T. Yoshimura, *J. Renewable Sustainable Energy* **1**, 033106 1-15 (2009). Copyright 2009, American Institute of Physics.

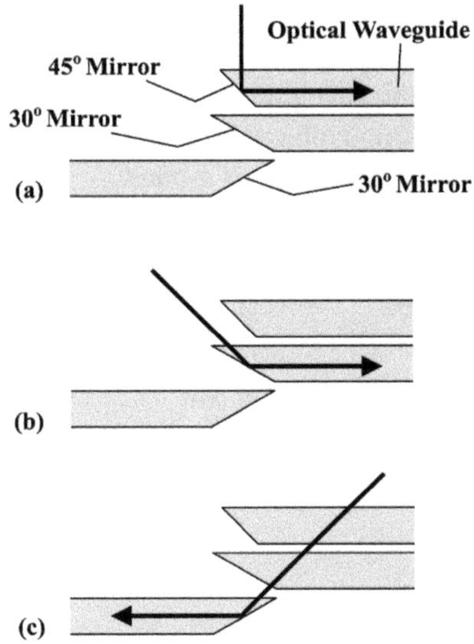

Figure 11.19. An example of the three-layer structure with 45° and 30° mirrors, denoted by 45°/30°/30°. (a) Incident light beams with vertical directions, (b) incident light beams from the left side directions, and (c) incident light beams from the right side directions. Reprinted with permission from R. Shioya and T. Yoshimura, *J. Renewable Sustainable Energy* **1**, 033106 1-15 (2009). Copyright 2009, American Institute of Physics.

left side directions, and the third layer is for beams from the right side directions. This structure is expected to provide wide-angle light beam collecting capability without solar beam tracking mechanisms.

Figure 11.20 shows models of the three-layer structure of 45°/30°/30° and the propagating light intensity for incident angles of 0°, 45°, and −45°. The distance between optical waveguides is 1 μm. The wavelength of incident light beams is 600 nm. It is found that the light beams are mainly collected by the first layer for vertical incidence, by the second layer for tilted incidence from the left, and by the third layer for tilted incidence from the right.

For comparison, light beam collecting characteristics of a light beam collecting film consisting of three-layer structure with only 45° mirrors, denoted by 45°/45°/45° are examined. The structure shown in Fig. 11.21 is the same as the model shown in Fig. 11.20 except for the mirror angles. While the collected light beams propagate in the optical waveguides smoothly for vertical incidence, for tilted incidence the collected light beams are reduced and propagate with considerably higher modes.

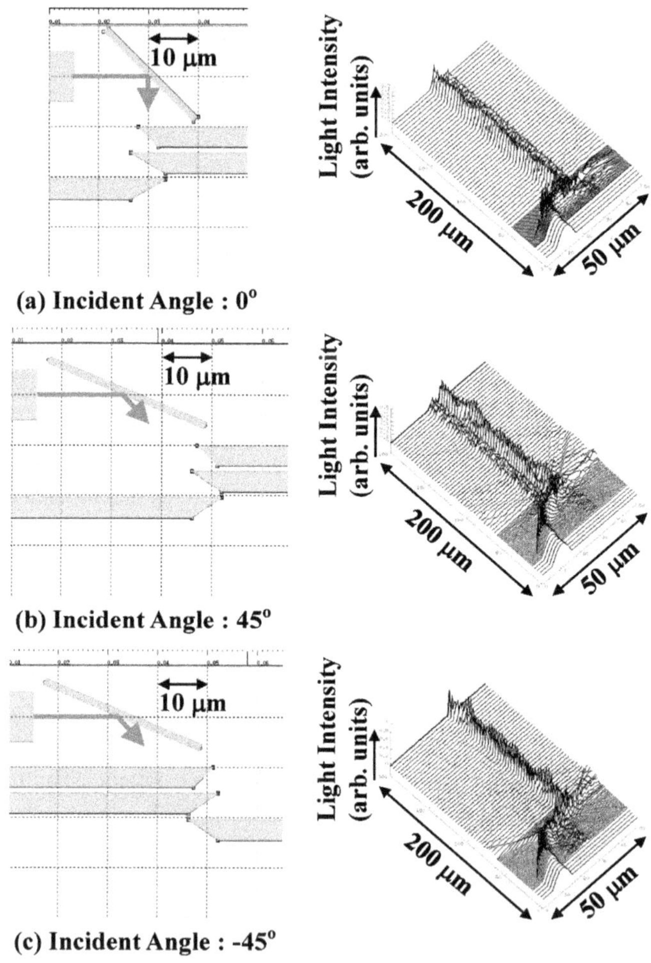

Figure 11.20. Models of the three-layer structure with 45° and 30° mirrors of 45°/30°/30° and the propagating light intensity. Reprinted with permission from R. Shioya and T. Yoshimura, *J. Renewable Sustainable Energy* **1**, 033106 1-15 (2009). Copyright 2009, American Institute of Physics.

Figure 11.22 shows dependence of light beam collecting efficiency on incident angles for the three-layer structures of 45°/30°/30° and 45°/45°/45°. 45°/30°/30° exhibits higher efficiency than 45°/45°/45° in whole incident angle regions. Especially, for incident angle of 45°, the efficiency is 1.5 times larger in 45°/30°/30° than in 45°/45°/45°. This indicates that the light beam collecting film consisting of multi-layer optical waveguides with mirrors of different angles provides wide-angle light beam collecting capability.

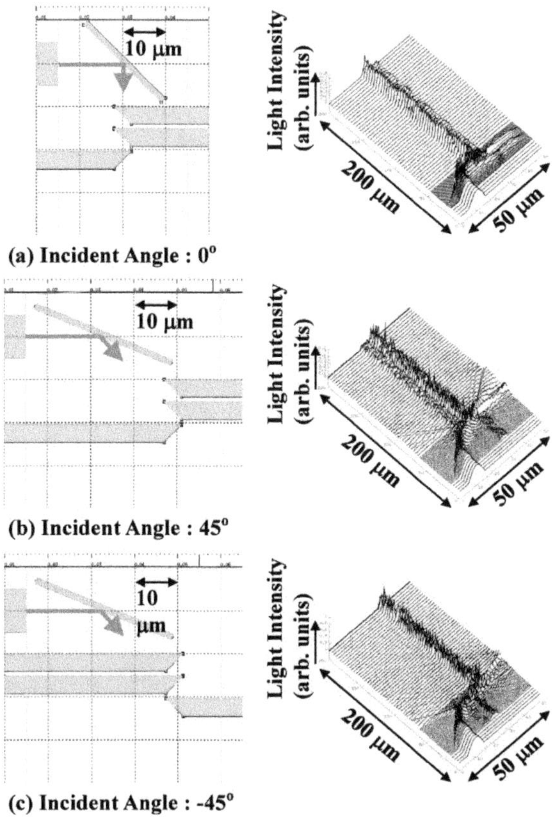

Figure 11.21. Models of the three-layer structure with only 45° mirrors, denoted by 45°/45°/45°, and the propagating light intensity. Reprinted with permission from R. Shioya and T. Yoshimura, *J. Renewable Sustainable Energy* **1**, 033106 1-15 (2009). Copyright 2009, American Institute of Physics.

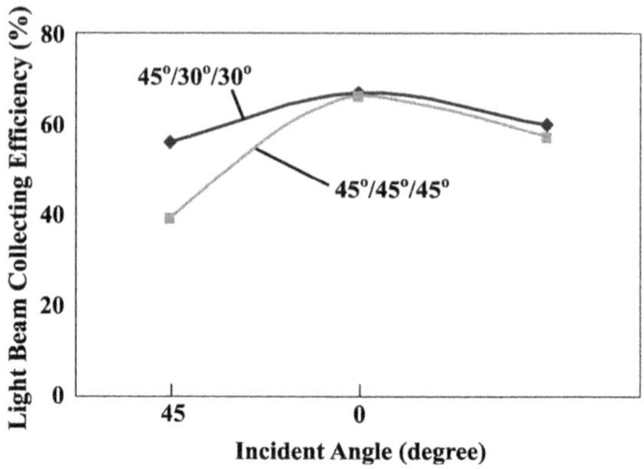

Figure 11.22. Dependence of light beam collecting efficiency on incident angles for the three-layer structure of 45°/30°/30° and 45°/45°/45°. Reprinted with permission from R. Shioya and T. Yoshimura, *J. Renewable Sustainable Energy* **1**, 033106 1-15 (2009). Copyright 2009, American Institute of Physics.

In Fig. 11.22, when incident angle is −45°, the efficiency of 45°/45°/45° is close to that of 45°/30°/30°. This might happen accidentally by prism effect of the 45° mirror part of the upper waveguide layers.

It should be noted that the model shown in Fig. 11.20 have 4 μm × 4 μm aperture in top views. This corresponds to aperture occupation for layer count of one in Table 11.2. So, when 3 × (4 μm × 4 μm) aperture, which corresponds to layer count of three in Table 11.2, is built by stacking two more film units of 45°/30°/30°, light beam collecting efficiency close to 80% is expected for incident angles of 45°, 0°, and −45°.

In addition the light beam collecting efficiency for incident angle of 45° in 45°/30°/30° was calculated to be 60%, 57%, and 65% for wavelengths of 400 nm, 600 nm and 800 nm, respectively. Thus, it is found that the wavelength dependence of the efficiency is relatively small, indicating that the proposed model could operate in wavelength regions from 400 nm to 800 nm.

11.2.2.4 Overall consideration

Overall light beam collecting efficiency $\eta_{\text{Over All}}$ in waveguide-type light beam collecting films is given as follows:

$$\eta_{\text{Over All}} = \eta_{\text{Mirror}} \times \eta_{\text{Merging}} \times \eta_{\text{Semiconductor}} \tag{11.1}$$

Here, η_{Mirror} η_{Merging}, and $\eta_{\text{Semiconductor}}$ are the light beam collecting efficiency via a vertical mirror, the light beam propagation efficiency via merging waveguides, and the coupling efficiency between an optical waveguide and a semiconductor flake, respectively.

η_{Mirror} of the tapered vertical/horizontal waveguide-type light beam collecting film (Model LH) and η_{Mirror} of the multi-layer waveguide-type light beam collecting film (three-film-unit structure of 45°/30°/30°) are supposed to be around 80%. For $\eta_{\text{Semiconductor}}$, when ZnO (refractive index *n*: 2.1) is used for the semiconductor and polymers (*n*: 1.52) for optical waveguides, reflection loss at the interfaces is ~16%. The coupling efficiency may increase near 100% by inserting appropriate AR coating between the optical waveguides and semiconductor flakes.

η_{Merging} was calculated by BPM. The model is similar to that shown in Fig. 11.13. The branching angle is 10°. The branching waveguides are tapered from a width of 4 μm at the vertical mirror to 0.1 μm at the joint, which is effective to increase the light beam propagation efficiency. Figure 11.23(a) shows the dependence of total propagation efficiency on the number of joints when light beams are introduced into all the vertical mirrors. The efficiency decreases with increasing the number of joints. For joint count of 8, the efficiency is 52%.

By substituting η_{Mirror} ~80%, η_{Merging} ~50%, and $\eta_{\text{Semiconductor}}$ ~100% into Eq. (11.1), expected overall light beam collecting efficiency is calculated to be ~40% for eight-step merging, that is, for the joint count of 8.

In the case of the eight-step merging, light beams from 16 vertical mirrors are collected into one semiconductor flake. Therefore, semiconductor consumption reduction factor, denoted by F_{SCR}, is 1/16 for the integrated solar energy conversion system with the multi-layer waveguide-type light beam collecting film. For the system with the tapered vertical/horizontal waveguide-type light beam collecting film, as discussed in Section 11.2.1, semiconductor consumption reduction of 1/9 is obtained at the tapered waveguide part. Therefore, the eight-step merging gives us F_{SCR} of (1/9) × (1/16) = 1/144. In the conventional solar cell, F_{SCR} is 1 because the semiconductor is placed all over the module.

As a measure of resource-saving ability, a ratio of $\eta_{\text{Over All}}/F_{\text{SCR}}$ is defined. This ratio is proportional to the amount of energy conversion achieved by semiconductors of unit area. In the conventional solar cell, since $\eta_{\text{Over All}}$ is nearly equal to 100%, $\eta_{\text{Over All}}/F_{\text{SCR}}$ ratio is $100/1 = 100$. For the integrated solar energy conversion systems with the multi-layer waveguide-type light beam collecting film, influence of the number of joints in the merging waveguides on $\eta_{\text{Over All}}/F_{\text{SCR}}$ ratio is shown in Fig. 11.23(b). The ratio increases with the joint counts, followed by saturation. In the case of eight-step merging, the ratio is $40/(1/16) = 640$, which

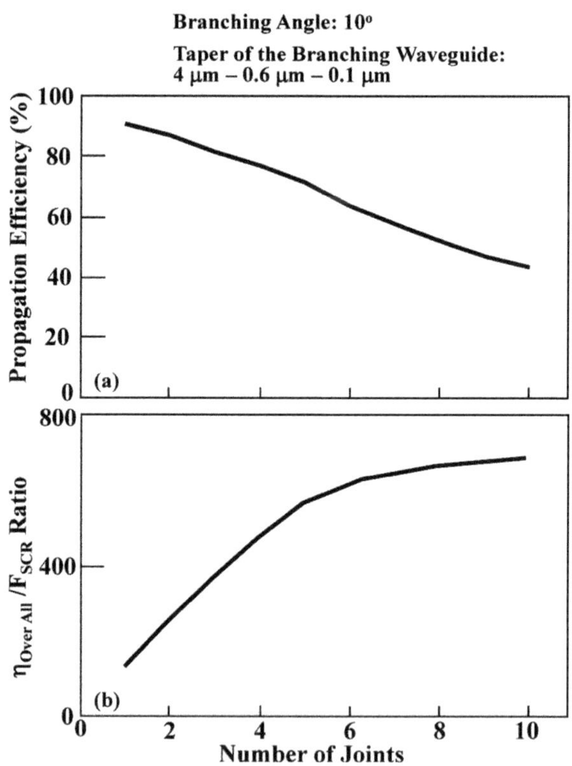

Figure 11.23. (a) Influence of the number of joints in the merging waveguides on total propagation efficiency. (b) Influence of the number of joints in the merging waveguides on ratio of $\eta_{\text{Over All}}/F_{\text{SCR}}$ for the integrated solar energy conversion systems with a multi-layer waveguide-type light beam collecting film.

is 6.4 times larger than the ratio in the conventional solar cell. For the system with the tapered vertical/horizontal waveguide-type light beam collecting film, the $\eta_{\text{Over All}}/F_{\text{SCR}}$ ratio is $40/(1/144) = 5760$, which is 58 times larger than the ratio in the conventional solar cell. Thus, it is concluded that the proposed integrated solar energy conversion systems are expected to realize drastic material saving.

In the multi-layer waveguide-type light beam collecting film in order to achieve aperture occupation of 100%, 39 layers are necessary for one angle of the vertical mirrors when the branching angle and the joint counts are respectively 10° and 8 in the merging waveguide. This 39-layer structure forms a module unit. To construct a multi-layer structure consisting of optical waveguides with 45° and 30° mirrors, one module unit for 45° mirrors and four module units for 30° mirrors are necessary to be stacked in order to collect light beams with vertical incident angles and tilted incident angles (from left side, right side, front side and back side). When the waveguide core thickness is 4 μm and the distance between waveguides is 1 μm, the module unit thickness is around 195 μm. Then, total thickness of the stack of five module units is 975 μm.

The estimated efficiency of the waveguide-type light beam collecting films is lower than that of Fresnel lens/mirror concentrators. So, at present the light beam collecting films seem not to be suitable for largescale solar energy conversion centers. However, as depicted in Fig. 11.3, they have an outstanding feature of flexible light-weight films, enabling us to use them as curtains, sheets and clothes, and to put them on cars, robots, buildings, roofs of fragile houses, surfaces of pavements/roads, walls, tables, various kinds of equipment, and everywhere we want. They also have a unique feature of color appearance as illustrated in Fig. 11.8. Since very high efficiency is not always necessary in these applications, light beam collecting film may sometimes be useful.

In the present design, we assumed 4 μm waveguide cores. In practice, larger cores with several tens-μm sizes can be used. On the contrary, when the tapered vertical/horizontal waveguide-type and the multi-layer waveguide-type are combined, namely, a number of the tapered vertical/horizontal waveguide-type light

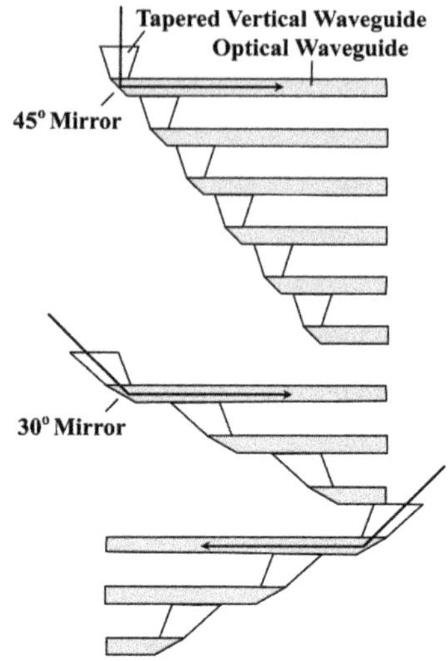

Figure 11.24. A light beam collecting film combining the tapered vertical/horizontal waveguide-type and the multi-layer waveguide type.

beam collecting films with 45° mirrors and 30° mirrors are stacked as illustrated in Fig. 11.24, the waveguide core sizes should be reduced to ~1 μm or less in order to prevent the total thickness of the whole module from increasing abnormally.

For waveguide fabrication, the built-in mask method as well as imprinting and stamping can be used. By precise investigation on the merging structures in the horizontal waveguides, the further improvement of the collecting efficiency is expected.

11.2.3 *Possible Fabrication Process*

Figure 11.25 shows an example of the whole structure of the integrated solar energy conversion system with a tapered vertical/horizontal waveguide-type light beam collecting film. A tapered vertical waveguide film and an optical waveguide film, in which semiconductor flakes are embedded by PL-Pack with SORT described in Chapter 8, are stacked by the film stack process for 3-D optical circuits described in Chapter 7. The tapered vertical waveguide film can be built by using the built-in mask method described in Chapter 6 In the optical waveguide film, electrodes

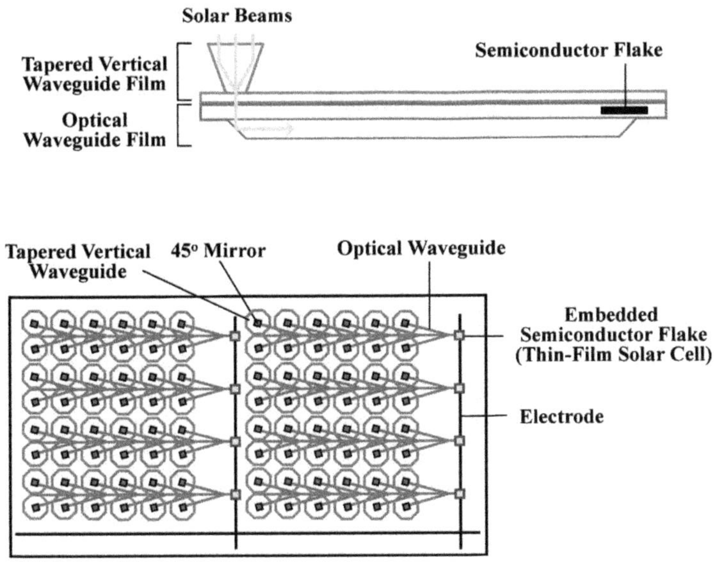

Figure 11.25. Whole structure of the integrated solar energy conversion system with a tapered vertical/horizontal waveguide-type light beam collecting film.

for electrical power transmission are formed. Thus, the integrated solar energy conversion system can be fabricated by the same technologies as those for the optical interconnects within boxes and 3D-MOSS.

An example of fabrication process of the tapered vertical/horizontal waveguide-type light beam collecting film by the built-in mask method is shown in Fig. 11.26. The tilted exposure of photo-definable materials makes cured regions having tapered walls. After development, a supporting substrate is attached on the top. By removing the removable layer, a tapered vertical waveguide film is obtained. The film is stacked on an optical waveguide film containing optical circuits of merging waveguides with embedded semiconductor flakes to form an integrated solar energy conversion system with a tapered vertical/horizontal waveguide-type light beam collecting film. For the system with a tapered vertical waveguide-type light beam collecting film, the vertical waveguide

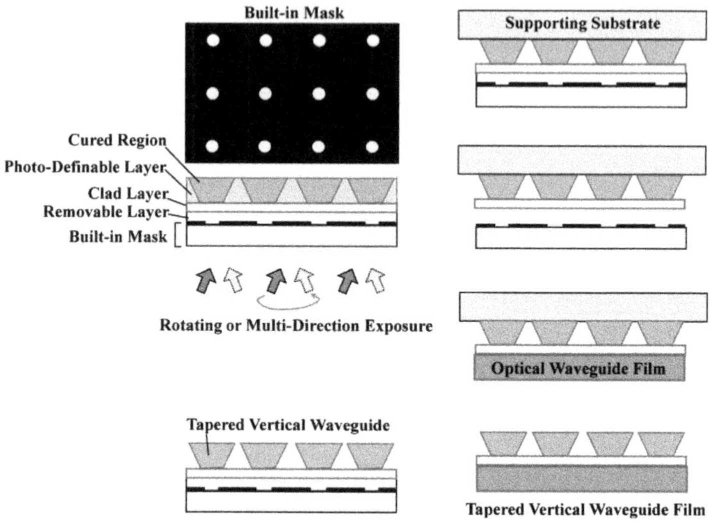

Figure 11.26. Fabrication process of the integrated solar energy conversion system.

film is stacked on a film, into which semiconductor flakes are embedded.

For the system with a multi-layer waveguide-type light beam collecting film, optical waveguide films containing optical circuits with merging waveguides are stacked each other with sliding by one vertical mirror aperture. The film separation can be kept by inserting spacers between films. From a viewpoint of light beam confinement, all-air-clad waveguides are preferable. The all-air-clad structures, as described in Section 8.5, can be constructed by core transfers from a substrate to another substrate by using PL-Pack with SORT which enables us to place waveguide cores and spacers onto films at one time.

Figure 11.27 shows photographs of an array of straight vertical waveguides and an array of tapered vertical waveguides fabricated by the built-in mask method. Vertical waveguides were formed on a clad film, which is placed on PVA layer coated on the built-in mask. The light beam collecting efficiency of the straight and tapered vertical waveguide was evaluated by introducing 633 nm light beams from the front side and measuring the power of the

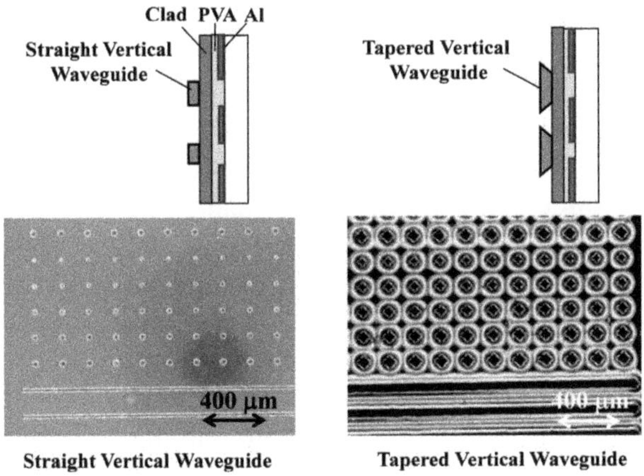

Figure 11.27. Photographs of a straight vertical waveguide film and a tapered vertical waveguide film.

light beams that transmit through the built-in mask window. The efficiency in tapered vertical waveguides was 1.5 times higher than that in the straight vertical waveguides, confirming the light collecting capability of the tapered vertical waveguides The efficiency in the tapered vertical waveguides will be improved by optimizing the design and fabrication processes as well as measurement setup.

11.3 Novel Structures of Photo-Voltaic and Photo-Synthesis Devices

Figure 11.28 shows examples of thin-film photo-voltaic and photo-synthesis devices [6, 9], which will be integrated in light beam collecting films. In the thin-film photo-voltaic device, collected light beams are introduced onto a semiconductor film through an optical waveguide with a vertical mirror. This structure enables us to replace an expensive and electrically resistive transparent electrode with low-cost metal one with low resistivity.

In the thin-film photo-synthesis device, using 3-D optical circuits with embedded wavelength filters, collected light beams branch into two optical paths for wavelengths of λ_1 and λ_2, which are respectively introduced onto two semiconductor films in an electrolyte, generating H_2 and O_2. The photo-synthesis process using two semiconductors is developed by Sayama et al. using the solution in which semiconductor powders are dispersed. Since the control of relative positions of the semiconductors is important in order to improve the photo-synthesis performance, a thin-film structure such as the one shown in Fig. 11.28 seems promising. Figure 11.29 shows several examples for positional configuration of thin films in the thin-film photo-synthesis devices. For the optimization of H^+ diffusion in the solution, dividing the thin films or opening windows in the thin films might be effective.

Possible fabrication processes for the thin-film photo-synthesis devices are shown in Fig. 11.30. In process I, thin film 1 and thin film

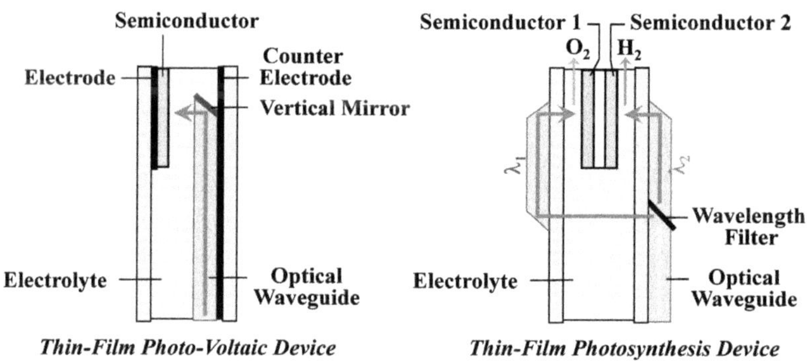

Figure 11.28. Examples of thin-film photo-voltaic and photo-synthesis devices.

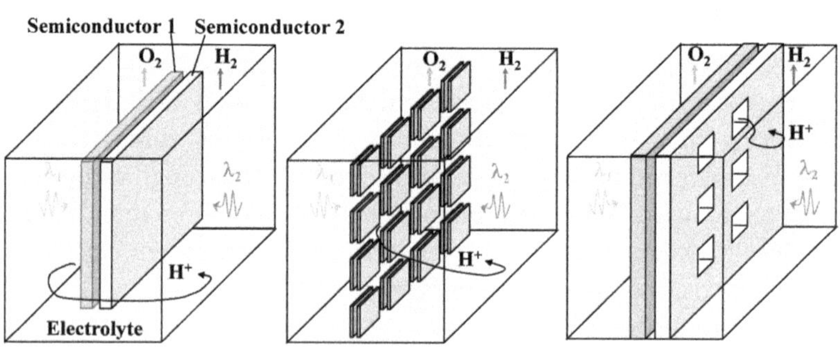

Figure 11.29. Whole structures of thin-film photo-synthesis devices.

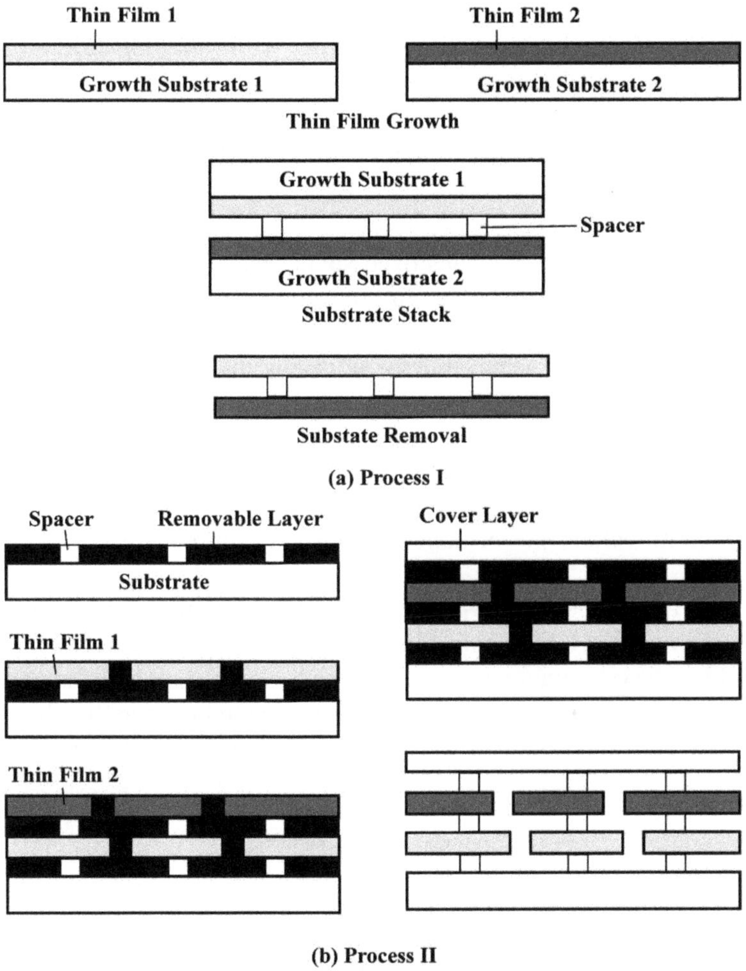

Figure 11.30. Possible fabrication processes for thin-film photo-synthesis devices.

2 are grown on growth substrate 1 and substrate 2, respectively. The films are attached to each other with spacers between them. Then the growth substrates are removed to obtain stand-alone coupled thin films. In process II, thin film 1 is deposited on a removable layer of a substrate with spacers. Then, another removable layer and spacers are formed on thin film 1, and thin film 2 is deposited

on them. After a cover layer is formed, all the removable layers are removed by etching to obtain a structure with thin films located in designated distributions three-dimensionally.

11.4 Waveguide-Type Photo-Voltaic and Photo-Synthesis Devices

In the conventional dye-sensitized solar cells, in order to increase the number of adsorbed dye molecules on semiconductor surfaces, porous semiconductors are used. In this case, although sufficient light absorption is achievable for normally incident light onto the semiconductor surfaces, the crystalline quality of the semiconductors tends to be degraded, suppressing electron mobility. Furthermore, the porous structure narrows the electron transporting channels. Consequently, the internal resistivity increases to degrade the photo-voltaic performance in the conventional photo-voltaic device.

In the waveguide-type dye-sensitized photo-voltaic device shown in Fig. 11.31, a thin film semiconductor with a flat surface and high crystalline quality, namely, high electron mobility, is used [2–5, 9]. So, the internal resistivity decreases to improve the photo-voltaic performance. However, if normally incident light is used for

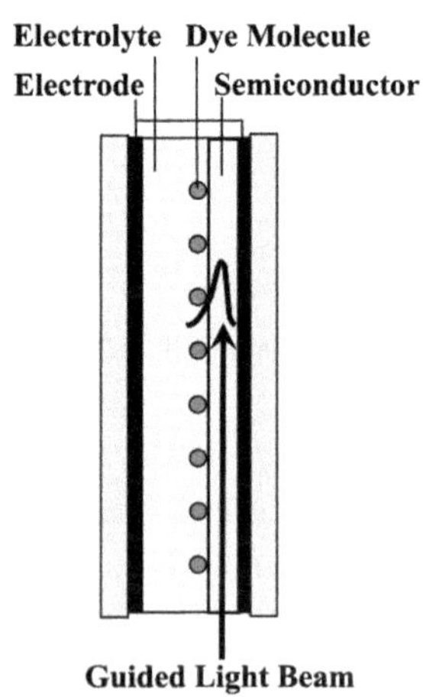

Figure 11.31. Waveguide-type dye-sensitized photo-voltaic device.

light absorption, the light passes through only mono-molecular layer of dye, resulting in very small light absorption. By using guided light beams in the semiconductor thin films, this problem can be overcome. In the guided beam configuration the light beams pass through a lot of dye molecules to enhance light absorption [2–5, 9]. Thus, high-performance dye-sensitized photo-voltaic devices are expected to be realized.

It should be noted that, if necessary, transparent electrodes can be replaced with metal electrodes, which reduces material costs and series internal resistance due to electrodes. The sensitized spectral regions can be widened by the two-dye sensitization with self-assembled double layers consisting of p-type dye molecules and n-type dye molecules [3, 9, 12]. The nano-structures can be fabricated by the liquid-phase MLD [3, 9].

Figure 11.32 shows a novel waveguide-type photo-voltaic device with the polymer MQD [2, 4, 9]. In this device, the dye molecules

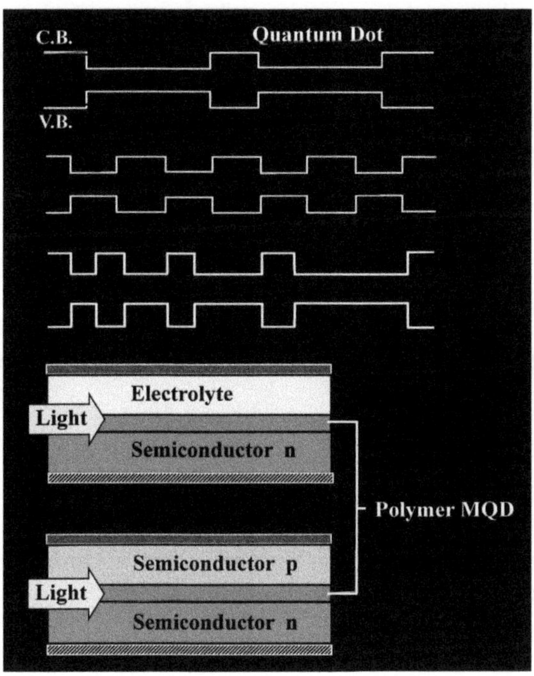

Figure 11.32. Waveguide-type photo-voltaic device with polymer MQD.

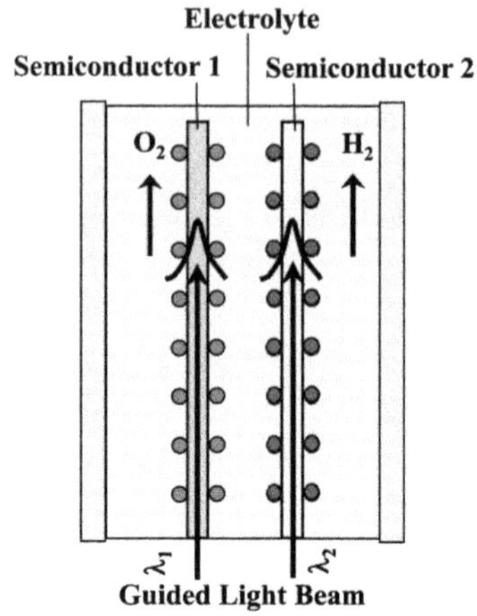

Figure 11.33. Waveguide-type dye-sensitized photo-synthesis device.

in the device shown in Fig. 11.31 are replaced with the polymer MQD. The polymer wires are expected to act as efficient and robust light absorption layers. By controlling quantum dot lengths, the quantum confinement effects are optimized to provide appropriate sensitization spectra with high efficiency. The polymer MQD can be fabricated by MLD. Details of the polymer MQD are described in Section 12.2.2.

Examples of waveguide-type dye-sensitized photo-synthesis devices are shown in Fig. 11.33. Instead of normally incident light beams, guided light beams in the semiconductor thin films are used. The light beams pass through a lot of dye molecules, which enhances the light absorption in the semiconductors to generate gases efficiently.

References

1. T. Yoshimura, *Molecular Nano Systems — Applications to Optoelectronic Computers and Solar Energy Conversion*, Corona Publishing Co., Ltd., Tokyo (2007) [in Japanese].
2. T. Yoshimura, A. Oshima, D. Kim, and Y. Morita, "Quantum Dot Formation in Polymer Wires by Three-Molecule Molecular Layer Deposition (MLD) and Applications to Electro-Optic/Photovoltaic Devices," *ECS Trans.* **25**

(4) "Atomic Layer Deposition Applications 5," from the Vienna, Austria Meeting, 15–25 (2009).

3. T. Yoshimura, H. Watanabe, and C. Yoshino, "Liquid-Phase Molecular Layer Deposition (LP-MLD): Potential Applications to Multi-Dye Sensitization and Cancer Therapy," *J. Electrochem. Soc* **158**, 51–55 (2011).
4. T. Yoshimura, R. Ebihara, and A. Oshima, "Polymer Wires with Quantum Dots Grown by Molecular Layer Deposition of Three Source Molecules for Sensitized Photovoltaics," *J. Vac. Sci. Technol. A* **29**, 051510-1-6 (2011).
5. T. Yoshimura, "Integrated Photonic/Electronic/Chemical Systems, Solar Energy Conversion Systems, Light Collectors, and Optical Waveguides," Japanese Patent, Tokukai 2009-4717 (2009) [in Japanese].
6. T. Yoshimura and K. Asama, "Integrated Chemical Systems and Integrated Light Energy Conversion Systems," Japanese Patent, Tokukai 2007-107085 (2007) [in Japanese].
7. T. Yoshimura, "Proposed Applications of 3-D Optical Interconnect Technologies to Integrated Chemical Systems. 2007 Digest of the LEOS Summer Topical Meetings [Bio-Inspired Sensors and Application/Imprinting on Photonic Integrated Circuits], Portland, Oregon, 129–130 (2007).
8. R. Shioya and T. Yoshimura, "Design of Solar Beam Collectors Consisting of Multi-Layer Optical Waveguide Films for Integrated Solar Energy Conversion Systems," *J. Renewable Sustainable Energy* **1**, 033106-1-15 (2009).
9. T. Yoshimura, *Thin-Film Organic Photonics: Molecular Layer Deposition and Applications*, CRC/Taylor & Francis, Boca Raton, Florida (2011).
10. T. Yoshimura, K. Ogushi, Y. Kitabayashi, K Naito, Y. Miyamoto, and M. Miyazaki, "Optical Waveguide Films with Two-Layer Skirt-Type Core End Facets for Beam Leakage Reduction at 45° Mirrors," *Proc. SPIE* **6899**, from Photonics West 2008, San Jose, California, 689913-1-11 (2008).
11. T. Yoshimura, M. Ojima, Y. Arai, and K. Asama, "Three-Dimensional Self-Organized Micro Optoelectronic Systems for Board-Level Reconfigurable Optical Interconnects — Performance Modeling and Simulation," *IEEE J. Select. Topics Quantum Electron.* **9**, 492–511 (2003).
12. T. Yoshimura, K. Kiyota, H. Ueda, and M. Tanaka, "Mechanism of Spectral Sensitization of ZnO Coadsorbing p-Type and n-Type Dyes," *Jpn. J. Appl. Phys.* **20**, 1671–1674 (1981).

Chapter 12

Future Challenges

In optical electronics, the important future issue is the development of high-performance functional materials such as electro-optic (EO) materials and semiconductors for photovoltaic devices. By developing high-performance EO materials, size, driving voltage, and power consumption of optical interconnects and optical switching systems will be drastically reduced. By developing high-performance semiconductors, high-efficiency and low-cost solar energy conversion systems will be realized.

So far, many kinds of EO materials have been developed, for example, LN, PLZT, and III-V compound semiconductors. Organic materials with π-conjugated systems — although their development is in a primitive stage — attract interest because they have both "large optical nonlinearity" and "low dielectric constant" characteristics. The low-dielectric-constant materials enable high-speed and low-power operation.

Organic EO materials can be classified into poled polymers, molecular crystals, and conjugated polymers, as summarized in Fig. 12.1 [1, 2]. For molecular crystals, the author found that the styrylpyridinium cyanine dye (SPCD) thin-film crystal exhibits an EO coefficient r of 430 pm/V [3], about 14 times that of LN, demonstrating the prospect of organic materials used in EO devices. The processability and reliability are the concerns in the molecular crystals.

Optical Electronics: Self-Organized Integration and Applications
Tetsuzo Yoshimura

ISBN 978-981-4310-90-1 (Hardcover), 978-981-4364-08-9 (eBook)
www.panstanford.com

EO Materials	$LiNbO_3$	Poled Polymer	Molecular Crystal	Conjugated Polymer
		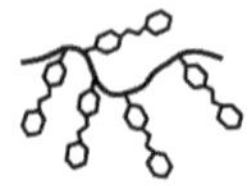	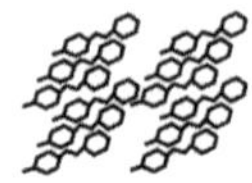	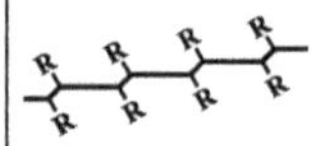
r (pm/V)	~30	~50	~400	>1000 (Simulation)
Processability	Good	Good	Poor	Good
Development Phase	Commercially Available	Test Vehicle	Material Preparation	Material Design

Figure 12.1. Organic EO materials.

Poled polymers have an advantage that they can easily be thin films of high optical quality by spin-coating although *r* of poled polymers is usually smaller than that of molecular crystals. In principle, by improving the molecular alignment, it might be possible to achieve an *r* of the same level as that of the molecular crystals, that is, ~10 times *r* (LN).

As mentioned above, the EO coefficient of about 10 times *r* (LN) seems to be possible using the molecular crystals and poled polymers. For the further improvement of the EO effect, another material, namely, conjugated polymers, will play an important role as remarkable candidates. The wires of conjugated polymers are regarded as natural one-dimensional systems (quantum wires) and are more favorable to induce a larger oscillator strength than two-dimensional or three-dimensional systems [1, 2]. When MQD structures are introduced into the wires, the wires have long controllable wavefunctions offering a greatly enhanced Pockels effect that is estimated to be about 100 times *r* (LN) [1, 2, 4] by calculations based on the molecular orbital method. We call the conjugated polymers with MQD as "polymer MQD."

The polymer MQD is also promising for photo-voltaic devices [2, 5, 6], as mentioned in Section 11.4, due to their controllability on electron confinement, which is effective to enhance quantum efficiency and to widen spectral responses of photocurrents.

In this chapter, molecular layer deposition (MLD) [2, 5–12], which is the key technology for polymer MQD fabrication, is reviewed. Possibility of the polymer MQDs as EO materials is predicted by simulations using the molecular orbital method.

As additional subjects, molecular circuits and thin-film bio/medical photonics are briefly described.

12.1 Molecular Layer Deposition

In polymer functional devices, such as optical switches, electroluminescent (EL) devices, thin-film transistors (TFTs), and photovoltaic devices, conjugated polymer wires play important roles due to their extended π-electron wavefunctions. In order to improve the performance of devices with conjugated polymer wires, self-organized growth of polymer wires is essential, which enables us to form polymer wires with (1) designated molecular sequences, (2) designated locations, and (3) designated orientations. These are achieved by molecular layer deposition (MLD).

In the present section, the self-organized growth of polymer wires based on MLD is reviewed.

12.1.1 *Concept*

The concept of MLD is shown in Fig. 12.2 [2, 7, 8, 12]. Different kinds of molecules, Molecules A, B, C, and D, are prepared under the following guideline:

- A molecule has two or more reactive groups.
- Groups in a molecule cannot react with groups in the same kind of molecules. For example, groups in Molecule A cannot react with groups in Molecule A.
- Groups in a molecule can react with groups in other kinds of molecules. For example, groups in Molecule A can react with groups in Molecule B.

Then, the same kind of molecules cannot be connected, and different kinds of molecules can be connected through the reactions between groups. When Molecule A is provided onto a substrate surface, a monomolecular layer of Molecule A is formed. Once the surface is covered with Molecule A, the deposition of the molecules is automatically terminated due to the small reactivity between the same kinds of groups, giving rise to thickness saturation. When

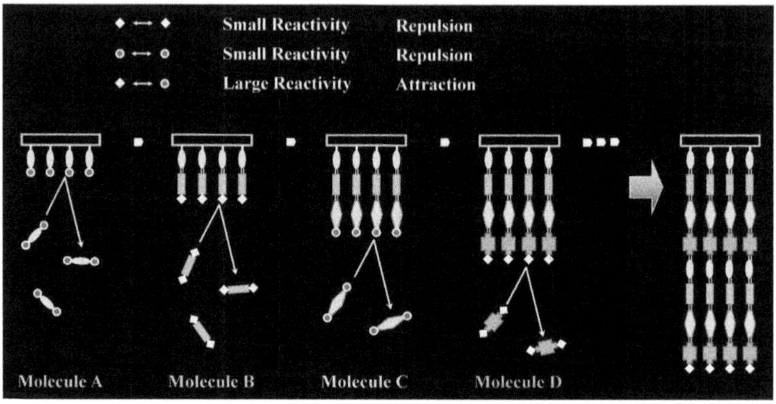

Figure 12.2. Concept of MLD. Reprinted from T. Yoshimura, *Trends in Thin Solid Films Research* (ed. A. R. Jost) Ch. 5 "Self-Organized Growth of Polymer Wire Networks with Designed Molecular Sequences for Wavefunction-Controlled Nano Systems." Nova Science Publishers, New York (2007). With permission from Nova Science Publishers, Inc.

Molecule B is provided onto the surface, Molecule B is connected to Molecule A to induce a rapid thickness increase followed by saturation, exhibiting monomolecular-step growth. This enables us to carry out sequential self-limiting reactions like the atomic layer deposition (ALD) [13] by switching molecules. By repeating the switching with a designated sequence, a polymer wire with a molecular sequence like ABCD... can be obtained.

In MLD, source molecules are supplied onto a substrate surface in gas phase or liquid phase. In K-cell type MLD, molecular gases are provided onto a surface in vacuum from low-temperature K-cells, in which source molecules are loaded. Molecular gas switching is performed by shutters or valves. In carrier gas type MLD, molecular gases are provided with carrier gas onto a surface. Molecular gas switching is performed by valves. In fluidic circuit type MLD, source molecules are carried with solvent to be supplied onto a surface. Fluidic processes using capillary circuits are effective to reduce the volume of solutions required for the MLD as well as solution switching duration. Figure 12.3 shows a photograph of the carrier gas type MLD equipment.

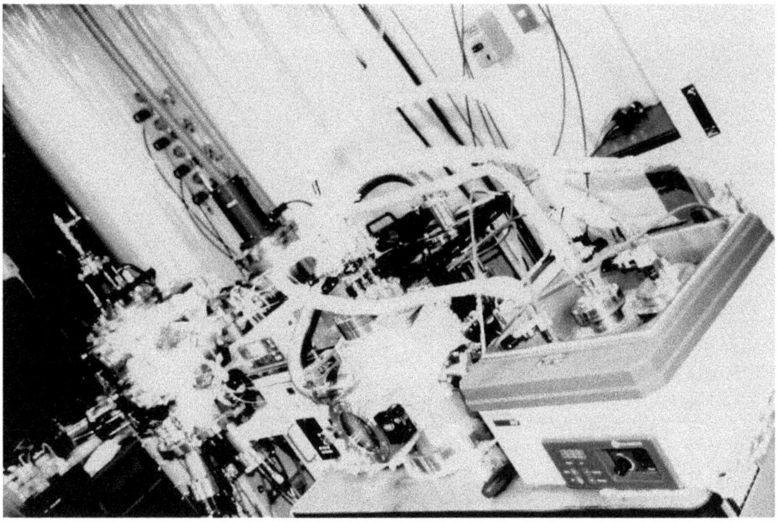

Figure 12.3. Photograph of carrier gas type MLD equipment.

12.1.2 *Experimental Demonstrations*

Figure 12.4 shows an example of growth reaction for conjugated polymer wires [2, 8, 12]. By using terephthalaldehyde (TPA) for Molecule A and *p*-phenylenediamine (PPDA) for Molecule B, conjugated polymer wires of poly-azomethine (poly- AM) [TPA/PPDA] are grown. TPA and PPDA react through –CHO in TPA and $–NH_2$ in PPDA to form –C=N– inducing π-conjugated systems. When oxalic acid (OA) for molecule A and oxalic dihydrazide (ODH) for molecule B are used, conjugated polymers of poly-oxadiazole (OXD) are grown with the assistance of heating in vacuum.

As shown in Fig. 12.5, when PPDA is provided on TPA with a molecular gas pressure of 1×10^{-2} Pa to 8×10^{-2} Pa, step-like growth of ~0.7 nm, which is close to a molecule size, is observed at substrate temperature $T_S = 25°C$. After removing PPDA from the vacuum chamber, by providing TPA on PPDA, step-like growth is observed again. These results indicate that monomolecular-step growth with self-limiting is achieved by MLD. It is also confirmed that the thickness is proportional to the molecular gas switching counts just like ALD. If more kinds of source molecules are used,

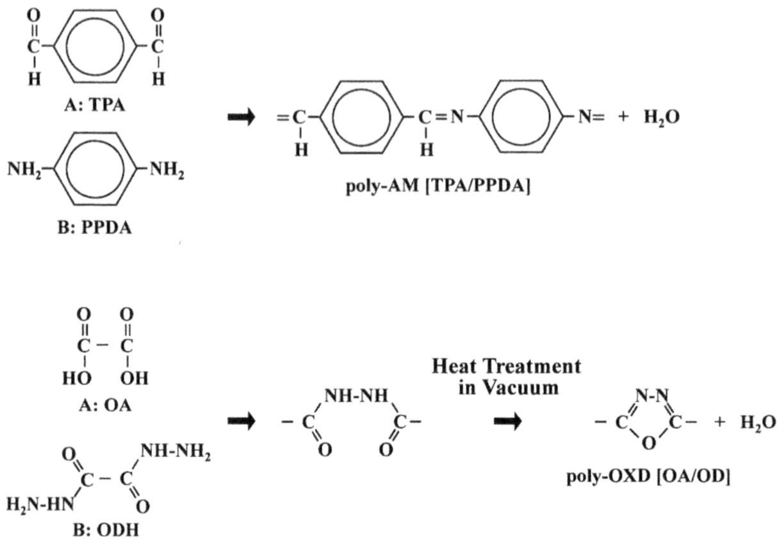

Figure 12.4. Reactions for conjugated polymer wire growth.

a variety of structures as illustrated in Fig. 12.2 can be fabricated. This enables us to realize molecule-by-molecule polymer wire growth with designated molecular sequences to control π-electron wavefunction shapes, optimizing the polymer wire performance.

MLD for polyamic acid, which is precursor of polyimide, was demonstrated [2, 7] using the reactions shown in Fig. 12.6 at molecular gas pressures ranging from 1×10^{-3} Pa to 5×10^{-3} Pa. Polyamic acid [PMDA/DDE] is obtained by using pyromellitic dianhydride (PMDA) for Molecule A and 4,4′-diamino-diphenyl ether (DDE) for Molecule B. As shown in Fig. 12.7, monomolecular-step growth is realized at a substrate temperature $T_S = 80°C$. A 10 nm growth is observed with 15 steps of molecular gas switching.

It should be noted that in addition to the reactions shown in Figs. 12.4 and 12.6, MLD can be performed by other reactions using, epoxy ring, –NCO, –OH, –COOH, and so on.

12.1.3 *Location/Orientation-Controlled MLD*

As one of the future challenges, we proposed location/orientation-controlled MLD, which uses seed cores. The concept is shown in

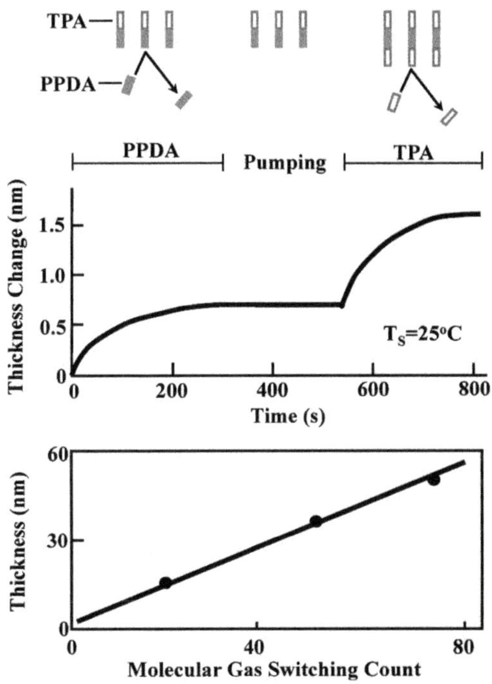

Figure 12.5. Thickness changes with time, and thickness versus molecular gas switching count characteristics in MLD of poly-AM [TPA/PPDA]. Reprinted with permission from T. Yoshimura, S. Tatsuura, W. Sotoyama, A. Matsuura, and T. Hayano, "Quantum Wire and Dot Formation by Chemical Vapor Deposition and Molecular Layer Deposition of One-Dimensional Conjugated Polymer," *Appl. Phys. Lett.* **60**, 268–270 (1992). Copyright 1992, American Institute of Physics.

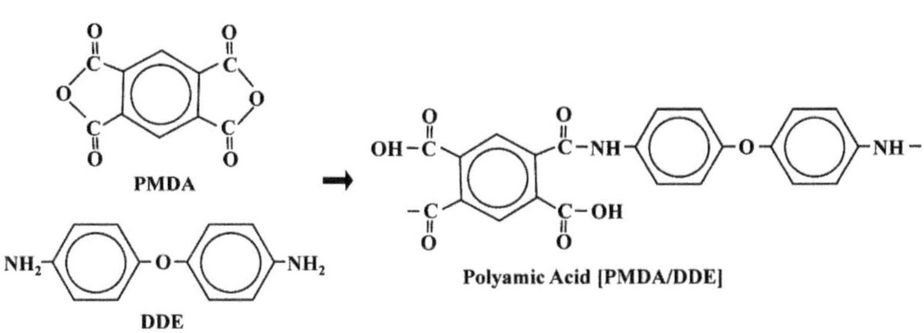

Figure 12.6. Reactions for polyamic acid wire growth.

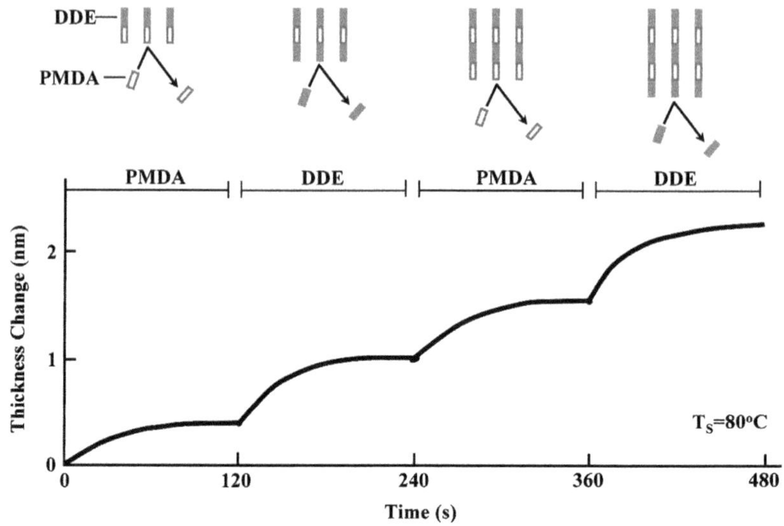

Figure 12.7. Thickness changes with time in MLD of polyamic acid [PMDA/DDE]. Reprinted with permission from [T. Yoshimura, S. Tatsuura, and W. Sotoyama, "Polymer Films Formed with Monolayer Growth Steps by Molecular Layer Deposition," *Appl. Phys. Lett.* **59**, 482-484 (1991). Copyright 1991, American Institute of Physics.

Fig. 12.8 [2, 9, 10]. On the top surfaces and/or sidewalls of the seed cores, self-assembled monolayer (SAM) is formed selectively. The seed cores are used for controlling polymer wire growth locations and orientations. For vertical growth, SAM is put on the top of the seed cores, while for horizontal growth it is put on the sidewalls. The regions where SAM should not be formed are covered with, for example, SiO_2 films. By distributing the seed cores with designated patterns, polymer wires are expected to grow with the designated configurations. Using the location/orientation-controlled MLD, a kind of polymer wire networks will be constructed.

Figure 12.9 shows a process for location/orientation-controlled MLD of poly-AM [TPA/PPDA]. In step 0, amino-alkanethiol SAM is formed on patterned Au films deposited on a glass substrate. In step 1, TPA molecules are provided onto the surface to connect TPA to the SAM. In step 2, after removing TPA molecular gas, the surface is exposed to PPDA molecular gas to connect PPDA to TPA.

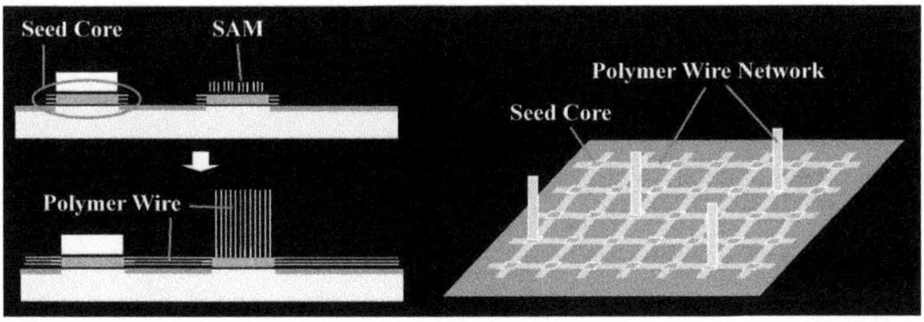

Figure 12.8. Concept of location/orientation-controlled MLD. Reprinted from T. Yoshimura, *Trends in Thin Solid Films Research* (ed. A. R. Jost) Ch. 5 "Self-Organized Growth of Polymer Wire Networks with Designed Molecular Sequences for Wavefunction-Controlled Nano Systems." Nova Science Publishers, New York (2007). With permission from Nova Science Publishers, Inc.

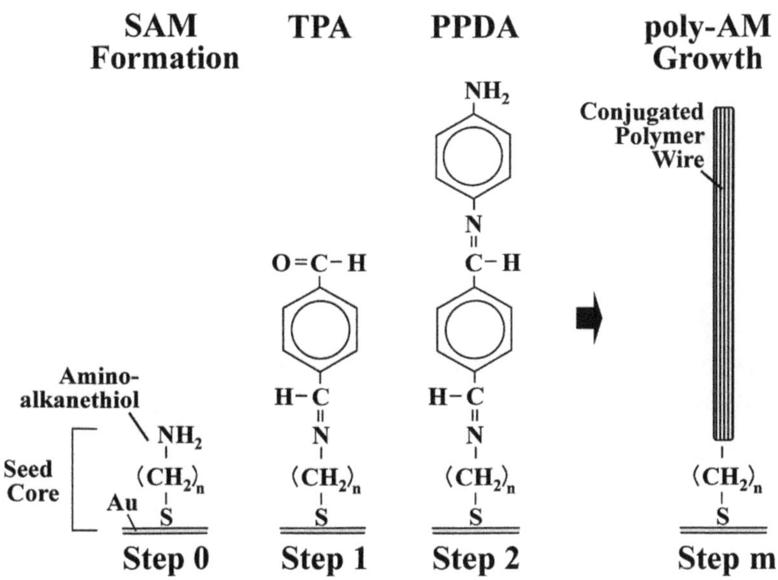

Figure 12.9. A process for location/orientation-controlled MLD of poly-AM [TPA/PPDA].

The aforementioned procedure can be repeated to grow poly-AM wires, reaching step m. Thus, by sequential molecular gas switching to perform MLD, poly-AM wires are grown vertically from the seed core selectively.

The Fourier transform infrared reflection absorption spectroscopy (FTIR-RAS) measurement on the absorbance ratio of "in-plane-direction vibration" to "surface-normal-direction vibration" in benzene rings revealed that polymer wires grow along nearly vertical directions from the seed core with SAM until step $m = 20$.

12.1.4 *Molecular Nano Duplication (MND)*

In order to reduce the fabrication cost of polymer devices fabricated by MLD, a mass production process is required. To do this, we proposed the molecular nano duplication (MND). Figure 12.10 shows the concept of MND [2, 14]. On a growth substrate, surface treatments such as SAM formation are applied in designed nano-scale patterns. On the SAM, removable molecules that have

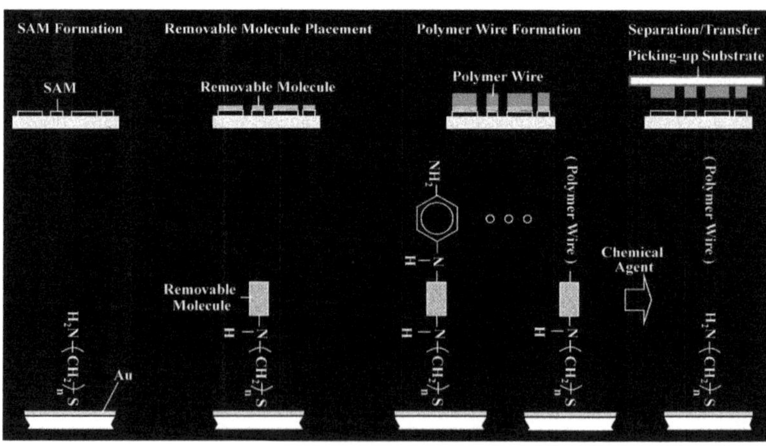

Figure 12.10. Concept of MND and process for poly-AM [TPA/PPDA]. Reprinted from T. Yoshimura, *Trends in Thin Solid Films Research* (ed. A. R. Jost) Ch. 5 "Self-Organized Growth of Polymer Wire Networks with Designed Molecular Sequences for Wavefunction-Controlled Nano Systems." Nova Science Publishers, New York (2007). With permission from Nova Science Publishers, Inc.

removable bonds connected to the SAM are placed by MLD, followed by polymer wire growth. After attaching a picking-up substrate on the surface, the removable bonds are cut by a chemical agent to separate the polymer wires from the SAM. Thus, nano-scale patterns consisting of polymer wires are transferred onto the picking-up substrate.

Since the growth substrate with patterned SAM is reusable, further nano lithography is not required for nano-scale pattern duplication. By combining MND with SORT, which is described in Section 8.7, SORT for molecular scale devices can be carried out, achieving heterogeneous integration of devices fabricated by MLD. The MND-combined SORT will achieve low-cost mass production processes for heterogeneous integration of nano-scale optical circuits.

Although MND is not demonstrated experimentally, we believe that the process will contribute to the cost reduction of systems containing nano-scale devices.

12.2 Polymer MQDs for EO Materials

As shown in Fig. 12.11, in order to enhance the EO effect, such as the Pockels effect and the Kerr effect, of the conjugated polymer wires, molecular-sequence arrangements in the wire should be controlled to optimize electron wavefunction shapes. Furthermore, for EO waveguide construction, the wires should be formed at designated locations with designated orientations. Therefore, to obtain high-performance EO waveguides, the self-organized polymer wire growth by MLD described in Section 12.1 is effective.

In the present section, the enhancement of the Pockels effect by controlling wavefunction shapes is predicted [1, 2, 4] using the molecular orbital calculations. The Pockels effect, which is an EO effect originating from the second-order nonlinear optical effect, induces refractive index changes proportional to applied electric fields. The possibility of MQD construction in polymer wires is also predicted. Experimental demonstrations of dimensionality control in conjugated polymer wires by constructing MQDs are presented.

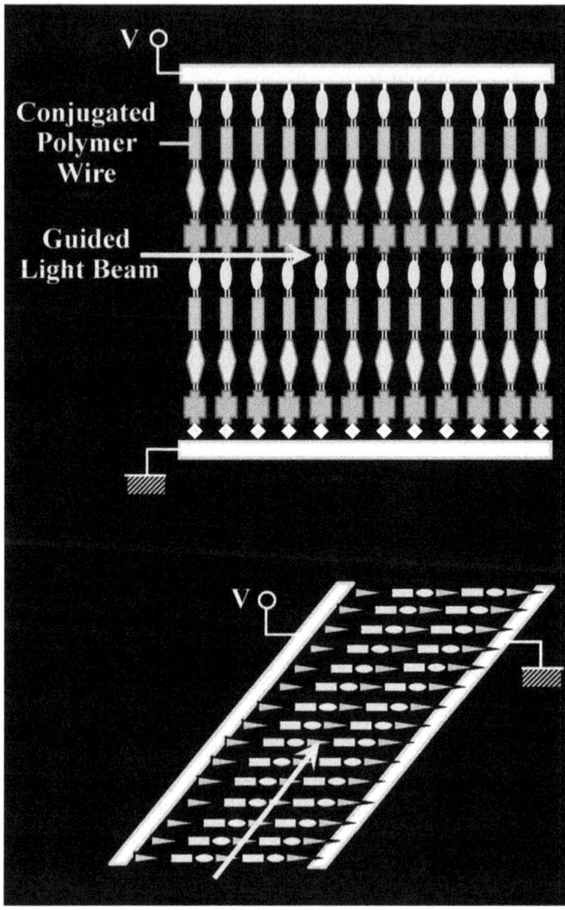

Figure 12.11. Structures of EO waveguides with conjugated polymer wires.

12.2.1 *Enhancement of the Pockels Effect by Controlling Wavefunction Shapes*

To measure the second-order optical nonlinearity of molecules, hyperpolarizability β is used. EO coefficient for the Pockels effect is proportional to β. In the two level model, $\beta \propto r_{\mathrm{gn}}^2 \Delta r$. Here, r_{gn} is transition dipole moment between the ground state and the excited state, and Δr is dipole moment difference between the ground

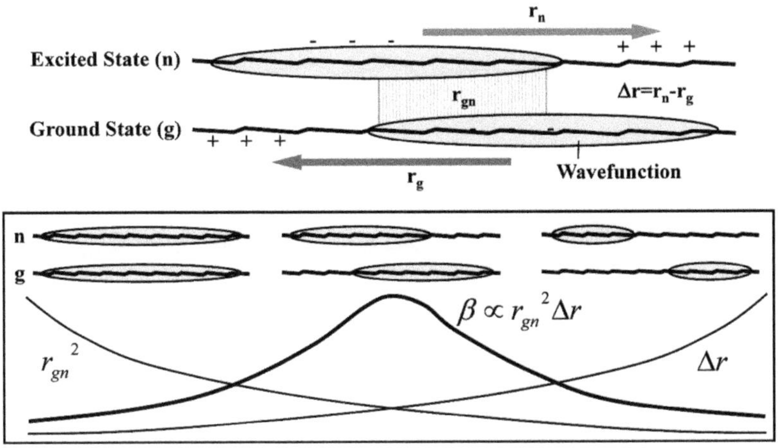

Figure 12.12. Guideline for enhancement of the Pockels effect.

state and the excited state. Therefore, the following guidelines for EO coefficient enhancement are derived as schematically shown in Fig. 12.12.

(1) Promote wavefunction overlap between the ground state and the excited state, increasing r_{gn}.
(2) Promote wavefunction separation between the ground state and the excited state, increasing Δr.

For three wavefunction shapes shown in Fig. 12.12, from left to right, the wavefunction separation increases while the wavefunction overlap decreases, i.e., Δr increases and r_{gn} decreases.

Δr and r_{gn} are not independent but are in a trade-off relationship, which indicates that the balance of the wavefunction overlap and the wavefunction separation must be considered simultaneously in optimizing β. Furthermore, Δr and r_{gn} depend on the dimensionality of the conjugated systems. Therefore, it is concluded that β is expected to have its maximum in a wavefunction shape of intermediate wavefunction separation with an appropriate conjugated system dimensionality existing somewhere between zero and one, i.e., between a quantum dot condition and a quantum wire condition.

Such optimized wavefunction conditions can be obtained by adjusting polymer wire lengths and donor/acceptor substitution

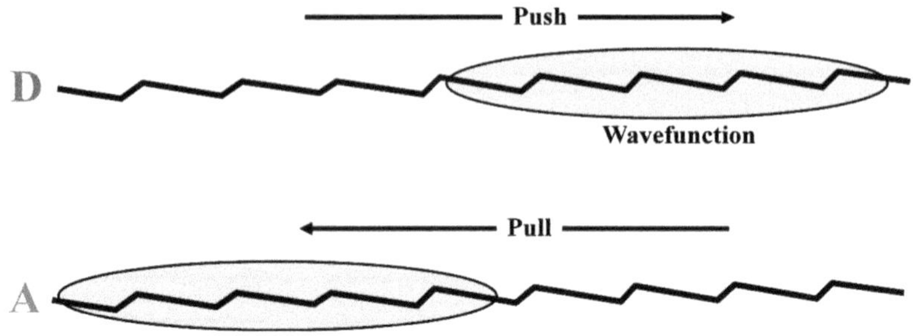

Figure 12.13. Push/Pull effect on wavefunction induced by donor/acceptor groups.

sites in the wires using the push/pull effects of the donors/acceptors shown in Fig. 12.13. Three types of models for conjugated polymer wires shown in Fig. 12.14 are considered. These molecules have poly-diacetylene (PDA) backbones. $-NH_2$ is the donor (D) and $-NO_2$ the acceptor (A). DA, DAAD, and DDAA represent the types of donor/acceptor substitution, and the numbers following them indicate the number of carbon sites N_C, which corresponds to molecular lengths. For these model molecules, the relationship between β and wavefunction shapes was simulated using the molecular orbital method. Details of the simulation are described in papers published elsewhere [1, 2, 4, 12]. The calculation revealed that wavefunction separations increase in the order of DA34, DAAD34, and DDAA30.

Figure 12.15 shows Δr, r_{gn}, and ρ_β of these model molecules for detuning energy of 0.2 eV. Here, ρ_β is β per 1 nm of molecular lengths. Δr increases and r_{gn} decreases in the order of DA34, DAAD34, and DDAA30, which makes ρ_β of DAAD34 in between maximum. This parallels to the tendency of the qualitative guideline shown in Fig. 12.12.

It is found from Fig. 12.16 that the molecular length affects ρ_β and that adjusting the length improves the second-order nonlinear optical effects. In DA and DAAD molecules, ρ_β reaches the maximum values near $N_C = 20$, corresponding to $\sim$2 nm in molecular length. Assuming PDA wire density of 1.3×10^{14} $1/cm^2$, the expected maximum EO coefficient of the DA and DAAD is about 3000 pm/V, which is 100 times larger than the EO coefficient r_{33} of LN.

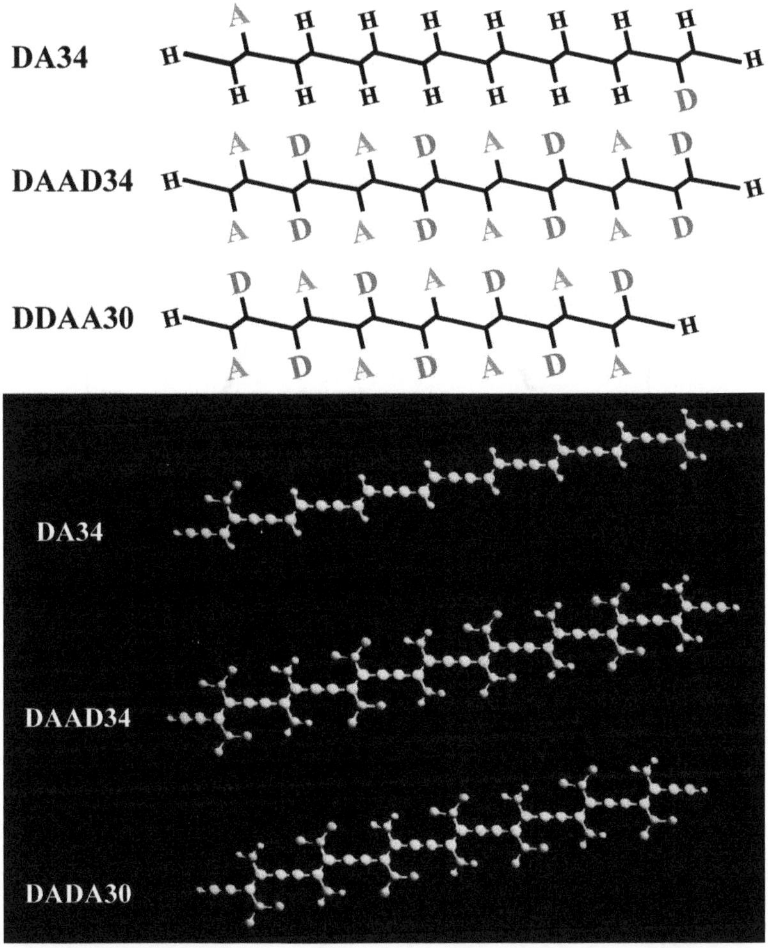

Figure 12.14. Models of conjugated polymer wires with donors and acceptors.

It is therefore concluded that a large EO effect will be obtained by controlling wavefunction dimensionality and separations, which can be realized by adjusting conjugated polymer wire lengths and donor/acceptor substitution sites.

As mentioned in Section 9.1.5, WPD optical switches with DAAD34 are expected to exhibit much larger deflection angles or much lower driving voltages, comparing with WPD optical switches

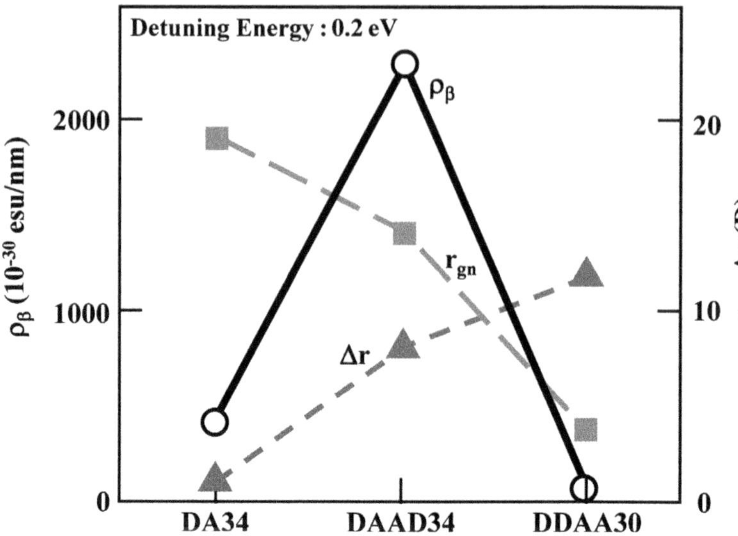

Figure 12.15. Calculated r_{gn}, Δr, and ρ_β for DA34, DAAD34, and DDAA30.

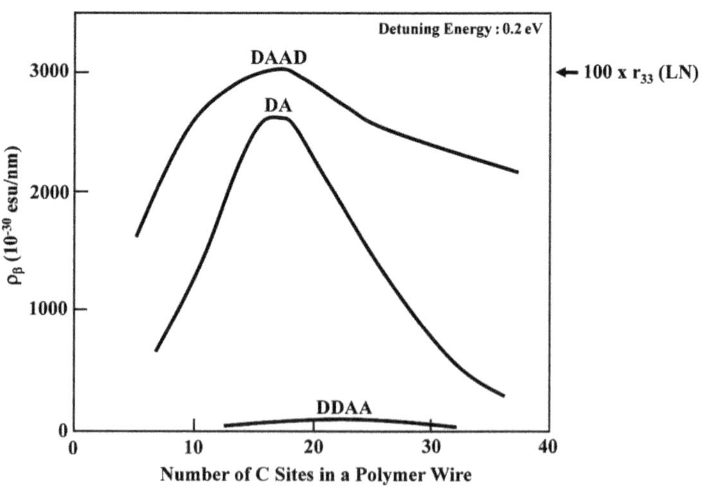

Figure 12.16. Calculated ρ_β of conjugated polymer wires as a function of the number of carbon sites corresponding to wire length. Reprinted with permission from T. Yoshimura, "Enhancing second-order nonlinear optical properties by controlling the wave function in one-dimensional conjugated molecules," *Phys. Rev. B* **40**, 6292–6298 (1989).

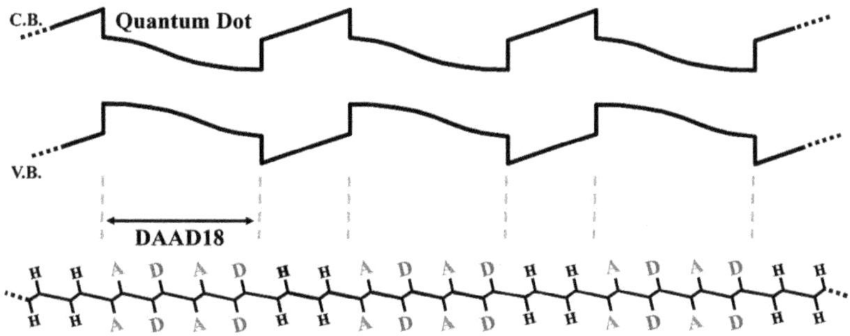

Figure 12.17. One-dimensional multiple quantum well (MQW) of a polymer super-lattice with corresponding electronic potential energy curves. DAAD18 molecules inserted in the polymer wire are regarded as quantum dots in the polymer MQD.

with LN. The dielectric constant for electrical signals might be relatively low in the conjugated polymer wire because the EO effect in the wires arises from electronic polarization. The low dielectric constant is suitable for high-speed operations with small RC delay and low power dissipation.

In the present simulation, PDA backbones are assumed. However, the same argument is applicable to other conjugated systems, for example, poly-AM and poly-OXD, etc.

The molecular orbital calculations revealed that energy gaps of DAAD molecules are smaller than those of PDA with no donor/acceptor substitution. Using the phenomena, it is possible to insert many DAAD molecules into a PDA backbone to construct a one-dimensional multiple quantum well (MQW), which corresponds to the MQD, of a polymer super-lattice as shown in Fig. 12.17 [1, 2, 4-6, 8, 12]. This structure enables the DAAD molecules, which are regarded as quantum dots, to be aligned perfectly along the wire direction. This is favorable for attaining large EO effects. In order to fabricate such molecular-order-controlled wire structures, MLD using source molecules with donors and acceptors is required.

12.2.2 *Fabrication of Polymer MQDs*

For experimental demonstrations of dimensionality control between zero and one in conjugated polymer wires, we constructed

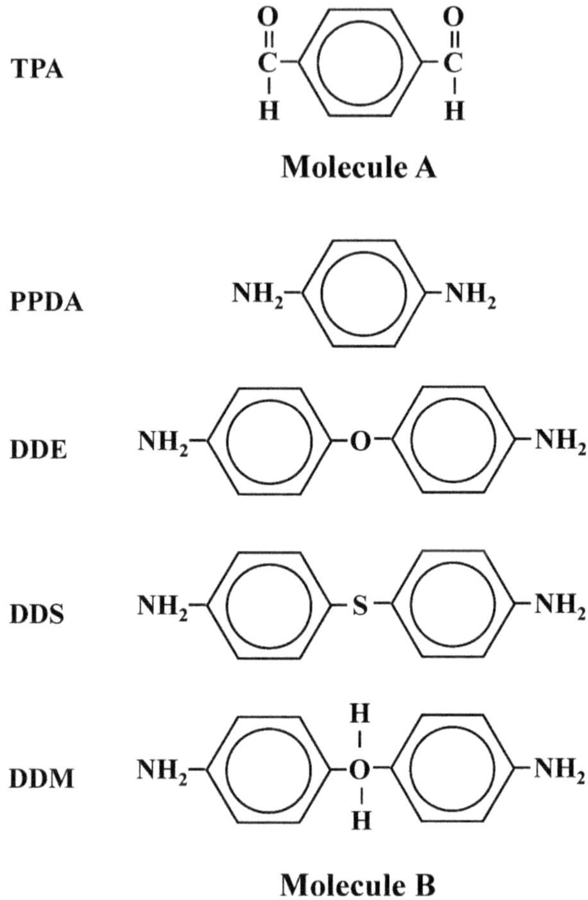

Figure 12.18. Molecules for making polymer MQDs based on poly-AM. Reprinted with permission from T. Yoshimura, S. Tatsuura, W. Sotoyama, A. Matsuura, and T. Hayano, "Quantum Wire and Dot Formation by Chemical Vapor Deposition and Molecular Layer Deposition of One-Dimensional Conjugated Polymer," *Appl. Phys. Lett.* **60**, 268–270 (1992). Copyright 1992, American Institute of Physics.

quantum dots in the wires using molecules shown in Fig. 12.18 [2, 8, 12]. The reaction between –CHO and $-NH_2$ is the same as described in Fig. 12.4 for poly-AM formation. Figure 12.19 diagrams molecules and polymer wires with their corresponding electronic potential energy curves. Benzene rings in TPA, PPDA, and DDE are

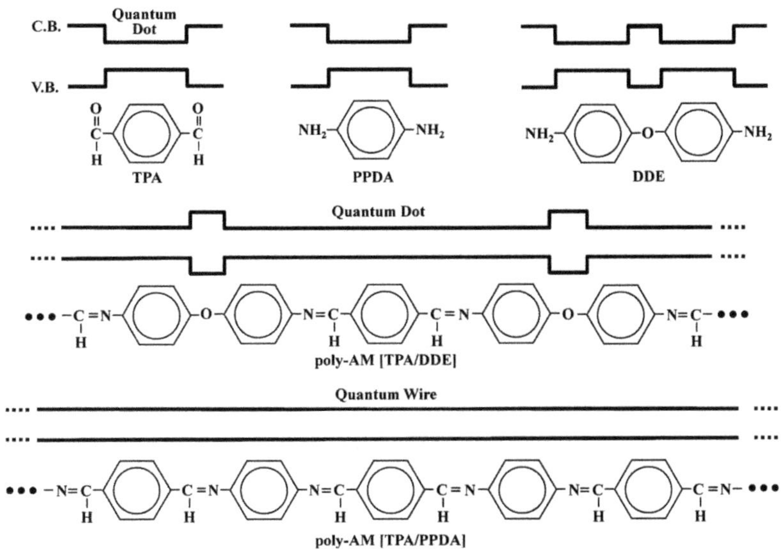

Figure 12.19. Molecules and polymer wires, where quantum dots and quantum wires are formed, with corresponding electronic potential energy curves. Reprinted with permission from T. Yoshimura, S. Tatsuura, W. Sotoyama, A. Matsuura, and T. Hayano, "Quantum Wire and Dot Formation by Chemical Vapor Deposition and Molecular Layer Deposition of One-Dimensional Conjugated Polymer," *Appl. Phys. Lett.* **60**, 268–270 (1992). Copyright 1992, American Institute of Physics.

regarded as short quantum dots of $\sim$0.5 nm length. In poly-AM [TPA/DDE] formed by combining TPA and DDE, π-conjugated part is regarded as a long quantum dot with barriers of –O–. The quantum dot contains three benzene rings and the length is $\sim$2 nm. In poly-AM [TPA/PPDA] formed by combining TPA and PPDA, π-conjugation is spread all along the wire to make a quantum wire.

Figure 12.20 shows the absorption spectra for TPA, PPDA, DDE, poly-AM [TPA/DDE], and poly-AM [TPA/PPDA]. In poly-AM [TPA/PPDA], a sharp peak of excitons is observed near 500 nm in wavelength. In poly-AM [TPA/DDE], the absorption band appears in the shorter wavelength side of that in poly-AM [TPA/PPDA] due to electron confinement in the quantum dot structure of $\sim$2 nm long. In TPA, PPDA, and DDE, the absorption band exhibits further

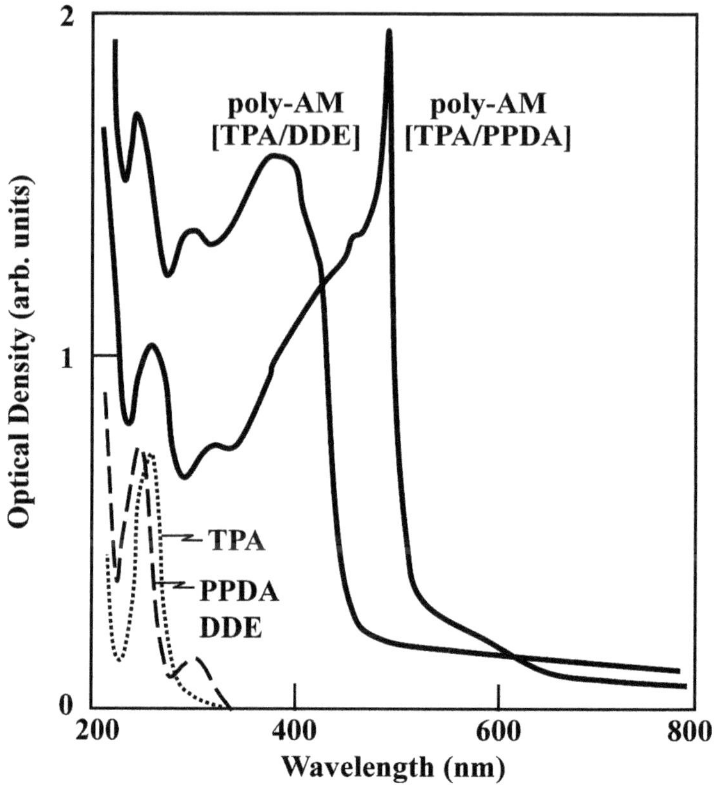

Figure 12.20. Absorption spectra of TPA, PPDA, DDE, poly-AM [TPA/DDE], and poly-AM [TPA/PPDA]. Reprinted with permission from T. Yoshimura, S. Tatsuura, W. Sotoyama, A. Matsuura, and T. Hayano, "Quantum Wire and Dot Formation by Chemical Vapor Deposition and Molecular Layer Deposition of One-Dimensional Conjugated Polymer," *Appl. Phys. Lett.* **60**, 268–270 (1992). Copyright 1992, American Institute of Physics.

shorter-wavelength shift, indicating strong electron confinement in the short quantum dots.

In poly-AM [TPA/DDS] grown form TPA and 4,4′-diaminodiphenyl sulfide (DDS), -S- is the barrier, and in poly-AM [TPA/DDM] grown form TPA and 1,4-diaminodiphenylmethane (DDM), $-CH_2$-is the barrier, just like –O– in poly-AM [TPA/DDE]. In the order of –S–, –O–, and $-CH_2-$, the absorption peak shifts to the shorter wavelength side, indicating that the barrier height increases to produce strong electron confinement. Energy of electrons confined in a quantum dot with infinite potential barrier height is expressed as $E_n = \hbar^2\pi^2n^2/\left(2m * m_0L^2\right)$, where, n =1, 2, ..., m_0 is electron mass, m^* effective mass in units of m_0, and L dot length. Assuming that the effective mass is 1 and the absorption peak energy for $L = \infty$ equals that of poly-AM [TPA/PPDA], the excitation energy for the lowest allowed transition is calculated as a function of L as shown in Fig. 12.21. Experimental and calculated results agree fairly well, confirming the electron confinement in the quantum dots.

Three-molecule MLD, which is carried out by using TPA, PPDA, and ODH, enables more precise control of quantum dot length in a range from ~0.5 nm to ~3 nm [2, 5, 6]. TPA and PPDA are connected

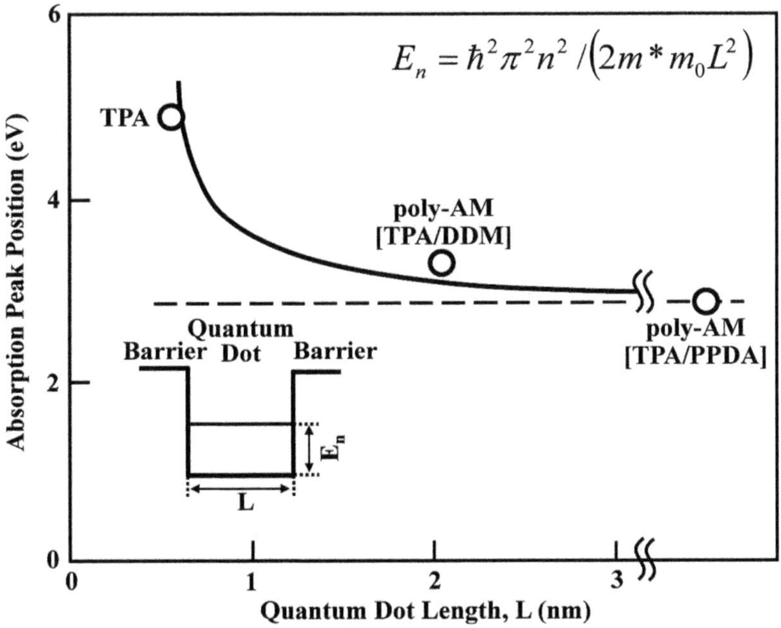

Figure 12.21. Dependence of absorption peak positions on quantum dot length. Reprinted with permission from T. Yoshimura, S. Tatsuura, W. Sotoyama, A. Matsuura, and T. Hayano, "Quantum Wire and Dot Formation by Chemical Vapor Deposition and Molecular Layer Deposition of One-Dimensional Conjugated Polymer," *Appl. Phys. Lett.* **60**, 268–270 (1992). Copyright 1992, American Institute of Physics.

with a double bond, allowing π-electron delocalization over the entire product molecule. The connection of TPA and ODH results in a series of single bonds, which severs the wavefunction at each bond. Therefore, by switching source molecules with a sequence of -ODH-TPA-ODH-TPA-PPDA-TPA-ODH-TPA-PPDA-TPA-PPDA-TPA-ODH—, as shown in Fig. 12.22(a), a polymer MQD containing three kinds

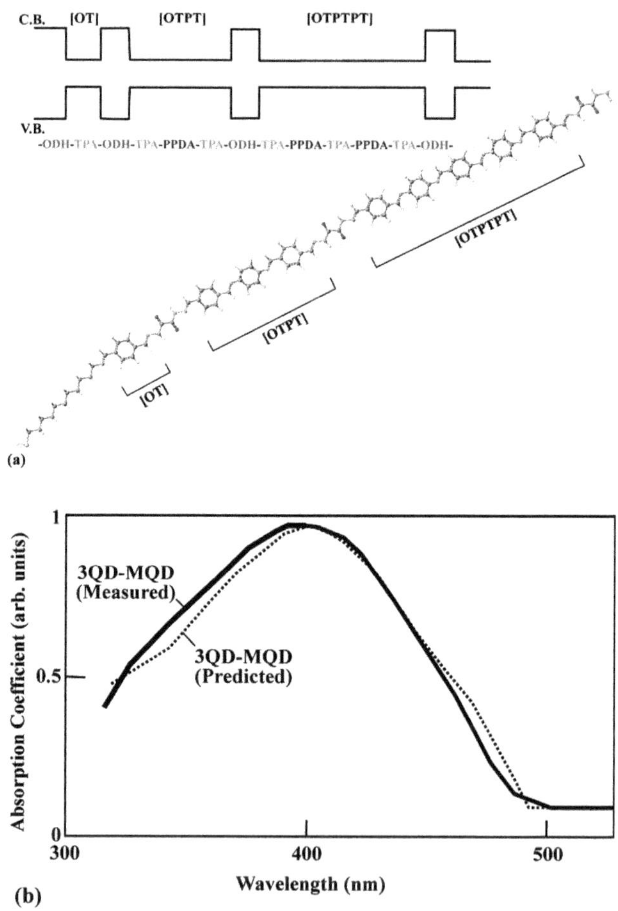

Figure 12.22. (a) Schematic energy level diagram and molecular structure of a polymer MQD, named 3QD-MQD, containing three different QDs of [OT], [OTPT] and [OTPTPT]. (b) Measured and predicted absorption spectra of 3QD-MQD.

of QDs, namely, [OT], [OTPTP], and [OTPTPT], can be grown [6]. The polymer MQD structure is named 3QD-MQD. As shown in Fig. 12.22(b), a measured absorption spectrum of 3QD-MQD is fairly coincident to the spectrum predicted by the superposition of component absorption bands of [OT], [OTPT] and [OTPTPT], indicating the capability of MLD to construct precisely-controlled polymer structures.

These results demonstrate the possibility of controlling dimensionality to enhance the EO effect in conjugated polymer wires. Future issue is to apply MLD to fabrication of new classes of quantum dot structures in conjugated polymer wires for high-performance EO materials by introducing donors and acceptors into the source molecules.

In addition, it is emphasized that polymer MQDs as well as molecular wires grown by MLD are also useful for sensitization of oxide semiconductors in photovoltaic and photosynthesis devices as described elsewhere in detail [2, 5, 6, 15].

12.3 Molecular Circuits

Conjugated polymer wires are expected to construct future molecular circuits [2, 12, 14] schematically illustrated in Fig. 12.23 due to the following reasons:

- High-carrier-mobility characteristics of their π-conjugated systems
- Small wire diameter, which is smaller than that of carbon nanotubes
- Molecular sequence controllability in the wire with mono-molecular scale, namely, precise wavefunction shape controllability, which is not available in carbon nanotubes and graphene

The molecular circuits consist of molecular LEDs, molecular PDs, molecular photo-voltaic devices and molecular transistors. Molecular LEDs, PDs, and photo-voltaic devices are made of conjugated polymer wires with pin junctions that are sandwiched by a pair of conjugated polymer wires for electrodes. In molecular transistors, a semiconductor conjugated polymer wire is aligned

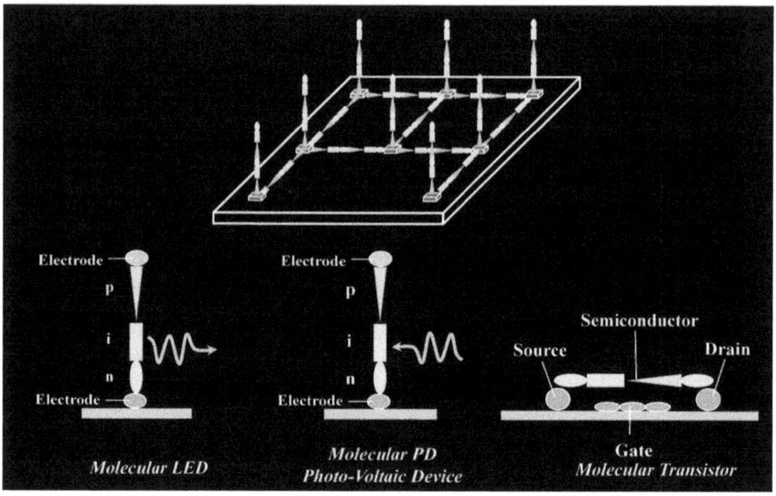

Figure 12.23. Concept of molecular circuits consisting of molecular LEDs, PDs, photo-voltaic devices, and transistors. Reprinted from T. Yoshimura, *Trends in Thin Solid Films Research* (ed. A. R. Jost) Ch. 5 "Self-Organized Growth of Polymer Wire Networks with Designed Molecular Sequences for Wavefunction-Controlled Nano Systems." Nova Science Publishers, New York (2007). With permission from Nova Science Publishers, Inc.

from a source electrode to a drain electrode, passing above a gate electrode. In order to construct the molecular circuits, the self-organized growth of polymer wire network by MLD described in Section 12.1 is effective.

Figure 12.24 shows tentative examples of conjugated polymer wires with pin junctions for LEDs, PDs, and photo-voltaic devices. For the donor/acceptor substitution-type pin junction, donors and acceptors are respectively substituted in the n-type region and the p-type region. For the backbone-type pin junction, an n-type backbone and a p-type backbone are connected. In Fig. 12.24, poly-AM and poly-OXD are assumed for the backbones to make a junction. Although p/n-type characterizations of poly-AM and poly-OXD are uncertain, if poly-AM and poly-OXD are formed within a polymer wire with molecular sequence arrangements like ABABABABABCDCDCDCD using the reactions shown in Fig. 12.4 with Molecule A: OA, Molecule B: ODH, Molecule C: TPA, and

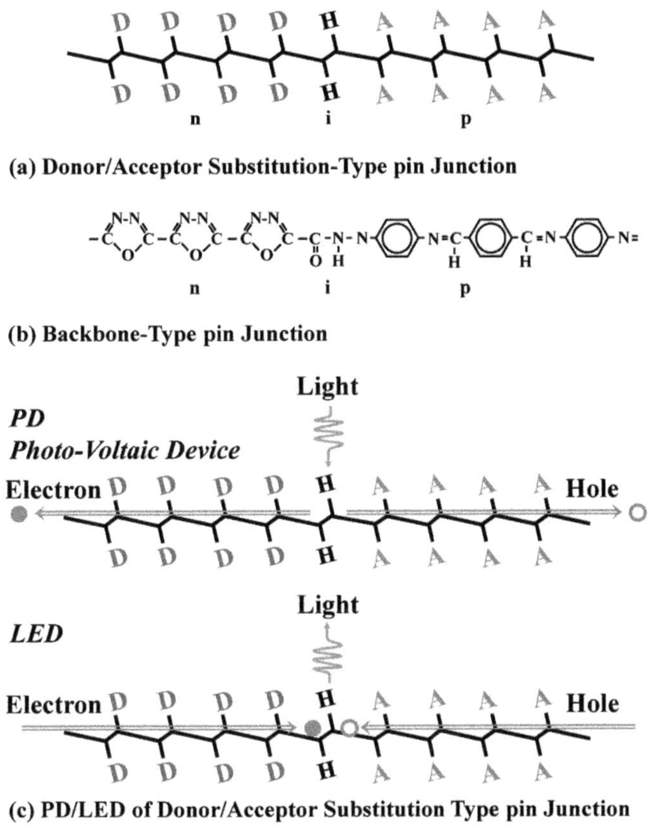

Figure 12.24. Tentative examples of pin junctions in conjugated polymer wires. Reprinted from T. Yoshimura, *Trends in Thin Solid Films Research* (ed. A. R. Jost) Ch. 5 "Self-Organized Growth of Polymer Wire Networks with Designed Molecular Sequences for Wavefunction-Controlled Nano Systems." Nova Science Publishers, New York (2007). With permission from Nova Science Publishers, Inc.

Molecule D: PPDA, some kind of "pin-like" junctions might possibly be formed in the poly-AM/poly-OXD interface region. For LED operation, electrons and holes are combined in the *i* region under a forward bias condition to emit light. For PD and photo-voltaic device operation, electrons and holes are generated by absorbing light in the *i* region to induce current/voltage.

12.4 Thin-Film Bio/Medical Photonics

12.4.1 *Integrated Photoluminescence Analysis Chips*

In integrated photoluminescence (PL) analysis chips [2, 16], MFCs, in which luminescent materials are contained, are inserted between two optical waveguide films as shown in Fig. 12.25. When excitation light beams with λ_{ec} in wavelength of a LD are introduced into MFC n from the upper optical waveguide, PL is generated. The PL is collected by the lower optical waveguide and is guided to PD passing through wavelength filter m, which transmits light beams of λ_m wavelength. By using arrayed MFCs, where different materials are injected, PL intensity measurements at λ_m can be done for the different materials at one time. By setting an array of optical circuits with embedded wavelength filters for different wavelengths, spectral analysis of PL can be performed for many kinds of materials at one time.

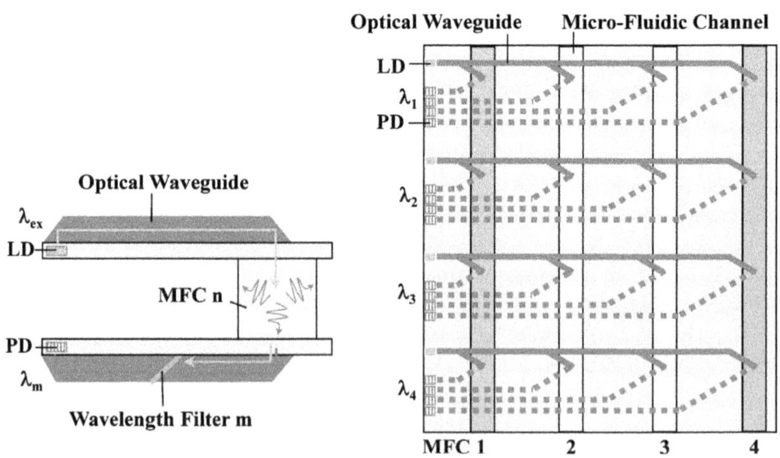

Figure 12.25. Integrated photoluminescence analysis chips with 3-D optical circuits. Reprinted with permission from T. Yoshimura, "Proposed Applications of 3-D Optical Interconnect Technologies to Integrated Chemical Systems," 2007 Digest of the LEOS Summer Topical Meetings [Bio-Inspired Sensors and Application/Imprinting on Photonic Integrated Circuits], Portland, Oregon, 129–130. © 2007 IEEE.

12.4.2 *Indicator for Reflective or Luminescent Materials using R-SOLNET*

Figure 12.26 shows an example of indicator for reflective or luminescent materials using R-SOLNET [2, 16, 17]. Between an optical waveguide and MFCs, PRI material is placed. From the optical waveguide, write beams are introduced into the PRI material. If Materials 1 and 3 are reflective or luminescent, the write beams and the reflected or photo-luminescence are superposed to increase the refractive index in the beam-overlapped regions since refractive index of the PRI material increases with beam intensity. This effect induces self-focusing to make SOLNET, namely, self-organized paths between the optical waveguide and MFC 1, 3, enabling us to indicate the locations of the reflective or luminescent materials as well as to guide the write beams, the reflected beams, photo-luminescence, and other light beams from the optical waveguide.

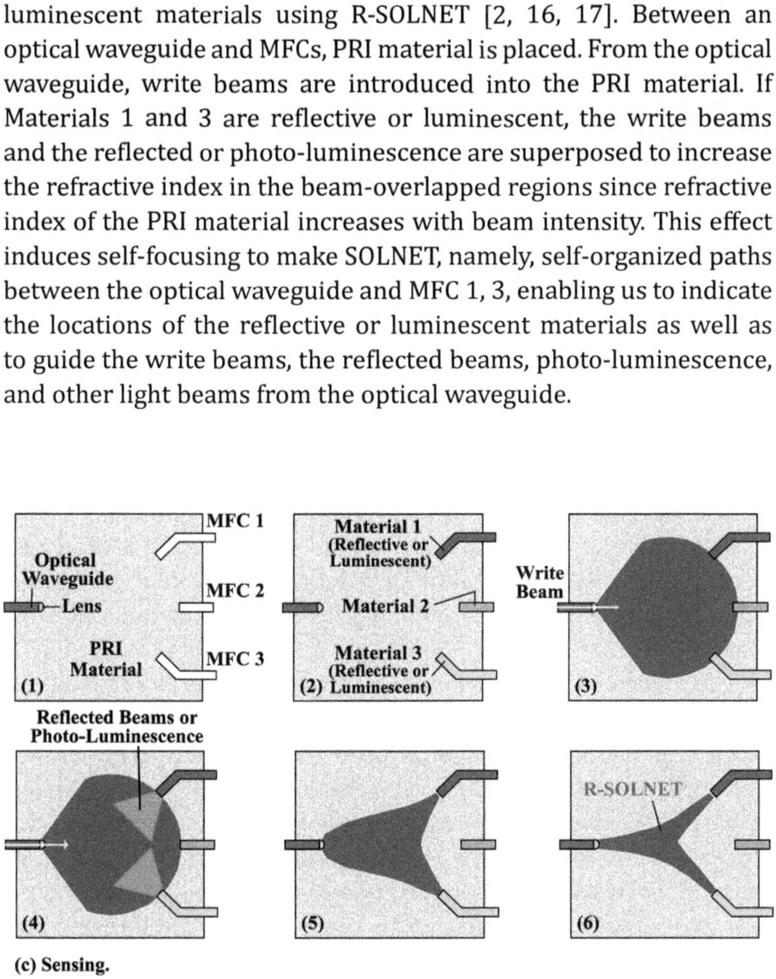

Figure 12.26. An indicator for reflective or luminescent materials using R-SOLNET. Reprinted with permission from T. Yoshimura, "Proposed Applications of 3-D Optical Interconnect Technologies to Integrated Chemical Systems," 2007 Digest of the LEOS Summer Topical Meetings [Bio-Inspired Sensors and Application/Imprinting on Photonic Integrated Circuits], Portland, Oregon, 129–130. © 2007 IEEE.

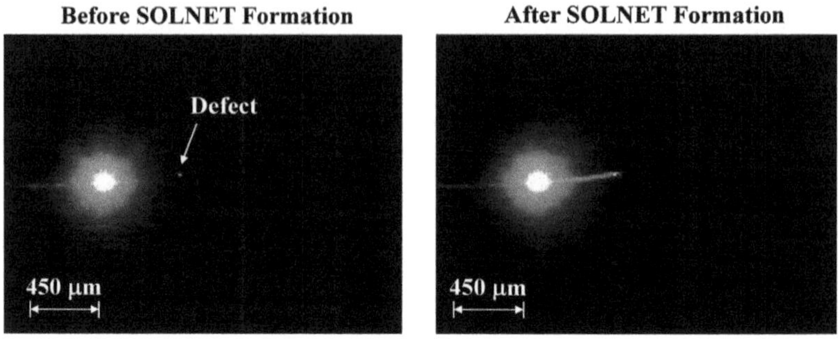

Figure 12.27. An example of the indicator for defects and dusts using R-SOLNET. Reprinted with permission from S. Ono, T. Yoshimura, T. Sato, J. Oshima, "Fabrication of Self-Organized Optical Waveguides in Photo-Induced Refractive Index Variation Sol-Gel Materials with Large Index Contrast," *J. Lightwave Technol.* **27**, 5308-5313© 2009 IEEE.

Figure 12.27 shows an example of the indicator for reflective materials [18]. In this case, the reflective material is a defect in a PRI sol-gel thin film described in Section 5.4. Write beams with power of 0.5 mW and wavelength of 405 nm were introduced into the thin film for 30 s under heating at 200°C. It is found that SOLNET is drawn to a defect in the PRI sol-gel thin film, indicating the locations of the reflective material as well as guiding the write beams and reflected beams.

The indicator described above, together with MLD, will be applicable to medical photonics [2, 16, 17]. Figure 12.28 shows the concept of the SOLNET-assisted laser surgery. First, luminescent molecules are attached to cancer cells by LP-MLD. After inserting an optical fiber and a PRI material into a region surrounding the cancer cells, a write beam is introduced from the optical fiber. Then, optical waveguides connecting the optical fiber and the cancer cells are constructed by the write beam and photo-luminescence from the luminescent molecules according to the mechanism of R-SOLNET with luminescent materials described in Section 4.4.6. By introducing surgery laser beams into the R-SOLNET via the optical fiber, cancer cells are destroyed selectively. By detecting the backward photo-luminescence emitted from the luminescent molecules, *in situ* monitoring of the degree of the cancer destroy might be possible.

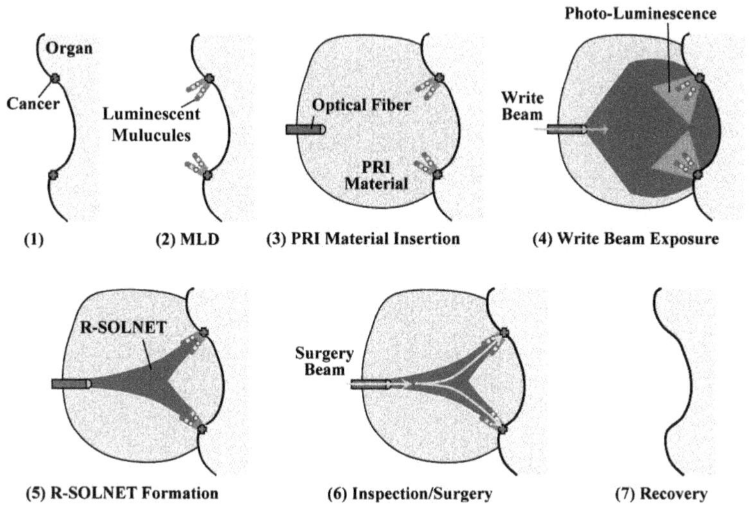

Figure 12.28. Concept of the SOLNET-assisted laser surgery. After luminescent molecules are attached to cancer cells by LP-MLD, R-SOLNET connecting an optical fiber and the cancer cells is constructed to guide surgery laser beams to cancer cell sites. See also Color Insert.

The optical fiber and the cancer cells with luminescent molecules in Fig. 12.28 respectively correspond to the optical waveguide and the targets of luminescent materials in the simulation results shown in Figs. 4.33 and 4.35. It is found in these figures that, with writing time, R-SOLNET is gradually constructed between the optical waveguide core and the targets. As a result, the probe beam, corresponding to the surgery beam, is guided to the target sites. The R-SOLNET construction between an optical fiber and an luminescent target is demonstrated experimentally as shown in Fig. 5.16. These results suggest that it is viable to construct R-SOLNET with luminescent materials connecting an optical fiber and cancer cells.

12.4.3 *Molecular Recognition Chip*

Molecular recognition chips consist of molecule-sensitive waveguides (MSWs) as shown in Fig. 12.29 [2, 16]. MSW is a waveguide ring resonator with receptors for particular molecules

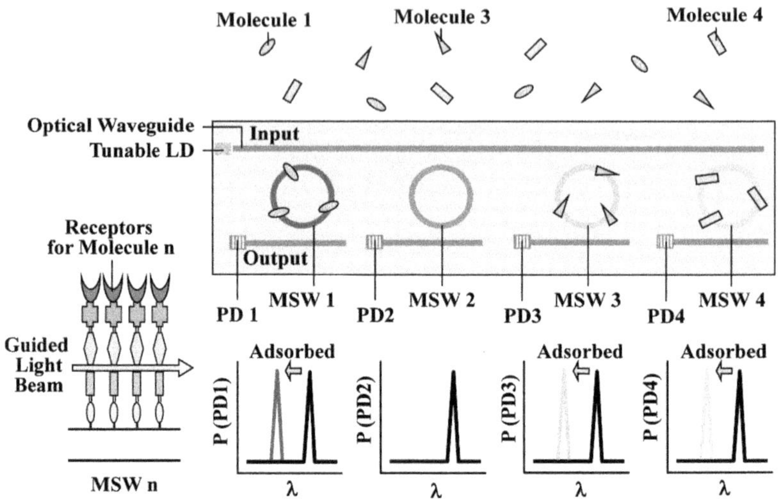

Figure 12.29. A molecular recognition chip with molecule-sensitive waveguide (MSW). Reprinted with permission from T. Yoshimura, "Proposed Applications of 3-D Optical Interconnect Technologies to Integrated Chemical Systems," 2007 Digest of the LEOS Summer Topical Meetings [Bio-Inspired Sensors and Application/Imprinting on Photonic Integrated Circuits], Portland, Oregon, 129–130. © 2007 IEEE.

for example. When wavelengths of light beams in input waveguide are scanned, optical power transmitted to PD at the edge of output waveguide exhibits a peak at the resonant wavelength of MSW. When molecules are adsorbed on MSW, refractive index of the MSW changes, inducing a resonant wavelength peak shift. For example, when Molecule 1, Molecule 3, and Molecule 4 are contained in the atmosphere, resonant frequencies of MSW 1, MSW 3, and MSW 4 shift. Thus, constituent molecules can be recognized at one time.

References

1. T. Yoshimura, "Design of Organic Nonlinear Optical Materials for Electro-Optic and All-Optical Devices by Computer Simulation," *FUJITSU Sci. Tech. J.* **27**, 115–131 (1991).
2. T. Yoshimura, *Thin-Film Organic Photonics: Molecular Layer Deposition and Applications*, CRC/Taylor & Francis, Boca Raton, Florida, (2011).

3. T. Yoshimura, "Characterization of the Electro-Optic Effect in Styrylpyridinium Cyanine Dye Thin-Film Crystals by an ac Modulation Method," *J. Appl. Phys.* **62**, 2028–2032 (1987).
4. T. Yoshimura, "Enhancing Second-Order Nonlinear Optical Properties by Controlling the Wave Function in One-Dimensional Conjugated Molecules," *Phys. Rev. B* **40**, 6292–6298 (1989).
5. T. Yoshimura, A. Oshima, D. Kim, and Y. Morita, "Quantum Dot Formation in Polymer Wires by Three-Molecule Molecular Layer Deposition (MLD) and Applications to Electro-Optic/Photovoltaic Devices," *ECS Trans.* **25** (4) "Atomic Layer Deposition Applications 5," from the Vienna, Austria Meeting, 15–25 (2009).
6. T. Yoshimura, R. Ebihara, and A. Oshima, "Polymer Wires with Quantum Dots Grown by Molecular Layer Deposition of Three Source Molecules for Sensitized Photovoltaics," *J. Vac. Sci. Technol. A* **29**, 051510-1-6 (2011).
7. T. Yoshimura, S. Tatsuura, and W. Sotoyama, "Polymer Films Formed with Monolayer Growth Steps by Molecular Layer Deposition," *Appl. Phys. Lett.* **59**, 482–484 (1991).
8. T. Yoshimura, S. Tatsuura, W. Sotoyama, A. Matsuura, and T. Hayano, "Quantum Wire and Dot Formation by Chemical Vapor Deposition and Molecular Layer Deposition of One-Dimensional Conjugated Polymer," *Appl. Phys. Lett.* **60**, 268–270 (1992).
9. T. Yoshimura, S. Ito, T. Nakayama, and K. Matsumoto, "Orientation-Controlled Molecule-by-Molecule Polymer Wire Growth by the Carrier-Gas-Type Organic Chemical Vapor Deposition and the Molecular Layer Deposition," *Appl. Phys. Lett.* **91**, 033103-1-3 (2007).
10. T. Yoshimura and Y. Kudo, "Monomolecular-Step Polymer Wire Growth from Seed Core Molecules by the Carrier-Gas Type Molecular Layer Deposition (MLD)," *Appl. Phys. Express* **2**, 015502-1-3 (2009).
11. T. Yoshimura, E. Yano, S. Tatsuura, and W. Sotoyama, "Organic Functional Optical Thin Film, Fabrication and Use Thereof," US Patent 5,444,811 (1995).
12. T. Yoshimura, "Self-Organized Growth of Polymer Wire Networks with Designed Molecular Sequences for Wavefunction-Controlled Nano Systems," in *Trends in Thin Solid Films Research* (ed. A. R. Jost), Nova Science Publishers, New York, Ch. 5 (2007).
13. T. Suntola, "Atomic Layer Epitaxy," *Mater. Sci. Rep.* **4** (7), 1989, Elsevier Science Publishers.

14. T. Yoshimura, Y. Suzuki, N. Shimoda, T. Kofudo, K. Okada, Y. Arai, and K. Asama, "Three-Dimensional Chip-Scale Optical Interconnects and Switches with Self-Organized Wiring Based on Device-Embedded Waveguide Films and Molecular Nanotechnologies," *Proc. SPIE* **6126**, from Photonics West 2006, San Jose, California, 612609-1-15 (2006).
15. T. Yoshimura, H. Watanabe, and C. Yoshino, "Liquid-Phase Molecular Layer Deposition (LP-MLD): Potential Applications to Multi-Dye Sensitization and Cancer Therapy," *J. Electrochem. Soc.* **158**, 51–55 (2011).
16. T. Yoshimura, "Proposed Applications of 3-D Optical Interconnect Technologies to Integrated Chemical Systems," 2007 Digest of the LEOS Summer Topical Meetings [Bio-Inspired Sensors and Application/Imprinting on Photonic Integrated Circuits], Portland, Oregon, 129–130 (2007).
17. T. Yoshimura, C. Yoshino, K. Sasaki, T. Sato, and M. Seki, "Cancer Therapy Utilizing Molecular Layer Deposition and Self-Organized Lightwave Network -Proposal and Theoretical Prediction," *IEEE J. Select. Topics Quantum Electron.* **18**, Biophotonics 1, May/June 2012 [to be published].
18. S. Ono, T. Yoshimura, T. Sato, J. Oshima, "Fabrication of Self-Organized Optical Waveguides in Photo-Induced Refractive Index Variation Sol-Gel Materials with Large Index Contrast," *J. Lightwave Technol.* **27**, 5308–5313 (2009).

Epilogue

In January 2009, I received offers from two publishers to write books on different subjects. One of the offers was from Pan Stanford Publishing for the present work. I was afraid that it might be very hard to complete two books simultaneously. So, I told my wife, Yoriko, "What do you think about writing two books at the same time in English?" She said to me, "You have accomplished anything you once told me." Her answer made me accept both offers. This book would not have been finished without Yoriko's encouragement.

While this book was being prepared, Yoriko passed away on September 27, 2009, after receiving warm and sincere care and encouragement from her friends of JOSHIGAKUIN, Musashino Academia Musicae and Morimura-Gakuen Parents, students of her private piano school, and her relatives and family. I thank Yoriko for her sincere support and encouragement for over 32 years since 1977.

Tetsuzo Yoshimura
January 1, 2012

Index

Color Insert

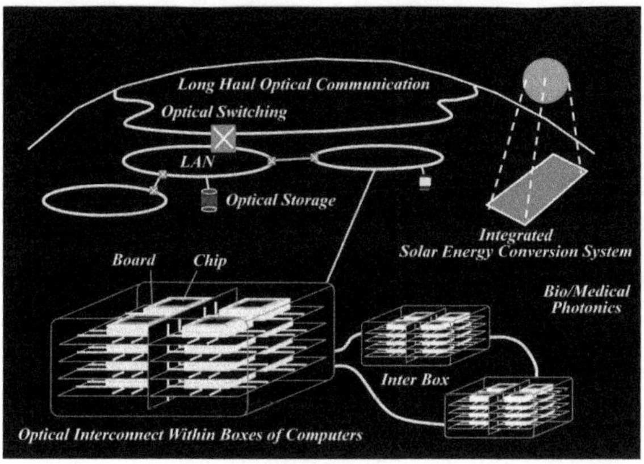

Figure 1.1

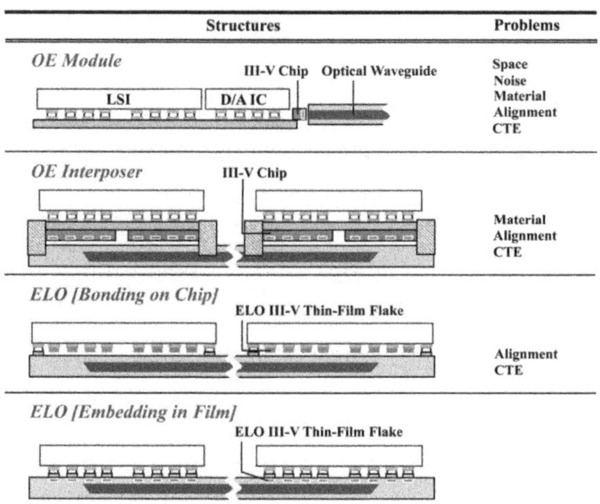

Figure 2.6

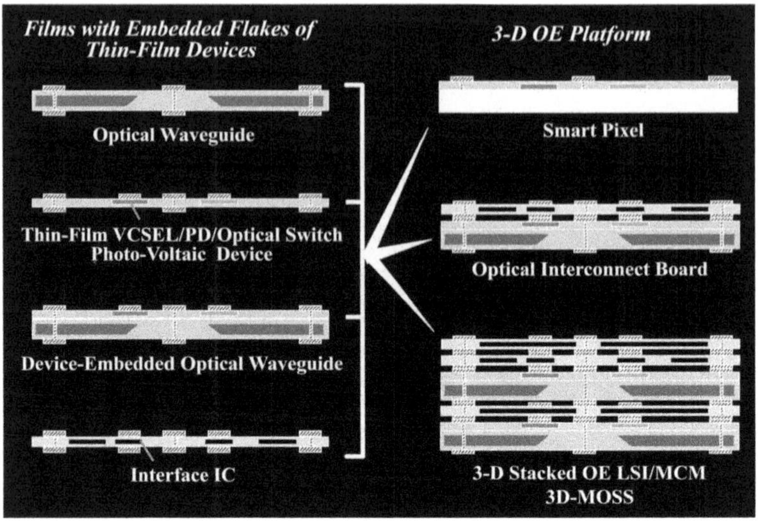

Figure 2.8

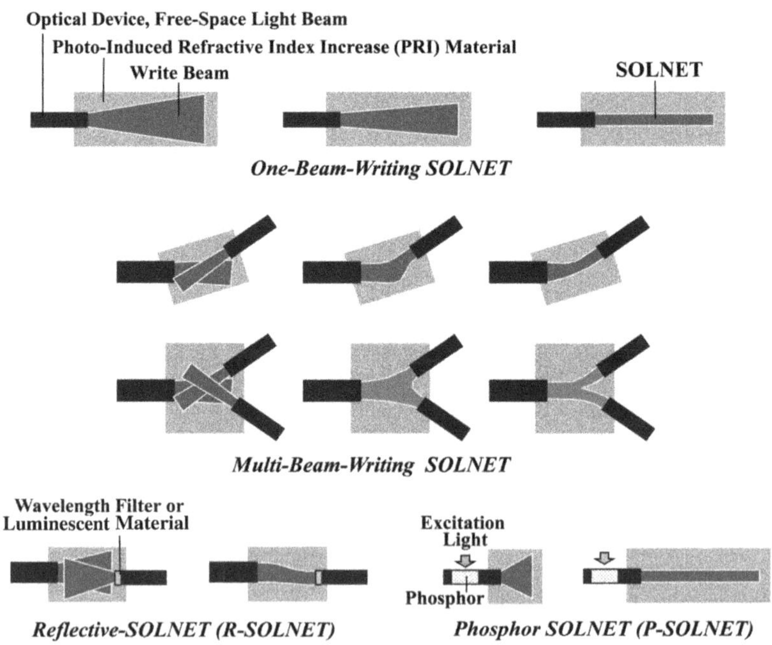

Figure 4.1

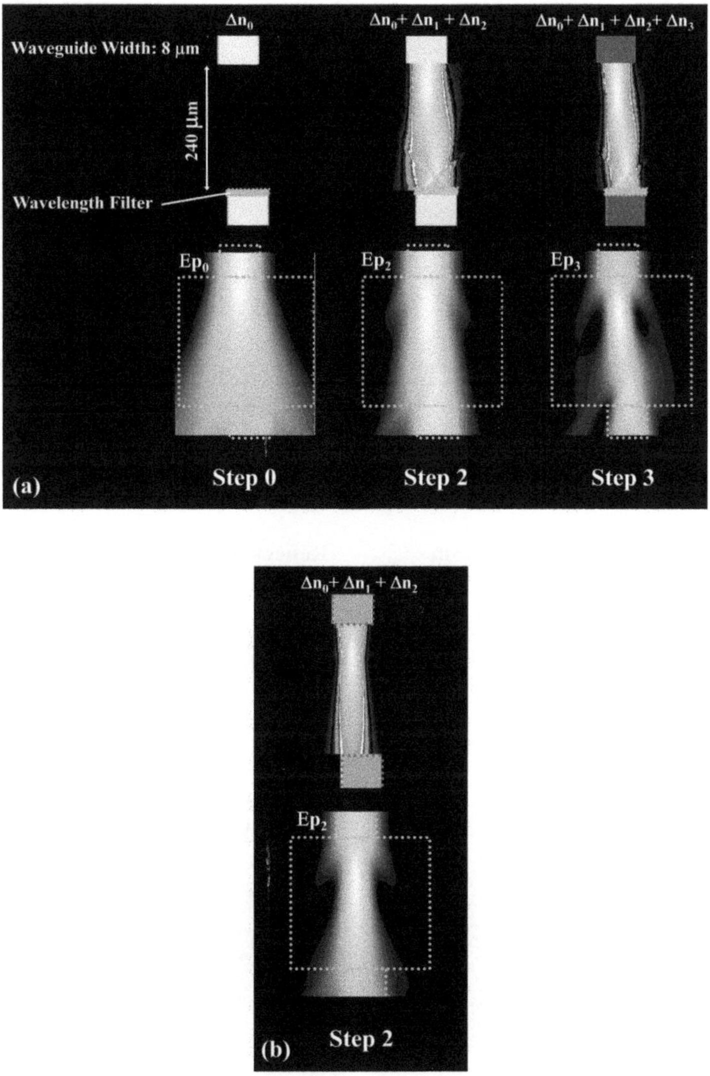
Δn_0
$\Delta n_0 + \Delta n_1 + \Delta n_2$
$\Delta n_0 + \Delta n_1 + \Delta n_2 + \Delta n_3$
Waveguide Width: 8 μm
240 μm
Wavelength Filter
Ep_0
Ep_2
Ep_3
(a)
Step 0
Step 2
Step 3
$\Delta n_0 + \Delta n_1 + \Delta n_2$
Ep_2
(b)
Step 2

Figure 4.13

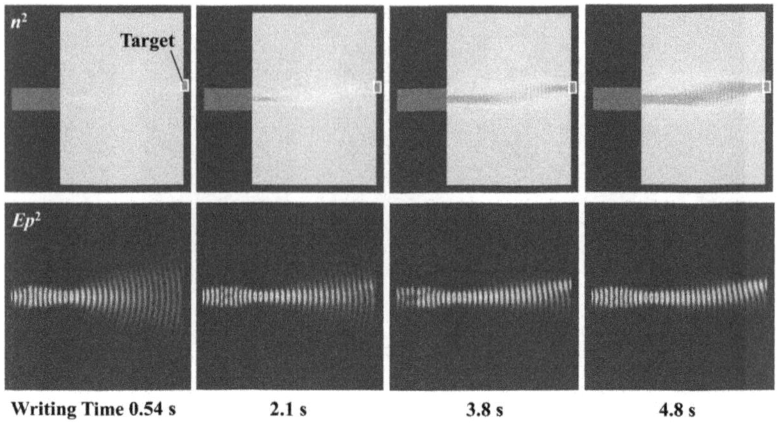

Figure 4.33

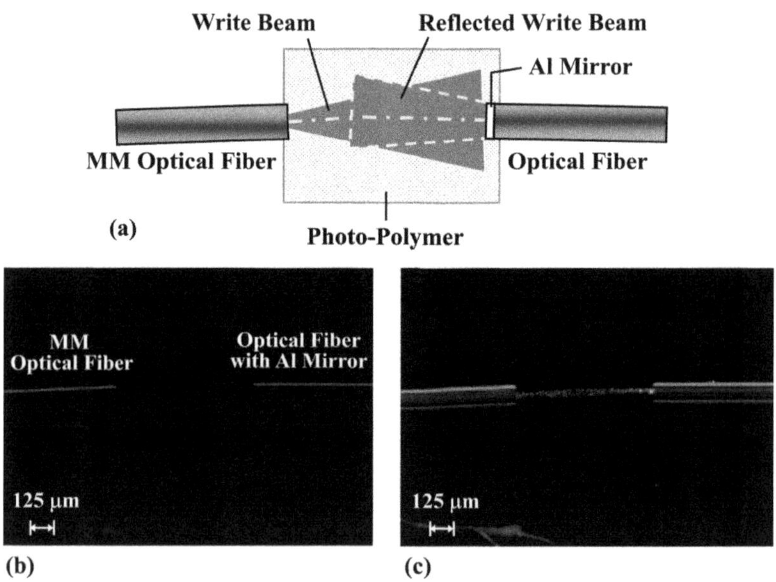

Figure 5.10

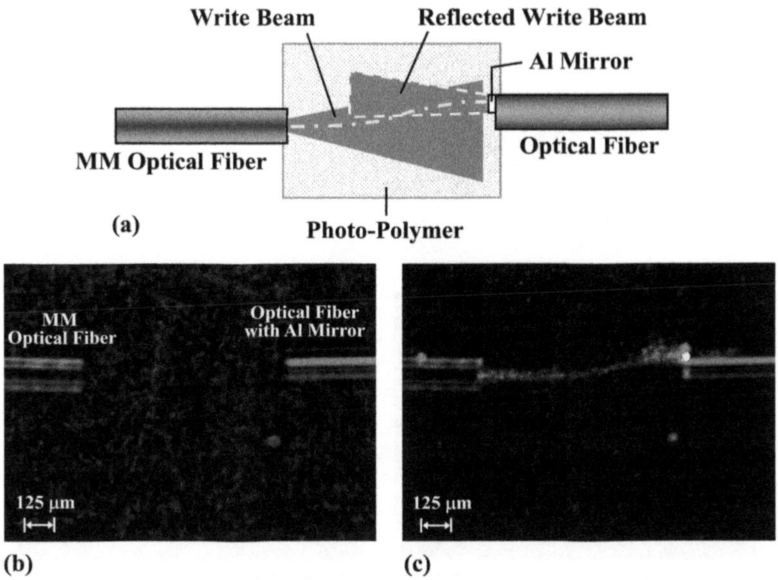
Write Beam
Reflected Write Beam
Al Mirror
Optical Fiber
MM Optical Fiber
(a)
Photo-Polymer
MM
Optical Fiber
Optical Fiber
with Al Mirror
125 μm
125 μm
(b)
(c)

Figure 5.12

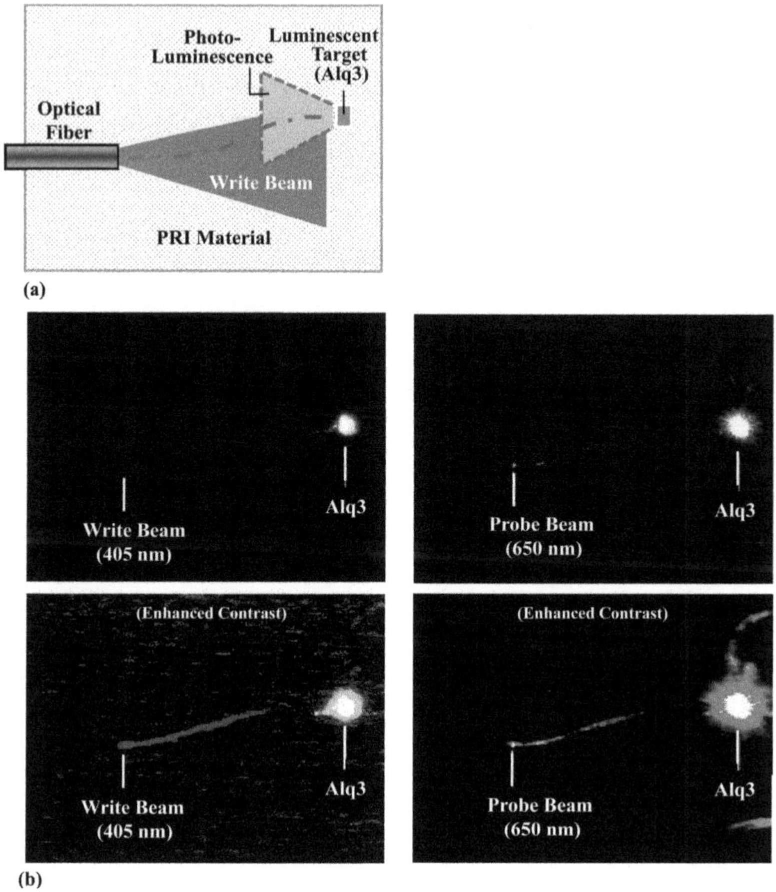
Photo-
Luminescence
Luminescent
Target
(Alq3)
Optical
Fiber
Write Beam
PRI Material
(a)
Write Beam
(405 nm)
Alq3
Probe Beam
(650 nm)
Alq3
(Enhanced Contrast)
Write Beam
(405 nm)
Alq3
(Enhanced Contrast)
Probe Beam
(650 nm)
Alq3
(b)

Figure 5.16

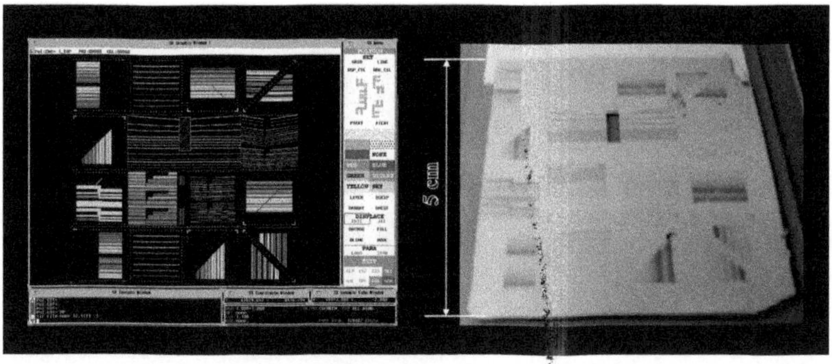

Figure 6.3

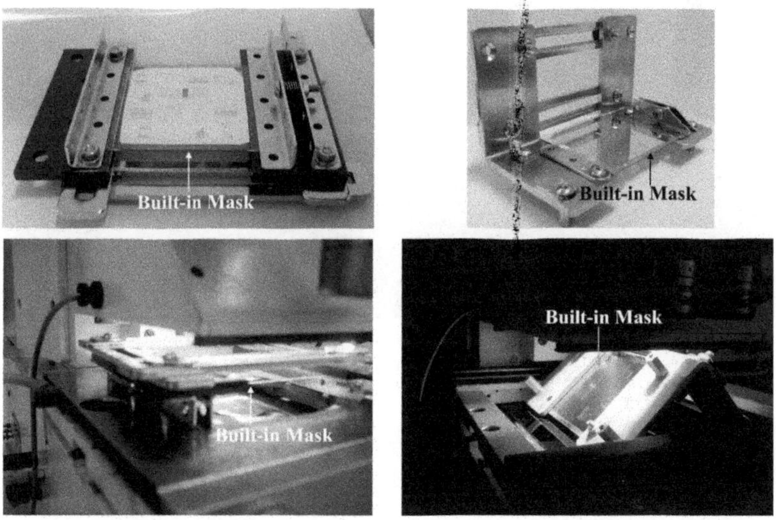

(a) Vertical exposure **(b) Tilted exposure**

Figure 6.4

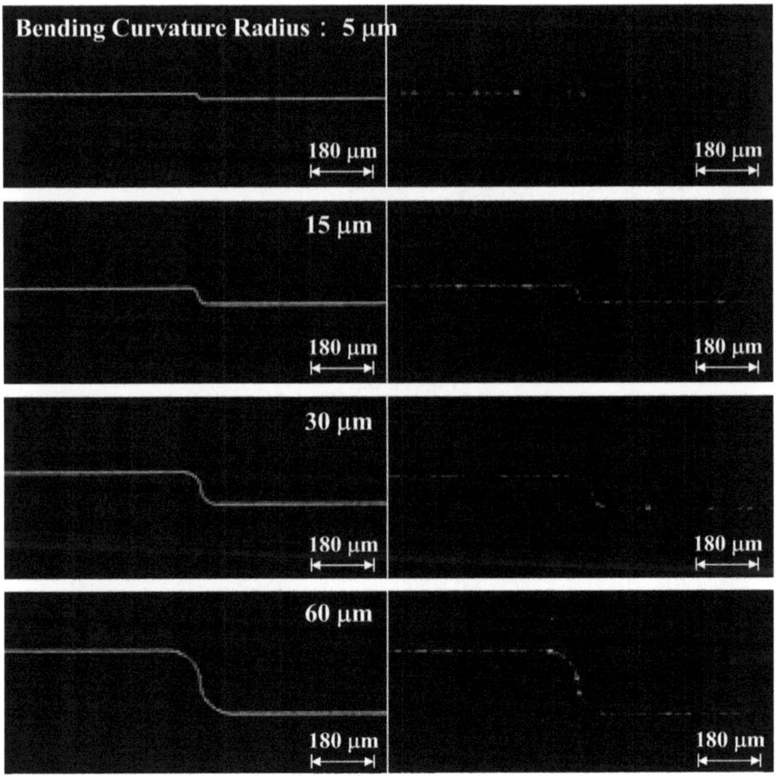
Bending Curvature Radius : 5 μm
180 μm
180 μm
15 μm
180 μm
180 μm
30 μm
180 μm
180 μm
60 μm
180 μm
180 μm

Figure 6.25

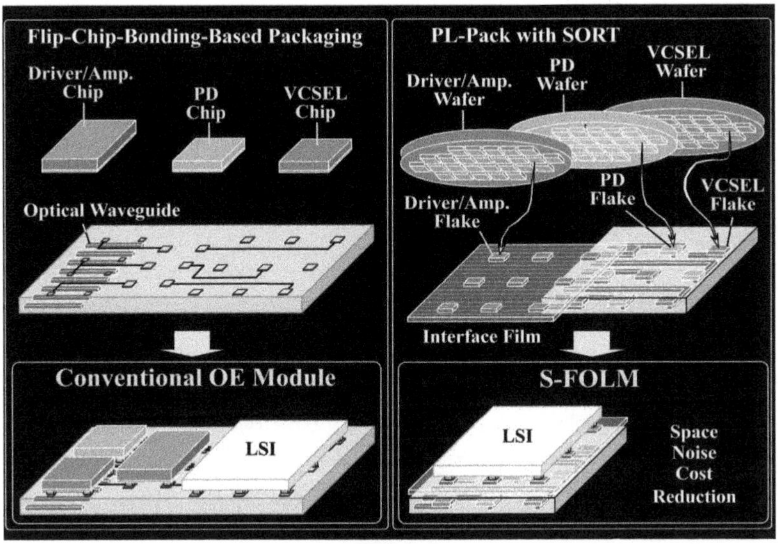

Figure 8.1

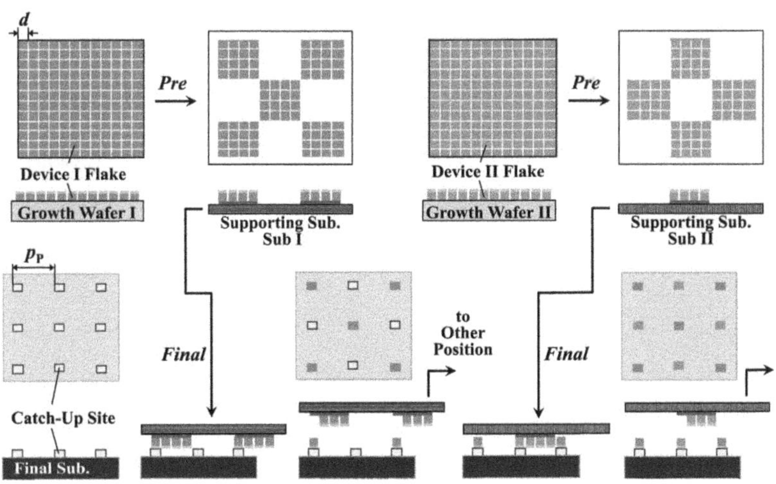

Figure 8.3

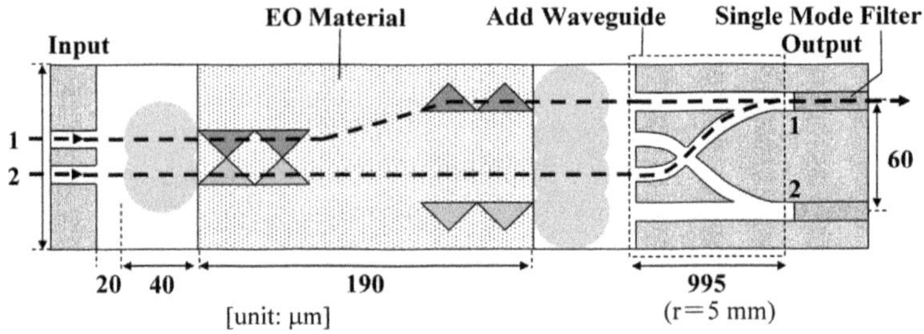

Figure 9.21

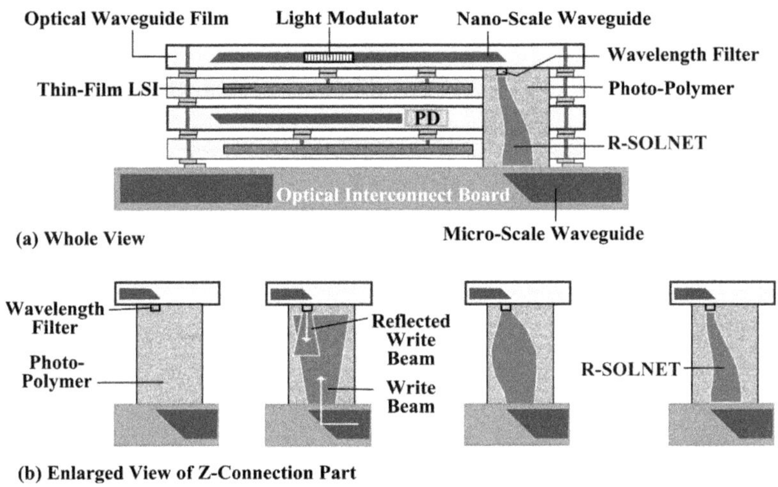

Figure 10.6

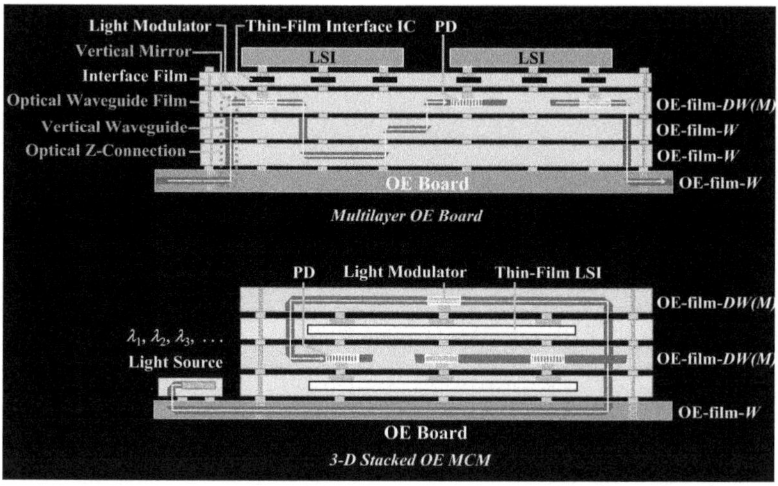

Figure 10.12

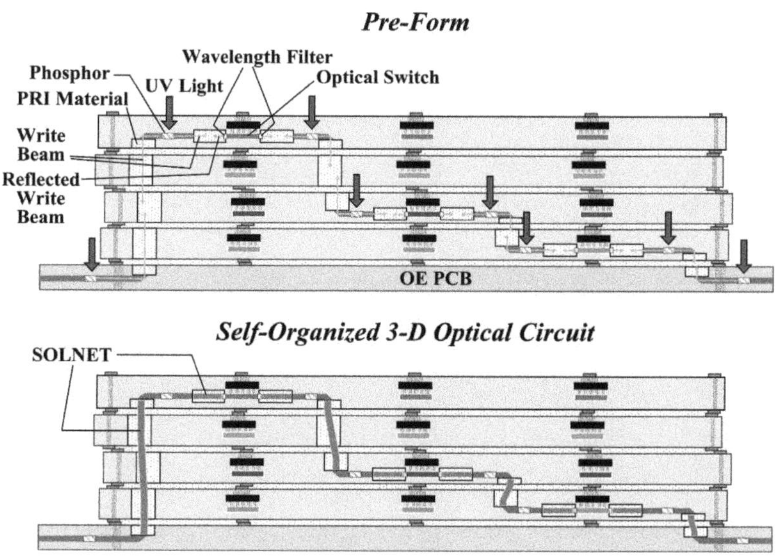

Figure 10.37

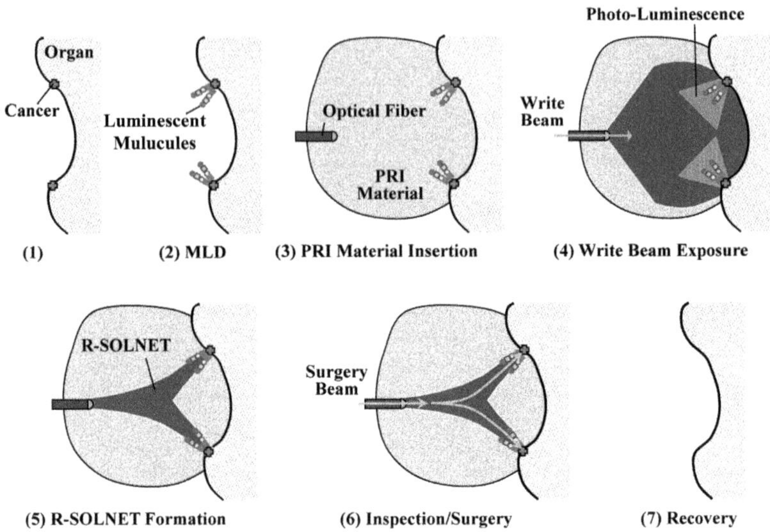
Organ
Cancer
(1)
Luminescent
Mulucules
(2) MLD
Optical Fiber
PRI
Material
(3) PRI Material Insertion
Photo-Luminescence
Write
Beam
(4) Write Beam Exposure
R-SOLNET
(5) R-SOLNET Formation
Surgery
Beam
(6) Inspection/Surgery
(7) Recovery

Figure 12.28

For Product Safety Concerns and Information please contact our EU representative GPSR@taylorandfrancis.com
Taylor & Francis Verlag GmbH, Kaufingerstraße 24, 80331 München, Germany

www.ingramcontent.com/pod-product-compliance
Ingram Content Group UK Ltd.
Pitfield, Milton Keynes, MK11 3LW, UK
UKHW021512250926
475702UK00043B/245
9789814310901